MATTER AND MIND

BOSTON STUDIES IN THE PHILOSOPHY OF SCIENCE

VOLUME 287

For further volumes:
http://www.springer.com/series/5710

MATTER AND MIND

A Philosophical Inquiry

by

MARIO BUNGE

McGill University, Montreal, Canada

Springer

Mario Bunge
McGill University
Dept. of Philosophy
Sherbrooke St West 855
H3A 2T7 Montreal
Canada
marioaugustobunge@hotmail.com

ISSN 0068-0346
ISBN 978-90-481-9224-3 e-ISBN 978-90-481-9225-0
DOI 10.1007/978-90-481-9225-0
Springer Dordrecht Heidelberg London New York

Library of Congress Control Number: 2010928855

Printed on acid-free paper

Springer is part of Springer Science+Business Media (www.springer.com)

I dedicate this book to
Nicholas Rescher,
the most learned, lucid and fair of us.

Preface

I am offered an avocado. *It is* nutritious – an objective statement; and *I like* it – a subjective sentence. Actually only a part of me likes it, namely my brain. *It* – one more thing, of course – is what makes *me* tick. Without *it* there would be no *me*. My brain is a material thing, though a living one, not just a physical one. And its mind, I mean mine, is a subset of my brain's functions, just as my smiles are contractions of my facial muscles – though not automatic but willed by my prefrontal cortex. No organ, no function. In short, there are material things, such as brains, as well as processes in them, such as thoughts and feelings. In other words, there are *its*, or material things, and there are *us*, our selves.

This is not a case of duality or bifurcation of reality, but of *distinction* between things, such as brains, and certain processes in them, such as thoughts. So, there it is: I am an unabashed monist. I belong in the same club as Democritus, not Plato's, and I feel impish glee at the great Aristotle's vacillation on this point, the intellectual source of all religions and philosophies.

I am a materialist but not a physicalist because, as a physicist, I learned that physics can explain neither life nor mind nor society. Physics cannot even explain phenomena (appearances), because these occur in brains, which are supraphysical things; nor can it fully explain machines, as these embody ideas, such as those of value, goal, and safety, that are nonphysical. Physics can only account for matter at the lowest level of organization, the only one that existed before the emergence of the earliest organisms some 3,500 million years ago. Hence physicalism, the earliest and simplest version of materialism, cannot cope with chemical reactions, metabolism, color, mentality, sociality, or artifact.

Our contemporary concept of matter is not Democritus', nor even Newton's, which is the one still held by most philosophers, and a reason that most people find it hard to believe that matter can think. They are right: A bunch of marbles cannot think. But brains are made of living tissue, which has peculiar properties that physical matter lack; and its constituent atoms are far more subtle and complex than the tiny marbles imagined by the ancient atomists. Whence modern materialism is not to be confused with physicalism, let alone mechanism, for it is inclusive rather than eliminative. And yet those confusions are rampant in the philosophical literature.

The orthodox mind/body duality is reflected in the chasm that separates the philosophy of mind from the philosophy of matter. Under the influence of Wittgenstein and Chomsky, the current philosophy of mind is often seen as an offshoot of the philosophy of language, and thus as utterly alien to the philosophy of matter. A consequence of this bifurcation is that a number of important philosophical problems are not being tackled in a fruitful way. For example, how can we ascertain whether the universe is mental, or whether the mental is physical, as some philosophers believe, in the absence of precise and up to date concepts of matter and mind?

Presumably, the world was viewed as one until religion split it into two: nature and the supernatural, the material and the spiritual, the lower and the higher, the accessible and the unattainable, the familiar and the mysterious. With the exception of the atomists, the ancient Greeks kept their gods tucked safely away in mount Olympus – save for Zeus' occasional escapades. And several Indian classical philosophies were naturalist as well as rationalist. This holds in particular for the important Samkhya school, which secularized the matter/spirit duality. This school anticipated by fourteen centuries Descartes's similar distinction between the *res extensa* or matter and the *res cogitans* or mind. However, neither of the two treatises that he refrained from publishing made much of that split. In particular, Descartes – following Galileo – suggested purely physical explanations of the phenomenal or secondary qualities; and eventually he located the soul in the pineal gland.

In the next century Berkeley attempted to eliminate the concept of matter altogether by reconceptualizing everything in terms of perceptions: he was the first modern phenomenalist. Hume and Kant followed suit, not realizing that phenomenalism is an aspect of anthropocentrism, the most primitive of worldviews. Holbach and other members of the French Enlightenment restored the unity of reality on the basis of materialism and scientism. In the following century the German idealists, in particular Hegel, rejected the Enlightenment and wrote cryptically about the *Geist/Natur* rift, though avoiding the subjectivism of Kant, Fichte, and Schelling. Marx and Engels rejected idealism but kept the dialectical mumbo-jumbo of that "mighty thinker."

Two generations later, the great scientist Ernst Mach unwittingly revived Berkeley, and decreed that the ultimate constituents of world are sensations. Still later, Alfred North Whitehead complained about the "bifurcation of nature" into the physical and the phenomenal, and attempted to restore unity by reinstating the subjectivism of Berkeley and Kant. At the same time his friend and coworker, Bertrand Russell, tried to overcome dualism by reviving Spencer's neutral monism, and by conceiving of a physical object as a possibility of sensations – just like John Stuart Mill. Rudolf Carnap reheated the old dish, seasoning it with the new logic. Closer to us, Karl Popper reinvented Descartes' mind-body dualism as well as the objective spirit described by Hegel and later by Dilthey. At the same time, many cognitive scientists, inspired by the computer revolution, claimed that humans are composites of hardware and software. There is nothing like forgetting the past to spew apparently new philosophies.

Throughout that long-drawn discussion between monists and dualists, most philosophers kept prescientific notions of matter and mind. For example, John Locke was not even aware of the revolution started by his contemporary Isaac Newton; David Hume did take notice of it, only to rejected it emphatically for going far beyond phenomena; and neither Locke nor Hume paid attention to medical psychology, which had kept alive the Hippocratic principle that mental processes and disorders happen in the brain. Even Bertrand Russell, possibly the most learned man of his time, repeated the myth that the peculiarity of matter is impenetrability – and this, two centuries after the birth of gas physics and chemistry, and long after his compatriots Michael Faraday and James Clerk Maxwell had fathered field physics. In our own time some famous philosophers, such as Bas van Fraassen, believe that quantum mechanics calculates phenomena (appearances) rather than physical properties inaccessible to the senses, such as atomic energy levels and scattering cross sections.

In 1888 Henri Bergson, despite his intuitionist stance, discussed the "immediate data of consciousness" in the light of experimental work by his famous contemporaries, Wilhelm Wundt, Wlliam James, and Pierre Janet. By contrast, most contemporary philosophers of mind are indifferent to psychology, or are remarkably misinformed about it, For example, when discussing the psychoneural identity hypothesis, they are likely to hold that, according to it, pain is identical to the firing of *C*-fibers – while as a matter of the fact nerve fibers do not fire, and those particular fibers cannot feel anything, because they only conduct signals that activate the dull-pain centers in the brain – which, incidentally, need no external stimuli to annoy us. This is no nitpicking, but a matter of medical interest because untreated pain may stay in the brain long after the injury that caused it: the memory of pain, just like the memory of a lost limb, can remain engraved for life in nervous tissue. So, a bad philosophy can give you chronic pain.

Nor are philosophers the only offenders. Even some of the founders of atomic and nuclear physics, notably Bohr, Heisenberg, Born, and Pauli, saw the quantum theory through the phenomenalism of Berkeley, Hume, Kant, Comte, and Mach. This metaphysics denies the existence of things-in-themselves – such as the atoms in the innards of the sun, which are beyond the reach of experiment – or at least the possibility of knowing them. Worse, some eminent physicists have claimed that the universe is created by observation. For example, they tell us that the observer has power of life and death over Schrödinger's unlucky cat, which, before the final observation, would exist half alive and half dead. Actually, this does not follow from the quantum theory: When analyzed, this theory is seen not to refer at all to organisms, let alone to observers and their mental processes.

The so-called observer effect invoked with reference to the quantum theory presupposes the ancient Egyptian view of vision, according to which light emanates from the observer's eye. Alhazen showed one millennium ago that that the arrow in the vision process points in the opposite direction. Stars are born from gases, not gazes. Of course, things are very different in experiment, an intervention that alters the object of study. But what causes such changes is the hand of the experimenter or his servomechanism, not his unaided mind. And presumably both the initial mental process (experimental design) and the final one (reading the dials) are brain processes. So, arguably all the links in the causal chain in question are material, though only the middle ones are physical. If thought could move matter without hands or neural prostheses, there would be no conservation of energy.

As for the other side of the it/us distinction, some of the most outstanding brain scientists of the last century, such as Charles Sherrington, Wilder Penfield, John Eccles, and Roger W. Sperry, clung to the old mind-body dualism while helping midwife cognitive, affective and social neuroscience. This young multidisciplinary field assumes and confirms the psychoneural identity hypothesis, that mental processes are brain processes, and it is currently blossoming, while brainless psychology is withering.

Thus, the old matter/mind duality is still firmly entrenched, not only in ordinary knowledge and in philosophy, but also in brainless (in particular computational) psychology. This book is one more attempt to reunite matter and mind, this time with the help of contemporary science, in particular quantum physics and cognitive neuroscience. The former gives us an unfamiliar but realist (observer-independent) concept of matter, and the new psychology overcomes the dualism inherent in brainless cognitive science. I suggest that we take seriously the two concepts in question, and that we look to science for answers to those old metaphysical questions.

We do not have to take physicists at their word when they make philosophical forays: we can avail ourselves of certain formal tools to find out what their theories are really about. For example, around 1930 Bohr and Heisenberg claimed that the task of physics is not to find out what nature is, but what we can say about nature. Any realist philosopher could have told them that the people who analyze what physicists say are philosophers and historians of science, whereas physicists like themselves study atoms, stars, and things, as shown by their formulas and experiments.

As for the mind, we have the advantage of living in the post-behaviorist and post-psychoanalytic period, when the most exciting psychological novelties are coming from studying mental processes in the living brain rather than in old books. Indeed, cognitive, affective and social neuroscience is gradually answering the old philosophical question "What is mind?" It has begun to explain how we feel, learn, and think, as well why caffeine keeps us awake, why we can get addicted to it, and why some persons suffer incapacitating mental disorders.

The philosophy of mind is one of the most active branches of contemporary philosophy. But few of its practitioners bother to keep up to date with the science of mind, in particular its cutting edge, cognitive neuroscience. This explains the large number of extravagant opinions, such as that it is dangerous to think that the mind is related to the brain (Wittgenstein); that the mind can guide the brain (Popper); that the mind is just a set of computer programs (Putnam); that the mind will forever remain unknowable (McGinn); or that we ignore what the physical is, whence it is wrong to try to reduce experience to the nonexperiential (Stoljar). In this field, like in the Romantic *Naturphilosophie* of two centuries ago, anything goes, particularly if it is scientifically outdated.

I submit that the arbitrariness that rules the contemporary philosophy of mind is largely due to its isolation from both the science of mind and the rest of philosophy, in particular ontology. I also submit that the right way to approach the problems of mind and matter is to make intensive use of the sciences of matter and mind, and to place those problems in an ontological system that embraces all the major categories that occur in the study of all the levels of reality, from the microphysical to the macro-social ones.

The reason for adopting this strategy is that all the Big Questions come in bundles, not one at a time – which is why they are big. They call for clear ideas about a number of inter-related categories, such as those of being and becoming, causation and chance, mind and society, meaning and truth, hypothesis and experiment. For example, to properly evaluate the ingenious if wild conjectures of the popular evolutionary psychologists, we need to know something about the ways the mind relates to brain and society, and how scientific hypotheses are evaluated.

In general, all the so-called Big Questions call for comprehensive and systematic philosophies, rather than a few clever aphorisms and thought experiments, such as imagining how people would behave in a dry twin of our planet. Particularity, fragmentation and unchecked fantasy are marks of philosophical improvisation. But of course breadth and system are not enough: We also want rigor, depth, and the promise of truth in tackling significant problems. In other words, we want to use the best extant knowledge to help solve important problems by placing them in a broad context and in relation to other knowledge items, even other disciplines if necessary, and handling them rigorously and in depth.

I believe that a philosophy is spineless without ontology, confused without semantics, acephalous without epistemology, deaf without ethics, paralytic without social philosophy, and obsolete without scientific support – and no philosophy at all with neither. All those branches of philosophy are treated in the nine tomes of my *Treatise* (1974–1989). The present book has a far narrower scope: it focuses on the modern conceptions of matter and mind. Incidentally, its comprehension does not require any specialized knowledge. Only the two appendices make use of some formal tools. My *Dictionary of Philosophy* (2003) may help elucidate some philosophical terms.

I thank Martin Mahner for pointing out several mistakes, as well as for attempting to restrain my bashings. I am also grateful to Graham Bell, Carlos F. Bunge, Silvia A. Bunge, Carmen Dragonetti, Bernard Dubrovsky, Alberto Galindo Tixeira, Antonio Lazcano, Jean-Marc Lévy-Leblond, Michael Matthews, Ignacio Morgado-Bernal, Andreas Pickel, Jorge Quintanilla, Arturo Samgalli, Dan A. Seni, Endel Tulving, Nicolás Unsain, Horacio Vucetich, as well as my late friends Dalbir Bindra, David Bohm, Andrés Kálnay, Raymond Klibansky, and Bruce G. Trigger. I owe all of them data, questions, suggestions, criticisms, or encouragement.

Montreal, QC, Canada Mario Bunge

Contents

Introduction

Most people believe, with Descartes, that the world is composed of entities of two radically kinds: material and spiritual – or bodies and souls. Materialists like Hippocrates, Democritus, Spinoza, Holbach, Diderot, and Engels, and idealists like Plato, Leibniz, Kant, Hegel, Bolzano, and Russell, have criticized this dualist worldview, holding instead that basically there is a single kind of substance or stuff. That is, they have advocated monism – materialist, idealist, or neutral – and have consequently rejected dualism. But of course dualism – the view that, whereas some things are material, others are spiritual – has always been the most popular metaphysics. By contrast, neutral monism – the view that the material and the ideal are only manifestations of an unknown neutral substance – is the least popular.

However, nowadays philosophy students are seldom told anything about materialism, whereas they are burdened with minutiae about Plato, Kant, Hegel, Dilthey, Husserl, and other idealists. Imagine faculties of science, engineering, or medicine, teaching that the world is composed of spiritual rather than material entities. Nothing would be discovered, and only the librarians and university bean counters would be happy, for laboratories, observatories, workshops, and experimental stations would become redundant.

This book argues that modern natural science, from physics to cognitive neuroscience, has tacitly adopted the materialist view that the universe is composed exclusively of concrete things, and that the social and biosocial sciences would benefit from following suit. This is not to deny that there are mental processes. Materialists claim only that there are no disembodied minds. They may add that there were no minds before the mammals and the birds arose. Further, materialists claim that qualia, feelings, consciousness, and even free will are real and within the reach of scientific research. And that laboratory results would be unreliable if it were assumed that freestanding spirits could interfere with measuring instruments.

This said, it must be admitted that materialism is still seriously underdeveloped, to the point that there is no generally accepted concept of matter, in contradistinction to the special concepts of materials handled by physicists, chemists, biologists, and engineers. Much the same holds for the concept of mind: We do not yet have a generally accepted materialist theory of mind. And the status of abstract objects, such as those invented by mathematicians, is even more precarious, for they seem to be neither material nor mental. This is why some of the main tasks of this book are

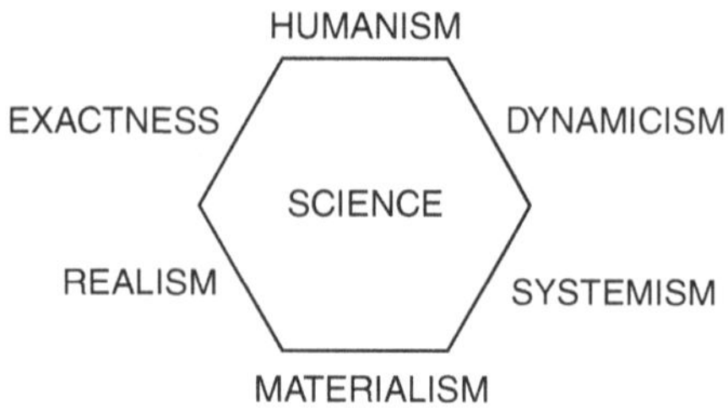

Fig. 1 Sketch of the author's philosophical system. Taken from Bunge (2009)

to elucidate the general concepts of matter and mind in the light of contemporary science. (The nature and status of mathematical objects and other transcendentals are discussed in Bunge 2006a.)

This feature suffices to locate this work outside mainstream contemporary metaphysics, which revolves around the concepts of possible world and counterfactual, with utter unconcern for the real world, and a fortiori for what could be done to improve it. I submit that, on the other hand, the main ideas in this work are congruent with the *matérialisme instruit* tacitly inherent in contemporary science (Changeux 2004, 8). But of course turning tacit and therefore inchoate ideas into explicit and articulate ones is part of the philosopher's job description.

The ideas expounded in this book are not stray, but form part of a comprehensive philosophical system whose components are shown in the following Fig. 1.

Part I
Matter

Chapter 1
Philosophy as Worldview

A worldview is a comprehensive conception of all there is, whereas a philosophy is a scholarly discipline divided into special fields, every one of which is usually cultivated independently of the others. For example, the typical philosopher of mind won't be interested in the philosophy of matter. As a consequence, he may find it hard to believe that matter can think. Or else he may be so radical a naturalist that he may believe that brains secrete cultures. I wish to restore the traditional unity of philosophy conceived of as an elaborate worldview or, if preferred, as a theory of everything. Such a unitary or integrated conception of philosophy should help place every philosophical problem in a network of knowledge items, instead of tackling it as an isolated puzzle.

Philosophy matters in all sectors of intellectual culture because a philosophical opinion may either encourage or obstruct the exploration of reality. For example, Joseph Needham, the great expert in Chinese culture, asked the question "Why was modern science not born in the most advanced civilization of its time, namely China?" His answer was roughly this: Because traditional Chinese intellectual culture was dominated by Buddhism, Taoism, and Confucianism. The Buddha had taught that all is appearance and illusion; Lao-tzu, that contemplation trumps action; and Confucius, that what matter are only peaceful coexistence and obedience to tradition. None of the three masters challenged people to go forth and explore the unknown, let alone alter the known.

There is no escaping philosophy. Yet, ironically, philosophy is the self-doubting discipline, the one that has no well-defined subject, and that is conceived differently by different scholars. In my view, any philosophy worth its name is an explicit and well-organized worldview rather than a collection of spotty opinions on this and that. I expect philosophers to tell us something interesting about the world, as well as about our knowledge of it or our place in it. That is, a philosophy proper is organized around an ontology or metaphysics: a theory of change and invariance, space and time, cause and chance, body and mind, person and society, and so on. I submit that a philosophy without ontology is spineless, just as it is confused without logic and semantics, headless without epistemology, and limbless without a practical philosophy. And a body of ideas lacking all five, as is the case with Wittgenstein's aphorisms and Heidegger's cryptic dicta, hardly qualifies as a philosophy. See Fig. 1.1.

M. Bunge, *Matter and Mind*, Boston Studies in the Philosophy of Science 287,
DOI 10.1007/978-90-481-9225-0_1,

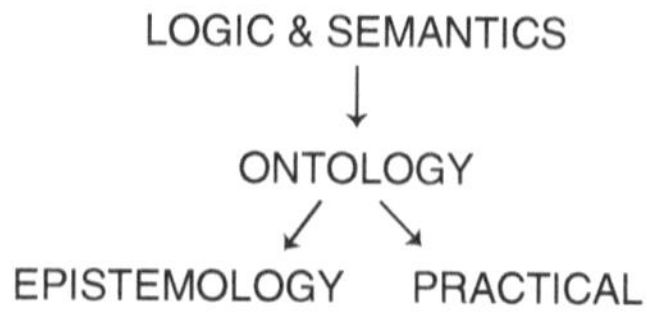

Fig. 1.1 Every philosophy worth its name is centered in an ontology

An ontology, cosmology, or worldview is much more than an information package: it is also bound to either inspire or inhibit research projects, whether timid or audacious. Just think of any of the following influential cosmological ideas, some purely speculative but none of them foolish: that the universe is infinite, or else finite, in both space and in time; that it is mostly void or totally full; that it is forever changing or unchanging, lawful or contingent, material or ideal; that reality is either a collection of individuals or a system; that the first organisms emerged from prebiotic materials by self-organization; that all biospecies are branches of a single tree of life; that the genome is either destiny or possibility; that mental processes are brain processes, or else changes in an immaterial soul; that human action is driven by passion or by interests, by environmental stimuli or by rational calculation; that humans are slaves or free; that we are altruistic, selfish, or half and half; that to live in society is to compete, cooperate, or do either in different circumstances; that social hierarchies are natural or artificial; that equality is a mirage or a doable ideal; that there are universal truths and moral norms; that morality trumps the law of the land or conversely; that art, philosophy, science, and mathematics are the highest spiritual activities; and that technology can be good or bad. People have argued passionately or even killed for some of these grand ideas, particularly those that have been included in religious or sociopolitical ideologies. And yet most contemporary philosophers manage to ignore those big questions, for they prefer to work on flowerpots rather than in open fields.

1.1 World and Worldview

It is hard to navigate across one's environment without having some ideas, however coarse, about it. Indeed, to face any situation we must know whether it is real or imaginary, profane or sacred, sensitive or insensitive to our actions, and so on. This is why even lowly organisms develop, if not worldviews, at least rough sensory maps of their immediate environment – as noted by ethologists from the start (e.g., von Uexküll 1921). But it is generally assumed that only humans can build conceptual models of their environments. And, except for some philosophers, humans distinguish maps from the territories they represent.

The notion of representation belongs in semantics (see, e.g., Bunge 1974a). It may be elucidated as follows. A set Σ of ideas in animals of kind K is said to *represent* a set Ω of objects for the Ks if the presentation, recall, or imagination of some members of Ω evokes a sensation, percept, or concept in some members of *K*. Note that different animal species are tacitly assumed to construct different representations, if any, of one and the same collection of objects.

With regard to the stuff of represented (Ω) and representing (Σ), we may distinguish three basic kinds: factual (things and processes), conceptual (concepts, propositions, classifications and theories), and semiotic (signs, figures, and sounds). See the following table.

Ω *(represented)*	Σ *(representing)*	*Example*
Factual	Factual	Scale model
Factual	Conceptual	Scientific theory
Factual	Semiotic	Scientific text
Conceptual	Factual	Computer operation
Conceptual	Conceptual	Numbers and points
Conceptual	Semiotic	Mathematical text
Semiotic	Factual	Text illustration
Semiotic	Conceptual	Text analysis
Semiotic	Semiotic	Translation

And with regard to form, in principle there are three possible representation relations: one-to-one, as in sound-musical note; one-to-many, as in the different cosmologies; and many-to-one, as in any general concept of man. One-to-one representations, the most faithful of all, are very hard to come by. For example, the point-number correspondence is one-to-one, but the numeral-number is not, because the overwhelming majority of real numbers are nameless, since whereas the former constitute a continuum, any list of names is countable. The word-fact, or concept-world correspondence is even more problematic, even though it underlies the correspondence theory of truth. In particular, the view that this correspondence is isomorphic, is untenable, because the isomorphism relation is defined only for sets, and the world happens not to be a set. In fact, the theory-world correspondence is roundabout (See Chapter 15).

The preceding presupposes a realist epistemology. Idealists have no use for the very concept of representation, as they hold that reality is constituted by ideas (objective idealism) or by signs (hermeneutics). In particular, the Pythagoreans thought that the world is made of numbers; philosophical hermeneutics asserts that facts are texts; and the contemporary "digital metaphysicians" claim that "its" (material entities) are made of bits (information units). An obvious objection to the Pythagoreans is that numbers have no physical properties, starting with energy. The its-from-bits crowd is wrong because bits are devoid of energy, and because the concept of information makes sense only with references to information systems, which are very special artifacts. Finally, hermeneutics too is false because the real world has neither grammar, nor phonology, nor style. Nature calls for a naturalistic approach, just as society must be described in terms of both social and biological categories, and ideas in conceptual terms as well as in neuroscientific terms.

1.2 Monism and Pluralism

The worldviews or ontologies can be classed into monistic and pluralistic, according as they postulate only one or more kinds of thing. There are three monism genera: materialist, idealist, and neutral. Each of these genera splits into two or more species. For example, materialism can be physicalist (basement-level) or emergentist (many-level). Idealism can be rationalist (everything is ideal), empiricist (in particular phenomenalist), semiotic (everything is linguistic), or informationist (every "it" is made from bits). And there are at least two versions of neutral monism: energetism (everything is energetic) and agnostic. According to the latter, the neutral substance is unknowable, but we can know its two manifestations, matter and mind. This is why this doctrine, held by Spencer and at one time by Russell, has also been called "double-aspect theory".

Materialism has been said to be the spontaneous ontology of scientists. This holds for experimental scientists. But a number of outstanding theoretical physicists have reinvented Plato's idealism and the phenomenalism of Buddha, Ptolemy, Hume, Kant, Comte, and Mach. And vitalism, though badly maimed in the nineteenth century by physiology and evolutionary biology, only died ca. 1920. Besides, idealism is still going strong in brainless psychology and in the outskirts of social science – interpretive (or hermeneutic) anthropology and phenomenological sociology.

Idealism and dualism have sometimes been regarded as secular versions of religion. This is particularly clear in Augustine's idealism and in the Aristotelianism of Thomas Aquinas and the neo-Thomists. But modern idealism, from Kant onwards, has been secular, and often atheistic. Actually idealism has ceased to fulfill a social function, except as an academic bulwark against Marxism and – in the case of the Neo-Kantian *Verstehen* school – an obstacle to social science.

Most primitive worldviews are dualistic, that is, they assume that reality is split into two non-intersecting strata: the worldly or profane, and the supernatural or sacred. In addition, the worldly is usually divided into two levels: matter and mind, or body and soul. In particular, animism postulates that everything, whether stone or star, plant or animal, is "animated," that is, inhabited by an immaterial spirit or soul. The spirit or soul is in principle detachable from its material carrier: it may be transferred to someone or something else, or it may survive the death of the body. Thus, the polite Brazilian for "dying" is "disincarnating."

Pantheism is nothing but monotheistic animism: it consolidates all the celestial and infernal powers into one, and postulates the identity of nature and God. The best-known modern pantheist was of course Spinoza. Another case was Ernst Haeckel, the great embriologist and Darwin's earliest German follower, as well as an influential popular science writer. Like Haeckel, Paul Carus held that all things are both material and spiritual, but their spirit was totally secular, and it amounted to the ability to know. And Einstein's Spinozism was only a ruse to defuse accusations of atheism: He emphatically rejected belief in a personal deity, and disliked all organized religions. No religionist was ever fooled by the naturalistic, secular, and timid deity of the pantheists: these were denounced as closet atheists. If

God is everywhere, and we hold Him in the hand every time we grasp something, then He is only it.

The main reason for the nature-supernature division may have been the wish to guarantee the immortality of the soul. And one of the reasons for upholding the Cartesian mind-body dualism was said to be that, whereas matter can only be studied from the outside, we have direct access to our own souls: we can introspect. Moreover, according to Augustine and Husserl, only introspection can yield deep knowledge of the world.

A few philosophers have needed more than two substances. For example, in his *Philosophical Notebooks*, Lenin (1981, 36: 182), infected by Hegel's enthusiasm for triads, wrote in 1914 that "there are *actually*, objectively, **three** members [sic]: (1) nature; (2) human cognition = the human brain (as the highest product of this same nature), and (3) the form of reflection of nature in human cognition, and this form consists precisely of concepts, laws, categories, etc." Note the absence of a fourth "member" between the first and second terms, namely, society. Presumably, Lenin included it in nature, thus unwittingly replacing his professed materialism with naturalism, maybe because the number 4 is unappealing to an admirer of Hegel's ubiquitous triads.

Little could Lenin suspect that half a century later a notorious anti-Leninist philosopher would adopt the very same three-worlds doctrine. In fact, Karl Popper (1967) divided reality into three "worlds": world 1 (physical), world 2 (mental), and world 3 (Hegel's objective spirit). He argued "for the (more or less) *independent existence of world 3*" (Popper's emphasis), as well as for the interaction among all three "worlds." Incidentally, these were not *worlds* proper (i.e., systems) but ragbags of heterogeneous items.

This triadic worldview is a piece of commonsense metaphysics. Indeed, it has always been common and convenient to distinguish between, say, brains, mental processes, and their "products," such as poems or theorems taken in themselves, that is, disregarding their origin. But such a distinction is not fundamentally different from that between a lump of clay, the potter's work, and the resulting vase; or between an organism, its metabolism, and the latter's consequences.

All such triads have a place in a materialist ontology provided their constituents are distinguished but not separated. For example, the brain is a material entity, ideations are brain processes, and pretending that the end product of any such process can be detached (in thought!) from its genesis is another mental process. But distinction does not necessitate separation. In general, the distinctions involved in description and analysis need have no ontological counterpart. For example, it is legitimate to distinguish bodies from their motions because, in fact, they belong to different categories, if only because different bodies can move in roughly the same ways. But this does not entail that movement can be detached from bodies or other material things.

Let us finally deal briefly with neutral monism. This view was upheld by Herbert Spencer, William James, Bertrand Russell, and energetists such as Wilhelm Ostwald. Some neutral monists proposed their doctrine as a compromise between materialism and idealism, and others as a bridge between science and religion.

Obviously, it found no takers. Compromise, so often desirable in technology and politics, is as disastrous in philosophy as in science and mathematics: here it is an indicator of shallowness.

The methodological partners of the various metaphysical principles are obvious. Materialists are likely to hold that the universe can only be known through the study of concrete things; idealists, that to know anything is to engage in a purely mental (rational or intuitive) exploration; and neutral monists, that the neutral substance is unknowable, even though its manifestations – matter and mind – can be known.

1.3 Metaphysics: Commonsensical, Speculative, and Scientific

Metaphysics, or ontology, is the study of the most basic and general problems about the universe and the mind. (This is the classical view. We shall disregard Quine's confusion between ontology and reference class or denotation.) There are four main attitudes to ontology: to deny its legitimacy (radical skepticism, positivism, and Wittgenstein); to build on folk physics (like Strawson) or folk psychology (like most philosophers of mind); to construct fanciful theories (like Leibniz's monadology, Hegel's philosophy of nature, or Whitehead's process metaphysics); and to work on Peirce's scientific philosophy project, which views ontology as general science. We shall adopt the latter stance (Bunge 1977a, 1979a, 1981).

Let us start by recalling some conventions ruling the use of a handful of key ontological categories. We shall do so by way of a familiar fact. Consider a pebble striking a pond and causing a ripple in it. The pebble and the pond are *things* sharing some *properties* (such as mass) while differing in others (such as composition and density). The impact of the pebble on the body of water is an *event* or, rather, a *process*, since it takes some time. (An event is an instantaneous change.) A process occurring in a thing may be conceptualized as either a string of events in the thing or, better, a sequence of *states* of the thing. And each state may be conceived of as a list of values of the relevant properties. A ripple in the pond is a process, one characterized by amplitude and frequency. (However, strictly speaking these are properties of the particles making up the liquid body.) Finally, impact and ensuing ripple are respectively *cause* and *effect*. Thus causation, or the causal nexus, relates events or processes, not things, properties, or states. (Hence to say, e.g., that a brain process causes a mental state, as Searle (2007, 40) does, is to violate a standard linguistic rule and to mess up the ontology.)

Thus far we have encountered six ontological categories: those of thing, property, state, event, process, and causation. Another two are definable as follows. *Change* can be defined as either event or process – in some concrete thing, of course. And a thing having a certain property, or being in a given state, or undergoing a certain change, is called a *fact*. Of course another two categories of the same kind have lurked in the background in the preceding: those of space and time. Note the logical order of the ontological categories mentioned in the above:

Thing < Property < State < Process < Causation.

This ordering is absolute, not contextual. That is, it would be wrong to treat an entity as a thing in one context and as a property, state, or process in another. There is a logical reason for this, namely that every one of the categories beyond the first one is defined in terms of a previous concept. True, usage is not consistent in philosophy or even in science. Thus, in solid state physics phonons are often treated as particles, although actually they are sound waves in solids – hence processes rather than things. True, phonons are similar to photons, in that their energy is quantized; moreover, like particles, they are scattered by incoming photons. Still, to state that phonons scatter light *as if* they were particles is not the same as affirming that they *are* particles. And the said analogy is only partial, since there are no free (solid-independent) phonons. Only fictionists and other truth-deniers deny the difference between fiction and reality: the rest of us are realists.

The preceding suffices to reject, as logically flawed, the definitions of a thing as either a bundle of properties (Russell 1914) or as a set of processes (Whitehead 1929). Indeed, every property is a trait of an entity; only entities (or, if preferred, properties thereof) are changeable; and all processes happen in things. Therefore it is mistaken to state that things are snapshots of processes (Lewontin and Levins 2007), for this is what states (of things) are. Likewise, it is wrong to assert that the world is the set of all facts (Wittgenstein 1922), or that it is composed of states (Armstrong 1997). This is so because, by definition, the composition of any concrete thing is a set of things; and also because there are neither facts nor states separate from things. (More in Chapter 14.)

I submit that in all of the factual sciences the real world is conceived of as composed of concrete things, from elementary particles and photons to persons and social systems. (Tadeusz Kotarbinski called this thesis *reism*.) The concept of a thing is so pervasive and overwhelming, that we often indulge in reification, that is, we call non-things, such as processes, concepts, and words, "things". At other times, particularly in the philosophical literature, the concept of a system, or complex thing, is missing, perhaps because it is only half a millennium old. Indeed, it emerged in the early days of modernity, when Copernicus conceived of the planets as members of the solar system, and Harvey hypothesized that the heart is the key component of the cardiovascular system.

Nowadays all the sciences and technologies deal with systems of many different kinds, from atoms to crystals to cells to multicellular organisms to machines to social systems , on top of conceptual systems such as vector spaces, and semiotic systems such as languages. The concept of a system has become so pervasive, that reality, or the universe, is increasingly being viewed as the system of all systems. This is the central postulate of *systemism* (Bunge 1979a). In turn, systemism implies *emergentism,*the thesis that every system has global or emergent properties that its components lack.

Think, e.g., of a water droplet and its component H_2O molecules; or consider a family and its members. A droplet has traits, such as surface tension and temperature, that its molecular constituents lack; likewise a family has properties, such as number of members and harmony (or its dual), that it members lack. The methodological counterpart of emergentism is the rule that recommends studying systems at two levels: the system's or macrolevel, and its constituents' or microlevel.

An inadequate study may miss a system, or mistake a mere assemblage for a system. For instance, at first the nervous, endocrine and immune systems were regarded as separate, whereas actually they constitute a supersystem; by contrast, the so-called limbic system is now known to be an assemblage of organs, only some of which are intimately connected.

Thus far we have taken the concept of a property as obvious, which it is not. Plato and is followers thought that "forms" (properties) precede "substances" (individuals, objects, things). Hence the familiar locution "thing *c* instantiates [exemplifies] property *P*". Aristotle corrected his erstwhile teacher on this crucial point, arguing that every property is a trait of some thing or other, that every thing has several properties, and that some of them are bound to change.

Science and technology followed Aristotle on this point: one takes it for granted that we are given things endowed with their properties. These may be intrinsic such as numerosity and composition, relational such as loving and betweeness, changeable like location and age, or invariant like laws. No property can be detached from its bearer. And many properties can be conceptualized in different ways. This forces us to distinguish properties from predicates: the former are ontological, whereas the latter are conceptual.

The preceding suggests that ontology should not be improvised. It also suggests that it is hardly possible to venture generalizations about the real world without using some ontological categories, such as those of system and emergent property: far from being disjoined from science, ontology belongs in the very marrow of science. And yet most scientists and technologists are likely to deny that all the accounts of facts in the real world are ontology laden. Even some philosophers have believed it possible and desirable to jettison the "metaphysical baggage" – e.g., by limiting science to "saving the phenomena" or appearances, as Ptolemy, Hume, Kant, Comte, and Mach had demanded, and as van Fraassen (1980) has recently reiterated. The only way to skirt metaphysics is to stick to particulars: all generalizations about facts, such as "All things change," and "No two concrete things are strictly identical," involve ontological categories. Hence keeping one's ontology tidy and up to date pays far more than damning metaphysics or allowing it to be mauled by wordsmiths.

However, let us face it: Metaphysics has earned a bad reputation among scientists and, until very recently, also among philosophers. The reason is of course that most of it has relied on obsolete knowledge or, worse, has consisted in either wild speculation (on, e.g., Saul Kripke's and David Lewis's possible worlds), or even in word-play (as in Heidegger's case). The popular "Twin Earth thought-experiment" devised by Hilary Putnam in 1973 is a case in point. Imagine a planet just like ours, and peopled by our identical twins. There is one difference, though: Twin Earth has no water: it is dry in one version of the game, and in another water is replaced with a liquid with an utterly different chemical composition. Stop right here: Philosophers are expected to know that life is impossible without water, which has unique properties – liquid in the "right" temperature interval, near-universal solvent, molecules joined by hydrogen bonds, very low electrical conductivity, etc. The joint possession of these properties makes water unique and indispensable to life, as Lawrence Henderson (1913) noted nearly one century ago. Obviously, a planet without water would be devoid of organisms, hence it would not be Earth's twin, whence it would merit a different name.

Philosophers are also expected to know that properties come in bundles – they are inter-related by laws – so that they cannot be replaced or removed arbitrarily. But of course the very notion of a law, which serves to distinguish real from merely imaginary possibility, is utterly alien to professors who play parlor games instead of tackling serious problems. We shall return to this theme in Section 11.5, when discussing the unconscious zombies imagined by Saul Kripke to refute psychoneural monism.

Charles S. Peirce (1935), in my opinion the deepest and most original and versatile of American philosophers, thought it possible to construct scientific metaphysics, that is, metaphysical theories using the vast storehouse of mathematics and factual science. I concur, and add that doing scientific metaphysics should be far more interesting than fantasizing about physically impossible worlds. Besides, whereas speculative metaphysics is groundless (or unjustified), the hypotheses of scientific metaphysics can be checked by their compatibility with current scientific knowledge. For example, the relational (or adjectival) theories of space and time are consistent with the theories of relativity, whereas the theories of absolute space and time are not. Again, whereas the philosophies of mind involving the psycho-neural identity hypothesis are consistent with cognitive neuroscience, the dualistic theories are not (see Bunge 1977a, 1979a).

Furthermore, I submit that every time scientists discover something, and every time technologists design a workable artifact, they confirm materialism. By contrast, no mountain of scientific and technological achievements is likely to satisfy the theologian, the pseudoscientist, the idealist philosopher, or even the scientist in love with the bizarre, or who wishes to *épater le bourgeois*. For example, the late John A. Wheeler, one of the most imaginative contemporary physicists, once claimed that the universe is composed of propositions. More recently, he and his followers asserted that "its are made of bits" (See Barrow et al. 2004). This idea, the core of the so-called digital metaphysics, originates in the fact that the laws of physics are expressible in terms of computer programs. This proposition is true, but it does not entail that pieces of matter are conglomerates of bits. To claim that they are is to conflate real things with our models of them – a piece of magical thinking. The "its from bits" fantasy is easily falsified by recalling that information clusters, such as sentences and texts, do not have physical properties, whereas pieces of matter do have them, even if they happen to encode or carry information. In short, there is no information without matter, whereas most material things encode no information. We shall return to this in Section 4.2.

Consider now the following items:

It is snowing. [1]
I am looking at snow falling. [2]
The proposition (or the sentence) "It is snowing." [3]

Few will dispute that these items belong in different categories: [1] is physical, [2] mental, and [3] conceptual (or cultural). The question is whether each of the above items belongs to a world of its own and in its own right: the physical world, the mental world, and the cultural world (or "world 3," as Popper called it). Common sense would answer in the affirmative. So did Lenin (1981) in the notebooks he kept in Zürich while studying Hegel's big *Logic* in 1914. Popper (1967), his formidable ideological foe, thought exactly the same, and expressed it in terms that looked as if he had cribbed them from Lenin – which of course he did not. All of which suggests once more the falsity of the dictum *Vox populi, vox dei.*

1.4 Determinism and Contingency, Causation and Chance

The twin concepts of determinism and contingency, as well as those of causation and law, are alien to spiritualism but central to naturalism and, a fortiori, to materialism. Indeed, according to spiritualism the soul and the spiritual beings, all the way from humans to the gods, are free, whereas matter, if it exists at all, is lawful. By contrast, modern naturalists and materialists deem every fact to be lawful. This, in a nutshell, is the *lawfulness principle.*

Nearly everyone agrees that there are man-made or conventional laws (or rather norms) on top of the natural laws. But there is no consensus on the meaning of "natural law". Let us therefore elucidate this notion. Realist will readily admit that "natural law" designates two different though intimately related concepts (Bunge 1959b): those of objective pattern, or law, and law statement, or conceptualization of a law. Naturalists and materialists hold laws to inhere in things; that they constitute the basic properties of things; and that one and the same law, or objective pattern, may be conceptualized differently in different theories. For example, the basic law of motion was "Force = constant × Velocity" for Aristotle, and "Force = Mass × Acceleration" for Newton.

The next pertinent distinction is that between causal and probabilistic laws. Again, Newton's second law is the classical exemplar of a law statement. But its corollary, the law of inertia "If *Force* = 0, then *Velocity* = constant," is obviously non-causal. Lavoisier's law of conservation of mass, "The total mass of an isolated chemical reactor is constant" is non-causal as well. So is the basic quantum-mechanical law for a free "particle": it concerns the probability distribution of its position, which is constant again. By contrast, the basic quantum-mechanical law for a "particle" subjected to forces will depend on these, and will thus involve chance and causation (external force) on a par.

Thus, there are causal laws, probabilistic laws, and laws that combine causation with chance. This suggests the convenience of broadening the concept of determinism to include probabilistic (or stochastic) laws. However, naturalists and materialists wish to keep miracles and magic out of science, so they will disown as non-scientific or pseudo-scientific any putative law violating Lucretius' principle *ex nihilo nihil*. This is why I have proposed (Bunge 1959a) redefining determinism as the conjunction of two logically independent principles: Lucretius' and lawfulness.

Nor is lawfulness restricted to facts: it also covers law statements. In fact, theoretical physics includes not only plenty of physical law statements, but also a handful of laws about them, or metalaws, such as Galileo's and Einstein's principles of relativity, and the PCT theorem in quantum electrodynamics (Bunge 1961). We shall not expatiate on this. Suffice it to note that, contrary to what possible-worlds metaphysicians assume, there are no arbitrary natural laws. Indeed, laws come in bunches, particularly in highly complex systems such as brains.

Were does determinism leave contingency? The answer depends on the definition of this ambiguous term. If Contingent = Lawless, then determinism disowns contingency. But if "contingent" is understood the subtle way it is used in developmental and evolutionary biology, then it may be accommodated in determinism. Indeed, biological contingency is just accident, as when a life history is severely deviated by a political catastrophe, or when a natural disaster causes the migration, decimation, or even extinction of a population – which may in turn open up the opportunity for its prey to expand and branch out. These are cases of chance as understood by Chrysippus, namely as intersections of initially parallel causal lines, such as those of continental plate drift and the phylogeny of ostriches, which branched out along with the separation of Africa from South America.

1.5 Epistemology: Skepticism, Subjectivism, Realism

The simplest view about learning is of course skepticism, which is an attitude and a practice rather than a theory. And the simplest attitude towards the theory of knowledge is to declare it impossible. However, note that skepticism comes in two strengths: radical or total, and moderate or partial. The radical skeptics, or Pyrrhonists, deny the possibility of knowing anything. It is doubtful that there are any radical skeptics outside academia, because survival calls for a modicum of self-knowledge as well as for exploring one's surroundings. By contrast, there are plenty of dogmatists, such as religious fanatics, incapable of making converts except by force; and market fundamentalists – so wealthy, that they can afford the free market, warts and all. It is generally admitted that serious humanists, scientists, and technologists, are moderate skeptics, that is, individuals who, while assuming lots of data, hypotheses, and rules, are willing to check them at the first suspicion of inadequacy.

To start any exploration or revision, one must make, if only provisionally, the following three assumptions: the autonomous existence of the external world; the possibility of knowing it at least partially and gradually; and the basic rules of logic. The latter are needed to reason cogently about anything; and the real world and its knowability must be assumed before any exploration of it is planned. For example, Columbus would not have planned his expedition had he and his sponsors not had indications of various kinds, that land would be found navigating westward. And physicists would not have spent years and billions trying to detect certain elusive entities, such as black holes and gravitational waves, if their existence had not been suggested by a theory with amazing track record – Einstein's theory of gravitation.

Note that all three above hypotheses are philosophical, and none of them is supposed to require proof. The hypothesis of the independent existence of reality is the main ontological postulate of realism, and the assumption of its knowabiliy is its epistemological partner. As for the assumption about the need for logic, it is central to rationalism, or the view that rational deliberation is possible and desirable provided some premises and rules of inference are admitted.

To be sure, one may doubt the existence of this or that thing, the possibility of getting to know items of a certain kind, or the reliability of a particular rule or theorem of logic. But such doubt can only be local or restricted, because every doubt must be stated against a background knowledge that is assumed if only for the sake of argument. Thus, one is justified in doubting the existence of "strings" because string theory contains hypotheses that contradict well-confirmed pieces of knowledge – e.g., that ordinary space has three dimensions, not 10. And one is justified in preferring classical to intuitionist logic because abidance by the latter impoverishes mathematics and renders mathematical proofs cumbersome.

How is logic related to ontology and epistemology? Logic is in their service, since it is the science of cogent argument. And, to be a good servant, logic must not owe them anything: it must hold regardless of the nature of things and our knowledge of them. In particular, logic must serve the materialist and the realist just as well as the immaterialist and the irrealist. If it did not, logic would not be a neutral umpire capable of moderating any debates about the nature of the world and theories about it. In particular, it would have no authority to clinch a debate by showing that a given thesis is untenable because it is self-contradictory, or because it leads to a conclusion that contradicts one of the initial premises.

But of course logic is, by construction, ontologically and epistemologically uncommitted: it presupposes nothing about the world or about the strategies to explore it. This is why it is called *formal*. For example, the tautology "*p* or not-*p*" holds for any proposition *p*, whether true or false, in chemistry or in alchemy. In other words, logic may refer to anything but it describes nothing in particular (Bunge 1974b). The contrary thesis, that logic is about the world, or about experience, has been held by a number of philosophers, though without proof. In short, formal logic underlies all cogent worldviews. Only irrationalists reject it.

Finally, how are ontology and epistemology related to science? A priori they can stand in either of three relations: independence, submission, or collaboration. Normally, philosophers do not listen to scientists. Thus, the most famous philosopher of his time, the Aristotelian materialist Cesare Cremonini, Galileo's contemporary and his colleague at Padua university (and the model for his Simplicio) refused to look through the scientist's telescope. Hume thought he refuted Newtonian mechanics – which he could not understand for want of mathematics. Kant, who could not understand it either for the same reason, claimed to have perfected it by adding a repulsive force. Hegel attempted to replace the whole of modern science with his own *Naturphilosophie*. Engels ridiculed some of the most eminent physicists of his time. Bergson, a Nobel laureate, criticized Einstein's concept of time.

And the Soviet philosophers during Stalin's rule denounced relativity, quantum mechanics, genetics, and more as idealist and bourgeois, while swallowing much of Hegel's nonsense.

The very idea of a scientific philosophy suggests that philosophy should always obey science. But once in a while philosophers will be able to correct science or, rather, scientists who go astray. Here is a short list of cases of this kind. Although Louis Pasteur is credited with have proved the impossibility of spontaneous generation, the biochemist Alexandr Oparin doubted this finding for purely philosophical reasons, and undertook to synthesize living beings from abiotic materials – a project that is currently in full swing. A few philosophers have shown that the Copenhagen interpretation of the quantum theory was a philosophical graft unwarranted by the theory's mathematical formalism, which made no room for any observers. Both the creationist and the steady-state cosmological theories have been attacked for violating Lucretius' ex nihilo nihil ontological principle: the former postulated the origin of the universe out of nothing, and the latter postulated the spontaneous creation of matter. Standard economic theory has been criticized for failing empirical tests and for presenting the so-called free market, aka unregulated capitalism, as nearly perfect and unavoidable. And statisticians have been criticized when adopting the Bayesian or subjectivist interpretation of probability theory, according to which probability values are credences, that is, belief strengths. In short, science should be critically examined to check whether it matches ontological principles and methodological rules with a distinguished track record in science itself.

This kind of criticism should be welcomed because it is constructive.

1.6 The Epistemology-Ontology Connection

At first sight, epistemology is logically independent of ontology. This is why realism has been held together with either naturalism or supernaturalism, and materialism has been combined with either realism or antirealism. For example, Kant was both a naturalist and a subjectivist; Karl Popper upheld realism and rejected materialism; by contrast, David Lewis, who called himself a materialist, shared Hume's phenomenalism and worked on many-worlds metaphysics, according to which all the imaginable universes are equally real. And many renowned thinkers, from Hegel to Heisenberg, identified positivism with naturalism or materialism, while actually positivism is just as subjectivist as Kantianism – from which it derived.

But in fact ontology and epistemology have often been strongly bonded to one another. For example ancient atomism, the earliest version of materialism, involved the slogan "Explain phenomena (appearances) in terms of imperceptibles." And subjectivism, from Berkeley, Hume, and Kant to the Copenhagen interpretation of quantum mechanics, has generated ontological phenomenalism. This is the view that, because we rely on perception for factual knowledge, "the world itself is a sum of appearances" (Kant). Accordingly, matter would be just possibility of perception (Mill) or even a collection of sensations (Mach, Carnap, and occasionally

Whitehead and Russell). The methodological partner of ontological phenomenalism is the positivist prescription to stick to appearances. In the case of psychology, this injunction reads: "Cling to overt behavior: do not neurologize."

Berkeley's is of course the most radical subjectivism, for it postulates that to be is to perceive or to be perceived. Today Berkeley's philosophy is found only in the writings of the students of the quantum theory who claim, contrary to all evidence, that this theory involves the postulate that quantum matter is a product of the observer or experimenter – while tacitly admitting that the latter, though composed of quantum entities, is not the child of laboratory operations. In other words, the standard or Copenhagen interpretation of the quantum theory involves this tacit contradiction: "The subject precedes the object, which in turn is generated by the object." Luckily neither the quantum-theoretical calculations nor the relevant laboratory operations depend on Berkeley's philosophy. This is why quantum mechanics can be formulated in an observer-free fashion (Bunge 1967b).

In recent times a different version of subjectivism has gained some popularity among students of science suspicious of disinterested research. This is constructivism-relativism, a sort of collectivist subjectivism. The best-known members of this school are Bruno Latour, David Bloor, and Harry Collins. The main thesis of this school is that everything, from molecule to star, is a social construction: scientists would create not only their ideas but also the objects they study. If this were true, there would be no differences between territory and map, nature and artifact, law and convention, universal and local, and so on.

Moreover, constructivism-relativism leads straight to the contradiction pit. For example, suppose that disease were a creation of the medical community, as the constructivists have claimed. Then either (a) the traces of tuberculosis found in Egyptian mummies could not be such, since the Koch bacillus was not identified until 1882, or (b) the ancient Egyptians are modern (More in Bunge 1999). Whoever challenges realism threatens the scientific endeavor, for this is nothing but the exploration of reality.

Realism, or objectivism, is likely to be the oldest and more robust of all epistemologies. Galileo's trial was the first serious challenge to realism. Indeed, the defendant argued that the Copernican or heliocentric model of the planetary system truly represented the latter. The Inquisition forced him to recant, and to reluctantly admit the Church's doctrine, that neither of the two rival models represented reality, and were mutually equivalent *façons de parler*. Three centuries later, Philip Frank and other logical positivists tacitly sided with the Inquisition, in defending the said equivalence. (In fact the equivalence is geometric but not dynamical: a planet's motion can be certainly be described in either Sun-bound or Earth-bound coordinates; but the smaller body will turn around the more massive one, not conversely.) In short, realism has been in the defensive from the moment it started to bear its most important fruit, namely modern science.

Ironically, positivism, which is practically extinct in the philosophical community, is still going strong among scientists despite its barrenness. For example, the physicist Mermin (1981, 397) once held that "[w]e now know that the moon is demonstrably not there when nobody looks." Presumably, the moon came into existence only when one of our remote ancestors looked up and said in a thundering voice: "Fiat luna!"; and its tidal effects did not emerge until Newton. In short, cognition would precede being, hence epistemology would be prior to ontology. In other words, empiricism (or positivism) can be seen as anthropocentric naturalism.

Though intimately bound to one another, ontology and epistemology should be distinguished, because they have different tasks: the former's is to understand the world, the latter's to understand understanding. Yet even self-styled materialists, such as Lenin (1908), confused realism, an epistemological stance, with materialism, a family of ontologies. And Roy Woods Sellars (1970) called his own philosophy "emergent[ist] realism," for he advanced non-reductive naturalism along with scientific realism.

Lenin and Sellars were not the first to conflate the disciplines in question. Kant (1780, 316) produced the most egregious and influential of confusions between them when stating that "all the things external to me are appearances, because the condition for determining their being is in me." That is, X would be subjective because I know X. In other words, phenomenalism entails either the confusion between metaphysics and epistemology, or the rejection of the former in favor of the latter. The positivists and neopositivists condemned metaphysics because they unwittingly adopted Kant's phenomenalist metaphysics, according to which "the world is a sum of appearances." What a retreat from Galileo's (1623) clear distinction between the primary or objective qualities, such as shape and motion, on the one hand, and the secondary or subjective ones, such as color and taste!

In my opinion, both realism without materialism, and materialism without realism, are weak, hence vulnerable. Realism without materialism invites speculating about disembodied minds, and materialism without realism has no use for the brains that it postulates. Only hylorealism, a synthesis of materialism and realism, is robust (Bunge 2006a). This is because hylorealism admits only the real or material world, and because it makes use of our knowledge of it instead of fantasizing about it: it attempts to be scientific.

The above-mentioned cases of Popper and Lewis are particularly instructive. Popper called himself a realist but, because he rejected materialism, he perpetrated or countenanced several violations of realism. Indeed, he claimed that the world of ideas, which he called "world 3", was just as real as both the physical world, or "world 1," and the "world" of mental events (Popper 1967). Accordingly, he defended mind-body dualism (Popper and Eccles 1977) and wrote about "knowledge without the knowing subject" (Popper 1967). Having detached mind from body, Popper had to admit the possibility of parapsychology, hence of paranormal modes of knowledge such as telepathy and precognition (Popper and Eccles 1977). And Popper (1961) chided me for criticizing the ex-nihilo creation fantasy included in steady-state cosmology.

As for David Lewis (1986), he believed uncritically in the real existence of all conceptually possible worlds, even those that violate all the known physical laws. But what kind of matter is that which has no energy, or which violates energy conservation? And how can we possibly get to know anything about imaginary worlds if they are assumed to be parallel to ours, hence isolated from ours? To access a thing or fact X it is necessary to receive signals from X or from a probe sent to explore X. But this is impossible if the putative world X is parallel to ours. Amazingly, such untestability does not bother the sympathizers of the many-worlds interpretation of the quantum theory proposed by Hugh Everett (1957).

Quantum physicists are particularly blind to the incompatibility between immaterialist metaphysics and the realist epistemology entailed by the scientific method. Thus Frank Wilczek (2008, 33–34), a Nobel laureate, asserts that quarks and gluons are "embodied ideas," and in general that "the its are the bits" – just because they, like the positron and the quarks, were theoretically predicted. Later in the same book Wilczek reports that those particles were actually produced by the Large Electron-Positron Collider. Now *this*, like any other artifact, is an embodied idea; but its constituent atoms existed long before the atomic hypothesis was invented. So much so, that most of them came from minerals that had to be extracted from mines, not from theorists' brains. Likewise, territories pre-existed maps, and people are born before their portraits are painted. In general, it is bits from its, not the other way round. We shall return to this subject in Section 4.2.

The reason that epistemology depends upon ontology is that obtaining some knowledge about object X depends not only on the knower and her tools but also on the nature of X. Thus, if X is readily accessible, anyone can claim to know it, and no special effort may be made to go deeper than appearances. By contrast, the knowledge of imperceptibles requires more ingenuity and effort. In short, "is knowable" is a three-place predicate: object X is knowable to subject Y with tools Z. Hence, any deep change in ontology may call for a change in epistemology. An example follows.

Electromagnetic field theory, born in the 1830s, changed not only the ontology of classical physics but also its methodology. Indeed, consider the problem of finding data about two very different universes: Newton's, constituted by corpuscles, and Fadaray's, filled with fields. The experimentalist hopes to be able to measure the position and velocity of every accessible point mass; but he cannot entertain the same hope with regard to Faraday's universe, because fields are continua, and these call for non-denumerable data sets. In short, the empiricist's dream is in principle realizable in Newton's universe but illusory in Faraday's.

The prediction problem is parallel. Laplace's demon, capable of measuring the position and velocity of every particle in the universe at a given instant, would be able to calculate the state of the universe at any future (or past) instant. But Faraday's universe would defeat the smartest demon, because no measuring instruments can measure all the parameters of a continuum, such as an extended body or an electromagnetic field. True, Cauchy's theorem assures us that, if we know the size, shape and velocity of a wave front at a given constant, then we can calculate the values of the same magnitudes at any later time. But the antecedent of this conditional statement is false: There is no way of producing a non-denumerable set of data: the most we can do is to obtain a finite sample.

This result dashed Laplace's dream – even though few if any philosophers noticed. Could computers do better? Of course not: being digital, they cannot handle continua. But we can and do help them, by approximating continuous manifolds by lattices, and replacing differential equations with finite-difference equations, and curves with staircases. Whereas theorizing often involves continua, computation always requires digitalization.

In summary, epistemology and ontology should match each another. In particular, a realist epistemology must go together with a materialist ontology if possibility is to be distinguished from actuality, discovery is not to be confused with invention, and the exploration of reality is not to be replaced with egology (Husserl's name for his phenomenology).

1.7 Practical Philosophy

Every worldview includes views on values, action, the right, and politics. There are several mutually incompatible practical philosophies: religious and secular, dogmatic and critical, submissive and independent, intuitionist and rationalist, humanist and anti-humanist, and so on. How to choose among them? I suggest evaluating every practical philosophy in the light of both philosophical and scientific considerations, and this for the following reasons. First, a practical philosophy should be consistent with all the other components of a philosophical system, namely logic, semantics, epistemology, and ontology (recall Fig. 1.1). Second, the norms proposed by a practical philosophy should be checked for fairness and efficiency, which calls for collaboration with the sciences of man, in particular psychology and social science. See Fig.1.2.

The particular practical philosophy I advocate (Bunge 1989, 2009) includes a value theory that is materialist, in the sense that it conceives of values as rooted in biological and social needs, instead of regarding them as above the material world. Thus I unabashedly perpetrate what idealists have called "the naturalistic fallacy", with the proviso that I share Hume's warning, that value judgments cannot be deduced from factual propositions without further ado. For example, the political imperative "Reduce social inequality" does not follow from the socioeconomic proposition "There is social inequality." But it does follow from the conjunction of the factual statement "Great social inequality harms both individual and society," and the moral norm "Abstain from harming unnecessarily."

My practical philosophy is also realist (or objectivist), in that it demands putting all moral and political norms to reality checks. More precisely, it adopts scientific realism, which is scientistic since it recommends crafting norms in the light of human biology and psychology, as well as the social sciences. Finally, I hope it is also consistent, and thus compatible with ordinary logic, as well as congruent with any semantics that enshrines both meaningfulness and truth. Consequently,

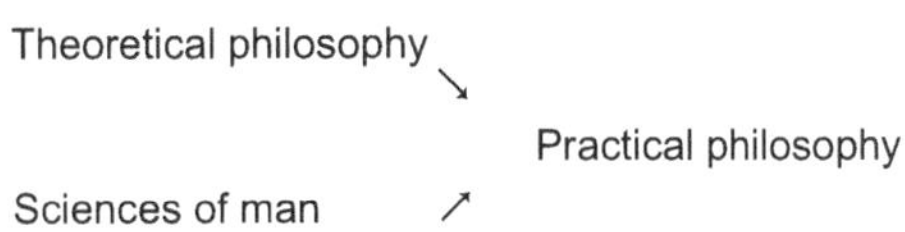

Fig. 1.2 The two inputs to practical philosophy

this practical philosophy opposes emotivism and intuitionism, both of which exalt gut feelings, ingroup loyalty, uncritical respect for authority, and tribal traditions, regardless of the harm they may cause.

My reason for advocating a scientistic approach to ethics is that I take this discipline to be concerned with moral dilemmas, all of which are social problems – the domain of social science. Indeed, I suggest that moral problems arise when scarce resources are handled by persons with unequal power. This is why the more divided a society, the more severe the moral problems that arise in it. I also suggest that the best way of resolving such conflicts in a fair and peaceful way is to discuss and bargain in the light of what is known about the social systems in question, so that the stronger party may compensate the weaker one, e.g., by offering it a share in a bunch of desirable items.

Obviously, the scientistic approach to morality opposes moral relativism as well as moral absolutism (or dogmatism). Contrary to the fashionable opinion, that we should tolerate all values and norms because they are only rooted in gut feelings or in custom, a science-oriented moral philosophy will promote fairness and reciprocity, for they favor individual welfare and social progress. The same approach will also favor close contacts between ethics and ontology, because moral behavior is a form of social behavior, the justification of which calls for a social ontology. Indeed, different social ontologies suggest different moral philosophies: Individualism underlies utilitarianism; holism undegirds deontologism; and systemism is the basis of any ethics according to which rights imply duties and conversely. An ethics without ontology, as demanded by Hilary Putnam (2004), is one for hermits.

Some philosophies are almost exclusively practical. This holds in particular for pragmatism and vitalism. Pragmatism flourished in America (Peirce, James, and Dewey), and vitalism in Germany (Nietzsche, Dilthey, and Simmel). Both schools were action-oriented and therefore anthropocentric, although Peirce, by far the deepest of them all, was seriously interested in ontology. These six men shared not only a practical orientation but also antirealism: to them, truth was the same as utility, and life took precedence over all the other values. On other matters they differed enormously. In particular, whereas the German *Lebensphilosophie* was antiscientific and antidemocratic, classical American pragmatism was scientistic and democratic. (By contrast, the neo-pragmatists, such as Goodman, Rorty, and Putnam, are antiscientistic.)

Those two philosophical schools had also different ends: whereas pragmatism is all but finished, vitalism survives in various post-modern sects, in particular constructivism and existentialism. But, ironically, constructivism is a metaphysical and epistemological doctrine, not a guide to life. And existentialism can hardly be said to be a philosophy, since Heidegger's peculiar sentences, such as "The world worlds," "Time is the maturation of temporality," and "The essence of truth is freedom," are unintelligible; and, like Nietzsche's teaching, existentialism is nihilistic. It is up to psychiatrists to find out whether it is lunacy, idiocy, or simulation.

In short, a nonsensical philosophy is bogus; and one that is only practical is at best useless, and at worst destructive. To be useful, a practical philosophy must make sense and it must be nourished by other branches of philosophy, as well as by the sciences of man. The reason is that successful human actions, unlike those of other animals, are planned in the light of the best available knowledge of the world and of people.

1.8 The Political Connection

Philosophers, like everyone else, are shaped by society, and in turn react on it. Their influence will be negligible if they only work on small problems; but it may be significant, and in such case healthy or not, if they tackle some of the so-called Big Questions, such as What is life?, What is mind?, What is human nature?, What is the good?, and Which is the good society? Consider briefly four turning points: the Scientific Revolution in the seventeenth century, the French Enlightenment in the eighteenth, the Counter-Enlightenment in the nineteenth, and totalitarianism in the twentieth.

The Scientific Revolution was part of the modernization process, one strand of which was secularization. A secular worldview leaves no room for gods, ghosts, or paranormal abilities: it involves a naturalist metaphysics. Nor does it admit mysteries: it adopts the realist thesis that the world is knowable precisely because it is composed exclusively of worldly things. At that time philosophers could adopt either of three stands with regard to the said revolution: to reject it, embrace it, or work out a compromise. For example, the heliocentric model of the solar system, revived by Copernicus, was rejected by both Catholic and Protestant theologians, and embraced by the likes of Galileo and Descartes.

In sum, far from being philosophical extravagances, materialism and realism were in the very marrow of the Scientific Revolution, as well as in that of its inheritor, the French Enlightenment. (The Scottish Enlightenment, by contrast, was philosophically commonsensical – Wittgensteinian *avant la lettre*.) The immaterialism and subjectivism of Berkeley and Kant, as well as the idealist fantasies of Hegel, Fichte, and Schelling, belonged in the Counter-Enlightenment. But it was hard to keep up an antiscientific stance amid the sensational achievements of nineteenth century physics, chemistry, and biology. So, a compromise was tacitly adopted: Allow natural science to follow its course, but stem the scientistic tide in the humanities and social sciences. That is, claim that the scientific method is fine to study nature, but mind and culture require a different approach – either intuition (Bergson) or *Verstehen*, aka "interpretation" (Dilthey, Rickert, Weber, Sorokin).

Better yet, recommended the later idealists: Ignore the big contemporary social questions, such as colonialism, militarism, poverty, and the advances of the democratic, labor, and feminist movements (Weber, Schütz). And dismiss all the attempts to philosophize scientifically: Go back to the hermeticism of the German Romantics, and claim that philosophy too calls for a method of its own, which consists in pretending that the real world does not exist, and catching the essences of things using a special intuition accessible only to the initiates (Husserl). An even simpler solution is to adopt "mysterianism," the view that certain important problems, notably the mind-body problem, are insoluble. This ancient obscurantist view was recently revived by Wittgenstein's sympathizer, Colin McGinn (2004), with Noam Chomsky's approval.

Most social scientists paid no attention to obscurantist philosophies. Or, like Weber, they claimed to adopt their methodology while actually proceeding in a scientific fashion (Bunge 2007a). Thus scientism – the scientific approach – made important inroads in anthropology, sociology, macroeconomics, and historiography. Even some scholars nominally wedded to dialectical materialism and historical fatalism have made important contributions to social science, particularly archaeology and history (see, e.g., Barraclough 1979; Hobsbawm 1997; Trigger 2006). True, the Soviet philosophers attacked all the great contemporary revolutions in natural science, particularly relativity, quantum mechanics, theoretical chemistry, genetics, evolutionary biology, and non-Pavlovian experimental psychology (see, e.g., Graham 1981). But their reactionary efforts succeeded only in biology, psychology, and social studies.

Ironically, the attempt to render social science scientific by "naturalizing" it has borne similar political fruits. Indeed, human sociobiology and its heir, popular Evolutionary Psychology, have reinforced the ancient opinion that human nature is uniform and invariable: that we are all basically aggressive and selfish brutes and that, whereas a few of us are to the manor born, the "herd" – as Nietzsche called the masses – are destined to be slaves.

In sum, philosophy and science are not immune to politics. The reason is that every political movement has an ideology, and every political ideology is a way of seeing, evaluating, and handling society. Since every modern society cultivates both science and the humanities, every serious political ideology involves science and humanities policies. If progressive, these will support those branches of culture and respect their freedom. But, however generous a political protector may be, it has claws, and may be tempted to use them. The practical lesson is obvious: Seek the patronage of political power but do not sleep with it. (More on the philosophy-politics connection in Barber 2003; Bunge 2009.)

Closing Remarks

Important philosophic-scientific problems, such as those about the nature of matter and mind, cannot be adequately discussed except in a broad philosophical framework. The philosophy used in the present work may be summarized thus:

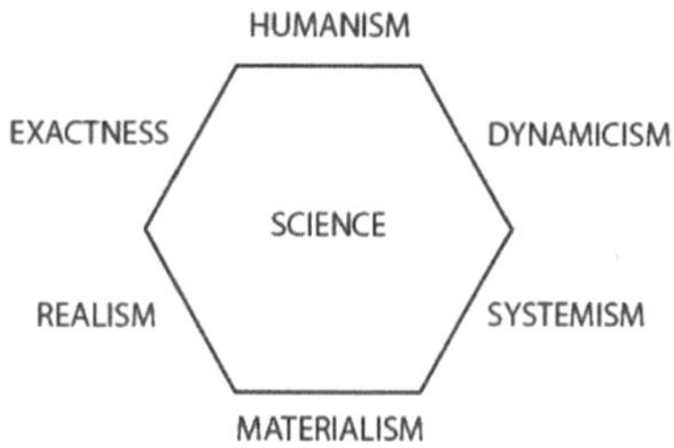

Chapter 2
Classical Matter: Bodies and Fields

Unsurprisingly, matter has been conceived of differently at different times. This chapter and the next are devoted to updating the concept of matter. We shall call *classon* any material thing that comes within the purview of classical physics, and *quanton* any thing describable, at least in principle, by a quantum theory. Specks of dust, organisms and galaxies are classons, whereas photons, electrons, and superconductors are quantons.

Under different names, matter has been a central concern of all ontological (metaphysical) systems, even those that denied its reality. And arguably all the factual ("empirical") sciences only study concrete (or material) entities, from photons to rocks to organisms to societies. Yet most contemporary philosophers have managed to ignore the modern concepts of matter. This is partly because many metaphysicians, following the lead of Saul Kripke and David Lewis, have preferred to speculate about simple conceptually possible worlds rather than studying the messy real world. (Their conception of possibility is so indigent, that it does not even distinguish between conceptual and physical possibility.) Unsurprisingly, their fantasies, such as that of the dry Twin Earth, have not helped science, let alone technology or politics. They have only succeeded in diverting attention from serious problems, both conceptual and practical. Tellingly, they have carefully avoided speculating about possible alternatives to our rather pathetic social world. Thir philosophies are just *jeux d'esprit*.

Many concepts of matter have been thought over the past five millennia. In particular, the rather counter-intuitive concepts of force field and quantum object have severely challenged Aristotle's and even Newton's concepts of matter. Yet, there are still philosophers who uphold Aristotle's hylomorphism, according to which physical objects are composites of matter (material or stuff) and form (shape or property), matter being the impenetrable and passive stuff upon which "forms" are impressed from the outside. Thus Patrick Suppes (1974, 49): "A body is matter endowed with a structure." Example: a potter's differently shaped products. Counter-examples: electrons, protons, photons, and neutrinos have no shapes of their own, and do not share a common material; they only share certain properties, mainly energy, momentum, spin, and the ability to interact with other material entities. Moral: An ontology built with the commonsensical categories of a two-millennia old world-view is bound to be utterly obsolete and therefore useless.

M. Bunge, *Matter and Mind*, Boston Studies in the Philosophy of Science 287,
DOI 10.1007/978-90-481-9225-0_2, © Springer Science+Business Media B.V. 2010

2.1 Traditional Concepts and Principles: Mechanism

The earliest concept of matter seems to have been that of stuff or material from which things are or can be made – stone for axes, clay for pots, bronze for swords, and so on. The ancient atomists, both Greek and Indian, enlarged this anthropocentric concept of matter to embrace all natural things, from water and air to celestial bodies and living beings. To them, every piece of matter was composed of atoms – simple, hard, self-existent, and imperceptible corpuscles moving in the void. And they invented not only atomism but also materialism, the earliest secular worldview, and the one according to which all there is, is material. The atomists also invented the accompanying methodology: Explain the perceptible (phenomenal) by the imperceptible.

Plato opposed vehemently both atomism and materialism: He built an idealist ontology according to which only ideas are self-existent and self-moving, everything else being but faint, ephemeral and passive shadows of ideas. Moreover, Plato held that only ideas can be known, and that no science of worldly things was possible. Thus he was the forerunner of modern idealism, except that – unlike most modern idealists – he argued cogently, understood mathematics, and wrote as clearly as was possible at his time.

Aristotle rejected both materialism and idealism, and replaced them with hylomorphism, a sort of compromise among them. Ironically, twenty-three centuries later the great Bertrand Russell (1954, 384) unwittingly revived hylomorphism under the name of "neutral monism". According to this doctrine, "a piece of matter is a logical structure composed of events." Whitehead (1929), his erstwhile co-author, hatched a similar view: his process metaphysics. Neither of them realized that, in the sciences, every event is understood as a change in the state of some concrete (material) entity.

At the same time, the logical positivist Rudolf Carnap (1928) too professed to stand above both idealism and materialism. Actually, he adopted the subjectivist view held clearly by Berkeley, less clearly by Kant, and again openly by Mill and Mach, that a concrete thing is "a possibility of sensations." Neither Mach – an eminent experimental physicist and psychologist – nor Russell, nor Whitehead, nor Carnap, realized that, if that were true, physics and chemistry would only employ psychological concepts, and scientists would be inspecting their own minds when claiming to study stars, fruit flies, or business firms.

Meanwhile physicists and chemists ignored psychology when crafting or using the concepts of mass, momentum, spin, energy, conductivity, valence, and the like, or when designing and operating experimental devices. And psychologists investigated other people's brains, behavior, and mental processes instead of indulging in introspection. Navel contemplation may be suitable for mystics and drug addicts, but it does not advance the knowledge of reality.

Luckily, scientists did not follow in the steps of Berkeley, Kant, Mill, Comte, Mach, Russell, Whitehead, or Carnap. Instead, they kept investigating matter by the most objective means, namely the scientific method. In particular, Galileo, Descartes, Boyle, and Huygens rejected Aristotelianism, and thoroughly vindicated the concept of matter; they held that physics is the science of matter in motion;

and they embraced a realist epistemology as well. (Descartes came closest to materialism in the two great works that he did not dare publish, but which were very influential after his death: his treatises on the world and on man, which were translated into English only three centuries later.)

Three influential philosophers, Hobbes, Gassendi, and Locke, adopted atomism and realism, as well as its principle (revived by Galileo and Descartes) that material things have only primary or objective properties, the secondary ones (like color, smell, and taste) being subject-dependent. And, of course, Newton's mechanics was the first thoroughly scientific version of mechanistic materialism (see Dijksterhuis 1986). Yet, with the exception of Hobbes, none of the above-mentioned thinkers dared dispute the existence of God or the immateriality of the soul: their materialism was prudently confined to physics and chemistry.

The earliest full-fledged and outspoken modern materialists and atheists were Helvétius, Holbach, and La Mettrie (see, e.g., Plekhanov 1967). However, Benedict Spinoza and John Toland too may be regarded as materialists even though they were deemed to be pantheists, since they equated God and nature. And the psychologist David Hartley, like the chemist Priestley, managed to hold materialist theories of mind along with Christian beliefs.

The founders of modern science kept religion separate from scientific research. And the Puritan scientists believed (or at least stated) that it behooved them to study Creation as a way of unveiling the grandeur of the Creator. As Merton (2001, 136) wrote in his doctoral dissertation, this came not so much from their theology as from their value system: "a complex of a scarcely disguised utilitarianism; of intra-mundane interest; methodical, unremitting action; thoroughgoing empiricism; of the right and even the duty of *libre examen*; of anti-traditionalism – all this was congenial to the same values in science." Thus, Puritanism – just as open-minded in England as narrow-minded when exported to America – unwittingly contributed to the advancement of science and the concomitant consolidation of the early modern materialist worldview.

The concept of matter inherent in the mechanistic materialism that prevailed between ca. 1600 and ca. 1850 was this. Nature is wholly material and lawful, and the physical laws hold throughout the world. Matter is one: there is no distinction between terrestrial and celestial matter. And material things are characterized by extension, shape, mass, and mobility. In the physics literature light was, and usually still is, counted as distinct from matter, although most materialist philosophers have always regarded it as a kind of matter.

True, classical mechanics employs the concept of a point mass (or rather massive point), which of course has no extension; but this was always understood as a simplified model of an extended body, just as the light ray of geometrical optics was rightly regarded as a stylized light beam. And from ca. 1750 on, liquids and gases, though not solid, were regarded as constituted by solid particles – atoms or molecules. The atomic theory built by the nineteenth century chemists kept the ancient view of atoms as tiny solid balls or systems of such. But in the early twentieth century experimental nuclear physics altered radically this simple picture: Rutherford showed that atoms are mostly hollow, hence penetrable. Only the lay and most philosophers kept thinking of matter as solid, impenetrable, and passive.

Throughout the reign of mechanism, space and time were assumed to constitute the immaterial and therefore unchangeable container of material things and stage of events. In particular, place, distance and duration were assumed to be absolute, in the sense that they did not depend upon material things and their changes. Galileo deprived place of the pride of place Aristotle had attributed it: places are exchangeable unless occupied by different things, and consequently the laws of motion are invariant under displacement (Galileo's principle of relativity). So much so, that motion is transformed into rest by a mere change in frame of reference. Also, rectilinear uniform motion is no real change; only accelerated motion, such as a freely falling body's, is real change. Therefore, rectilinear motion at constant speed can be neither cause nor effect. In other words, motion of this type is not a causal chain, where every effect is the cause of another event.

Newton speculated that space and time were the sense organs of the deity, but he abstained from attributing them any spiritual properties, and did not bother to inquire any further. Only Leibniz ventured to ask the hard ontological question, "What are space and time?" He also revived the ancient relational (or adjectival) view that, far from existing by themselves, space and time are relations among things and events. But Leibniz's metaphysics was just as unclear and marginal as his physics. Likewise, Kant's thesis, that space and time are subjective, was ignored by scientists. (Johann Heinrich Lambert, that amazing polymath, tried to persuade him that he was wrong on this point, but did not succeed.) The relational view was revived much later by Mach, and again, within physics, by the two theories of relativity.

The mechanists took it for granted that matter possesses some additional features, all of them intuitive; but these presuppositions became clear only when the quantum theory, born about 1925, questioned them. One such universal feature is that all physical quantities, with the possible exception of some properties of the universe as a whole, are finite. Consequently, any theory including infinities (singularities, "divergences") must be false.

The ban on physical infinities may have two exceptions: the size and age of the universe. Indeed, at the time of writing we do not yet know whether the universe is spatially finite or infinite, and there is no compelling argument for a beginning of time. Philosophy cannot help with the question of spatial infinity, but it is not indifferent to the question of temporal origin: any naturalistic ontology will demand that the universe has always existed.

What about the Big Bang, which is usually supposed to have occurred between 10 and 20 billion years ago? At present there are at least three possible answers to this question:

1. The Big Bang happened, and it was God's creation out of nothing. This answer is obviously unacceptable to any physical cosmology, for it invokes the supernatural and violates Lucretius's *ex nihilo nihil* principle.
2. The Big Bang is only a simplistic interpretation of the singularity occurring in the simplest of all cosmological models. This model assumes that the universe is the maximal balloon, and that there is a cosmic time in addition to the uncounted local times attached to all the possible reference frames. As Lévy-Leblond

(1990) has argued, even admitting this model does not force us to interpret the time at which the radius of the universe was nil as the origin of the universe. This is because the model is not defined for that time, just as the ideal-gas formula "Pressure × Volume = const." is not defined for the null volume, at which the corresponding pressure is infinite – a physically impossible value. Adopting this sober approach leads Lévy-Leblond to concluding that the Big Bang never happened.

3. The Big Bang did happen, but it was only the sudden and worldwide expansion of the universe that existed earlier in a state about which we know nothing. Nor will we ever discover anything about the pre-Big-Bang universe, because the explosion destroyed the records. One possibility is that the event consisted in the sudden emergence of ordinary matter (electrons, photons, etc.) out of the pre-existing electrodynamic vacuum, or space filled with "virtual" particles. But, given the scarcity of astronomical data, and the unrestrained fantasy of cosmologists, I suggest suspending judgment until more realistic cosmological models are crafted.

In any event, we should heed Tolman's (1934, 488) warning at the end of his massive treatise: "we must be specially careful to keep our judgments uninfected by the demands of theology and unswerved by human hopes and fears. The discovery of models, which start expansion from a singular state of zero volume, must not be confused with a proof that the actual universe was created at a finite time in the past."

Further famous infinities are the energies of a point electric charge and of a plane electromagnetic wave. These are unavoidable flaws of classical electrodynamics. Quantum electrodynamics contains clever tricks ("renormalization" procedures) that disguise such warts as beauty spots. While most physicists buy them, a few – the present writer included – regard them as stopgaps to be absent in better theories.

However, let us return to the task of listing the most common features of nature. Another supposedly universal trait is *continuity*: *Natura non facit saltus*. All changes are incremental, none are discontinuous: there can be no quantum jumps. In particular, the energy of a body increases or decreases in a continuous manner. (Mathematical counterpart: All the functions representing basic physical magnitudes are smooth, at best continuous, and at worst piecewise continuous.)

However, classical physics did admit several discontinuities, such as the abrupt change in the velocity of a solid body upon impact on a solid surface; uncounted threshold effects; the occurrence of a denumerable set of modes of vibration of an elastic body; phase transitions, such as liquid → gas, and paramagnetism → ferromagnetism; and the quantization of electric charge, which Faraday discovered in electrolysis.

Another allegedly universal property of things is *individuality*: Every thing has some properties that characterize or individuate it in a unique fashion. If two things had exactly the same properties, they would be one. But what if two things, though different, turned out to be equivalent and therefore exchangeable? In classical physics each of them would keep its individuality – not so in quantum physics. (See Section 3.2.)

A further trait traditionally attributed to matter, albeit only tacitly, is *separability*. Since all the classical forces weaken with distance, when a complex thing explodes, its fragments interact less and less strongly as they recede from one another, until they end up by being practically separate. Shorter: It was assumed that, when systems dismantle, their components become mutually independent. Again, quantum physics has limited the scope of this principle (see Section 3.3 on entanglement).

A third feature that used to be attributed to matter is that it exists in space and time, conceived of as the public and neutral receptacles or stages, and as impartial witnesses of all events. In other words, space and time were regarded as constituting an independent grid indifferent to the vicissitudes of matter.

It was taken for granted that vacua, or regions entirely devoid of material things, have no physical properties, and therefore are not material things. The ancient atomists put it this way: The universe is a collection of atoms hurtling in the void. Thus empty space would coincide with the vacuum: both would be immaterial and therefore inactive. Quantum electrodynamics was to demolish this hypothesis three centuries after Otto von Guericke built the first vacuum machine. His sensational experiment had a strong impact on physics and philosophy: it vindicated atomism and, by the same token, it discredited the plenum speculations of Aristotle and Descartes. Moreover, in vindicating atomism, Guericke's experiment weakened religion, since at that time "atomism", "Epicureanism", "materialism", and "atheism" were pragmatically synonymous. However, let us move on.

Passivity or inertia used to be regarded as a further trait of matter. That is, all the changes in a material thing were asumed to originate outside it: *Omne quod movetur ab alio movetur* ("Everything that moves is moved by another.") This assumption leads to either an infinite regress or to postulating an immaterial first mover. The first consequence satisfies nobody, and the second – Aristotle's solution – can please only theists. Consistent materialists cannot admit the passivity postulate, for they have learned from science about emergence, spontaneous self-organization, and pattern formation. Just look at snowflakes, crystals, flowers, or birds. As Philip Ball (2001) put it in a beautiful book, nature is the self-made tapestry.

(Caution: The Austrian school of economics and most neoconservative ideologists are great believers in spontaneity because they want the "invisible hand," not the state, to regulate the market, even at the expense of those who cannot afford free enterprise. They do not admit that all artificial systems are constructed and therefore anything but self-organized and self-regulating. Everyone admits that the worst two economic crises in recent times, those that started in 1929 and 2008, resulted from the laxity of market controls.)

A corollary of the passivity dogma is the principle that nothing except God can cause itself: only the top deity is *causa sui*. Naturalists reject this thesis: They assert that nature is self-caused. (Caution: This statement should not be understood literally, because events, not things, are the relata of the causal relation. The correct version of the thesis in question is that nature is autonomous, or self-sufficient.)

A second consequence of the principle that matter is passive, is the Aristotelian postulate (actually first stated by his Medieval followers) *Causa cessante, cessat effectus*, that is, effects cease with their causes. Newton tacitly negated this proposition when he stated his principle of inertia. But this did not prevent the behaviorists to hold on to it: they stated that all responses are extinguished when the corresponding stimuli cease. They overlooked such familiar counter-examples as afterimages, pains felt long after their causes, and so on. And they were bound to overlook such spontaneous mental events because they refused to study the brain, which spends nearly all of its energy budget in processes that are not elicited by external stimuli. (See Section 9.3.)

Another consequence of the passivity formerly attributed to matter was *stability*: Material things might certainly change as a result of either external forces or internal stresses, but they would not alter in the absence of either. That is, loss of mass and spontaneous radioactive and radiative decay were excluded; and so were gain of mass and spontaneous self-organization. By the same token, spontaneous neuron discharges and non-stimulus-bound mental processes were out of the question. The entire stimulus-response or behaviorist doctrine, as well as the assumption that the isolated brain must be inactive except for household functions, were instances of the Peripatetic maxim that nothing moves by itself. (More on brain self-generated activity in Section 9.3.)

A related but far more comprehensive principle is that of *conservation of matter*, first stated by Lucretius: *ex nihilo nihil*. This postulate was first formulated in a quantitative manner by Lavoisier as the conservation of *mass* in all chemical transformations. (In this sentence, "mass" means quantity of matter rather than inertia.) Half a century later the postulate in question resurfaced as conservation of *energy* in a closed system, or the first principle of thermodynamics. As will be seen in Section 9.3, mass is not conserved in certain changes, whereas energy is assumed conserved in all. (Actually, what are conserved are energy distribution and its average.) But both mass and energy are properties, not things. So, Lucretius' principle of conservation has been conserved. (True, in quantum electrodynamics there is talk of "virtual" processes, in which energy is not conserved. But it may be argued that such processes, like the movements to the past, are imaginary, whence the correct name "virtual".)

2.2 Further Features of the Classical Picture

Another apparently obvious feature of material things is that *the values of all their properties are sharp rather than blunt*. In contemporary metaphysical jargon: All tropes are well-defined. For example, all position and velocity values are exact. (Mathematically speaking, all physical variables can be represented by real-valued functions. Thus, the position of a point mass p, at time t, and relative to a reference frame f, is $X(p, t, f) = <x, y, z>$, a triple of real numbers.) In metaphorical terms: The world has sharp contours, and only our knowledge of it may be smudged. Quantum physics falsified this postulate: it showed that all dynamical properties are blunt rather than sharp (Section 3.2).

Two principles concerning causes and effects were assumed: those of *antecedence* and *causality* (see Bunge 1959a). The antecedence principle asserts that causes (or inputs) precede their effects (or outputs), or that the past determines the present. And the principle of causality states that every event is produced by some other change: there is no spontaneity or self-determination. To be sure, there are coincidences galore, but they are assumed to consist in the crossing of initially independent causal trajectories, as the Stoic Chrysippus had assumed.

In classical physics, chance or randomness is always thought of as resulting from causation at a lower level; and at the lowest level – the atomic level – everything was assumed to proceed causally. Shorter: there is no basic or irreducible chance. Quantum physics overthrew this principle of the mechanistic worldview: We now know that the exact opposite is true – that at the lowest level, irreducible randomness occurs along with causation, just as Epicurus and Lucretius had surmised. And we should also realize that the idea that chance is nothing but ignorance, is wrong too: chance is for real even to an omniscient being. For example, shuffling a new deck of cards transforms order into disorder, and this is nothing but chance, even if an omniscient being could certainly follow the trajectory of every card.

By contrast, the antecedence principle stays. And, far from being an idle metaphysical fantasy, it is used in sciences to weed out error. For example, the principle is used to reject as physically meaningless one half of the mathematically correct solutions to the equations of electromagnetic theory, namely the advanced potentials. According to these, there are electromagnetic fields coming from the future: these are ruled to be mathematically true but empirically false. Likewise, the precognition hypothesis, held by most parapsychologists, can be rejected out of hand because it involves the fantasy that the future, which does not yet exist, can act upon the cognitive subject. True, Richard Feynman and other theoretical physicists regarded the positive electron (positron) as a negative electron going to the past, just because certain formulas remain invariant upon the joint inversion of the time and electric charge signs. But the experimental physicists were never fooled: They knew that it is impossible to go back in time because the past is no longer – they stuck to the antecedence principle.

This antecedence principle, together with Lucretius's principle, can be invoked against cosmological and biological creationism. Indeed, both principles together rule out any assertion of the absolute beginning of anything material. In particular, the Big Bang may be understood as the start of a new phase in the infinite history of the world rather than as its absolute beginning; new biospecies are known to have evolved from their ancestors according to Darwin's Tree of Life; and the earliest organisms are assumed to have self-assembled from abiotic precursors. In short, there are no absolute beginnings. Thus, consistent naturalism, in particular mechanism, involves atheism or, at least, the deistic heresy that God abandoned the world to its mechanical fate once He created matter and endowed it with its immutable laws (Descartes 1664, 37).

Then, there is the principle of *least action*, according to which in all motions the action of a mechanical system is the least possible – or, more generally, either a minimum or a maximum. (In analytical mechanics the action of a thing is a global property of it defined as the time integral of the difference between the kinetic and the potential energy.) This principle is remarkable for various reasons: it concerns the entire history of a thing; it entails its equations of motion; it sharply distinguishes physical from conceptual possibility; and it occurs in all the branches of physics (see Lanczos 1949). In the eighteenth century, Pierre Maupertuis used it to argue for the parsimony of the Creator – which provoked Voltaire's hilarity; two centuries later, Max Planck used that principle to prove His intelligence.

Let us finally peek at chaos theory, better known as nonlinear dynamics. Whether this theory calls for any important epistemological changes, particularly with regard to predictability, is still an open question. What is beyond dispute is that chaos theory has broadened science, by studying instabilities, or cases of small causes-large effects, such as avalanches (see Glass and Mackey 1988).

Some of the most remarkable successes of chaos theory have been the discovery that some of the orbits of members of three-body systems are chaotic, and a first account of the periodic explosions and crashes of insect populations. By contrast, the speculations on the chaotic nature of political "turbulence" are just superficial analogies.

Note that any chaos-theoretic model of real processes is more than a set of nonlinear differential or finite-difference equations: These have got to be supplemented with a factual interpretation (of function symbols as properties). That is, unless some of the symbols are made to represent momenta (quantities of motion) or their rate of change (forces), the model will be purely descriptive, as explanation calls for mechanism, which in turn involves energy.

Finally, a couple of linguistic observations about chaos theory. Firstly, its successes in accounting for certain irregularities suggests giving up the traditional identity of "law" and "regularity." Secondly, it should be clear that "chaos" is a misnomer, since all the chaos-theoretical equations express lawful changes. Therefore the non-committal "nonlinear dynamics" is preferable.

In short, the mechanistic worldview postulates that the universe is at once objective, thoroughly material, and lawful. This cosmology is in stark contrast with the view of Berkeley, Hume, Kant, and their successors, in particular the positivists, logical positivists, intuitionists, and many-words fantasists – a view that is subject-centered (in particular phenomenalist), immaterialist, and lawless.

So much for the mechanistic worldview. Let us now look at its decline. We will see that mechanism was eventually refuted in details but not in its grand conception: Objectivity, materiality and lawfulness were kept even though the mechanistic restrictions on matter and law were removed.

2.3 The Decline of Mechanism: Fields

Mechanism reigned in physics until mid-nineteenth century. It declined irreversibly during the second half that century as a consequence of the emergence of four new ideas: those of force field, energy, chance, and the micro-macro distinction (see d'Abro 1939).

Michael Faraday introduced the idea of a radically new kind of matter: the field of force, in particular the electromagnetic field. A physical field is a region of space-time every point of which has one or more physical properties, such as energy density. The core of a field theory is a set of field equations that describe the distribution and change of the physical properties in question. A peculiarity of field theories vis-à-vis mechanics is that they do not contain position variables – that is, time-dependent coordinates – and consequently cannot be interpreted in terms of particles.

An electromagnetic field accompanies electric charges and currents, and it interconnects them, but – as became clear later on – it can subsist independently of its sources and of the medium where it exists. Electrodynamics studies fields that accompany electrically charged bodies, as well as fields that are generated by oscillating charges and electric currents, and that become detached from their sources, and may eventually be absorbed by bodies. Thus, unlike classical mechanics, electrodynamics studies certain qualitative changes in addition to merely quantitative ones.

Contrary to Ampère's theory, which it replaced, the electromagnetic theory sketched by Michael Faraday in 1831 and perfected by James Clerk Maxwell in 1865 rejected the idea of action at a distance: it postulated that the space between electrically charged bodies is filled by a field, and that all actions between such bodies were mediated by fields. Moreover, Maxwell's mathematical theory of these fields suggested the existence of electromagnetic waves, which Heinrich Hertz produced and measured a decade after Maxwell's death. It turned out that light rays and radio beams were nothing but bunches of electromagnetic waves. Later it was found that X-rays too are electromagnetic waves.

According to Einstein (1934, 213), the transformation of the conception of reality brought about by the Faraday-Maxwell theory was "the deepest and most fruitful experienced by physics since Newton." (Incidentally, Einstein himself perfected that theory by showing that it was in no need of the ether: that fields are freestanding things.) That theory also suggested an ambitious research project that fascinated Einstein all his life: that of reducing particles to fields. This program was partially realized by the quantum theory, which conceives of photons, electrons, mesons, and other elementary "particles" as the quanta, or units, of so many fields – one per kind of basic building block. However, let us now go back to classical physics.

Electromagnetic waves, unlike water and sound waves, can exist on their own, in free space, without any support: the ether invented in the earlier century to account for the wave aspect of light – obvious in diffraction and interference – turned out to be fictitious. (However, as we shall see in the next chapter, quantum electrodynamics involves a new kind of ether: the quantum vacuum.) The same holds for gravitational waves: these ripples in spacetime are assumed to subsist on their own once generated by moving bodies such as planets. Moreover, fields have no shape of their own; they propagate but do not move along neat trajectories (orbits); and they need not have a mass. The light ray is the closest field-theoretic analog to a particle trajectory. But, as Huygens showed four centuries ago, far from being basic and simple, light rays result from the interference of waves.

Unlike the equations of motion of corpuscles and extended bodies, the field equations describe the field intensity at every point in the region it occupies. In the electromagnetic case the field intensity has an electric and a magnetic component, whose changes in space and time determine one another.

The field equations for the vacuum involve only the basic field properties, that is, its electric and magnetic intensities. In particular, they do not involve the masses of the field sources (electric currents and magnets). The total energy of a field in a region is calculated by squaring those intensities and adding them up over the given region. This suggests regarding energy as the measure of matter, just as in classical mechanics the quantity of matter is mass. We shall return to this in Section 2.4.

Whereas corpuscles can move at different speeds, all the electromagnetic fields (or waves) move in a void at the same speed, namely the limit velocity c. However, inside a transparent material these fields can slow down to snail pace. And, unlike particles, electromagnetic fields have no inertia, because they have no mass. But of course they carry energy. Contrary to untutored intuition, the electromagnetic energy flux is perpendicular to the electric and magnetic field components.

Fortunately, the intuition that developed from our commerce with solid bodies can be corrected and enriched through studying fields. True, many physicists resist calling light beams material entities, so that they may occasionally say that fields are only intermediaries between particles; many of them also call "forces", and radiation "energy". But in this case philosophers, if materialist, know better: They suggest that the discovery of fields, and the invention of field theories, forced an extension of the concept of matter to include fields. In classical physics since Faraday's day there are then two kinds of material entity: body-like and field-like. As will be seen in the next section, the quantum theory overcomes this body-field duality, in the sense that the so-called particles turn out to be the quanta (elementary units) of the corresponding fields.

2.4 Aditional Decline: Thermodynamics

Thermodynamics developed at about the same time as electromagnetism. But the two theories are radically different from each other as well as from mechanics. Indeed, classical thermodynamics deals only with large systems, such as engines, and with bulk properties, such as temperature, entropy, and free energy; and it models every material entity as a black box endowed with volume, internal pressure, temperature, energy, entropy, and properties derived from these. Furthermore, thermodynamics distinguishes two kinds of energy, mechanical and thermal, so that the total energy of a thermodynamic system, such as a star or an engine, equals its mechanical energy plus its thermal energy. The first axiom of thermodynamics says that the total energy of a closed (or isolated) system is constant. Many cosmologists assert that an expanding universe loses energy whereas a collapsing one gains it. But actually the principle does not apply to the universe as a whole because this is not a closed system.

Moreover, heat and work can be converted into one another – though not in a completely symmetric fashion. Indeed, mechanical energy can be fully transformed into heat, but the converse transformation is never complete: there is always a thermal residue that can no longer be transformed into motion, for it stays bound to the system. This is what the second principle of thermodynamics says: that, although the total energy of an isolated system is constant, its quality tends to be degraded, in the sense that it gradually dissipates. Thus highly concentrated energy, such as that in an electric battery, tends to disperse, as when the battery is connected to a system composed by an electric bulb, a heating element, or a machine. Increase in order, as in self-assembly, can only occur in open systems, and at the expense of increasing disorder in their environment.

The difference between higher and lower energy forms is explained by statistical mechanics, which analyzes thermodynamic systems as systems of randomly moving particles, and attempts – alas, thus far not with complete success–to reduce all thermodynamic properties to mechanical ones. For example, internal pressure is reduced to the sum of the molecular impacts on the system's walls. And entropy is reduced to molecular disorder or, more precisely, to the number of microconfigurations compatible with a given macrostate. This is the famous formula "$S = k \ln W$", where S represents the entropy, W the number in question, and k a universal constant, that is, one that does not depend on the stuff the system is made of.

Because of the crucial role played by the molecular randomness (or disorder) hypothesis, the reduction in question is incomplete or partial, notwithstanding the dominant opinion (see Bunge 1973a). What statistical mechanics did accomplish was not the reduction of thermodynamics to mechanics but the bridging of the two. The above formula is the most obvious bridge between the two disciplines.

Thus thermodynamics, born from the desire to understand and perfect the steam engine, ended up by causing a quiet ontological revolution, for it introduced three ideas alien to the mechanistic worldview. These are those of level of organization (micro/macro, or particle/extended body), objective randomness (disorder, entropy), and the rundown (increasing disorganization) of all closed macrophysical things.

However, it is usually claimed that, unlike quantum randomness, classical randomness is just a matter of ignorance and therefore subjective. This is the gist of the interpretation of thermodynamics in terms information theory and Bayesian (subjectivist) probability championed by Jaynes (1967).

While it is true that, had He nothing better to do, God would be able to track down every particle in a gas, it is also true that molecular disorder is objective, whence entropy – a measure of such disorder – is just as objective a property as quantity of heat and temperature. (So much so, that the increase ΔS in entropy that accompanies a warming up by ΔQ at temperature T is $\Delta S = \Delta Q/T$.) Such increase in disorder upon heating is not confined to anyone's mind: it is objective. This is why different physicists or engineers will come up with roughly the same numbers when measuring with the same instruments the increase in the entropy of a system: Disorder increased in the system, not in their brains. Moreover, temperature measurements are possible only when the system attains equilibrium, the state at which its entropy is maximal – as would be the experimenter's uncertainty according to subjectivism. That is, knowledge and ignorance would be the same.

Classical statistical mechanics assumes that every component of a system has a precise though unknown position, as well as an equally precise and unknown momentum. Thus, since each position in ordinary space is specified by a position coordinate with three components, a system of *n* particles is assigned a 3*n*-dimensional configuration space, and a 6*n*-dimensional state space. These properties are assumed to be real but unknowable. Thus, paradoxically, in this case realism involves an epistemic pretense. Such explicit admission of ignorance of microscopic detail deprives place of the pride of place it held since antiquity: We assume that things move in space, but do not seek to discover their trajectories because in practice we cannot find them. Thus theory, far from reflecting practice – the way empiricism and pragmatism demand – ignores it. As we shall see in the next chapter, quantum physics does not even pretend that its microphysical referents have definite albeit unknowable positions, hence trajectories and shapes as well.

Back to ontology. Whereas mechanism had postulated that nature exists on a single level, classical statistical mechanics showed that there are at least two physical levels: macrophysical and microphysical – exactly what the ancient Greek and Indian atomists had surmised. Later on atomic, nuclear and particle physics increased the number of physical levels. And, of course, biologists and social scientists added a number of supraphysical levels (see Chapter 5).

2.5 Special Relativity

The next scientific revolution was the emergence of Einstein's special theory of relativity in 1905. Actually, in a sense it was a reform rather than a revolution, for it altered mechanics but not electrodynamics, and it affected the theory of spacetime as well as the theory of matter. In particular, the new theory showed that space and time, far from being mutually independent, fuse into one: spacetime. Special relativity also showed that the values of certain properties, such as distance, duration, mass, temperature, and electric field intensity, are frame-dependent, whereas others, such as space-time distance, electric charge, and entropy, are invariant with regard to changes in reference frames. In other words, whereas some physical properties are the same in (relative to) all reference frames, others are not. Thus, the relativization was partial, and it concerned the relation to the objective reference frame, not the inquiring subject.

For example, the distance $\Delta s^2 = \Delta x^2 - c^2 \Delta t^2$ between two points in spacetime is the same in all reference frames: it is an invariant under Lorentz transformations. By contrast, every body has as many masses and energies as there are inertial reference frames – that is, infinitely many if the universe is spatially infinite. And such differences do not result from any changes, hence they do not call for explanation in terms of mechanisms, in particular causal mechanisms. All changes cause differences, but the converse is not true. Incidentally, such relativity has often be misunderstood as unreality – whence the expression "apparent mass", by contrast to "proper mass", which is the mass relative to the rest frame. There is nothing unreal or apparent about frame-dependence, whether in relativistic or in classical physics. (Recall that position, velocity, and therefore kinetic energy as well are frame-dependent in classical mechanics too.) Invariance implies reality but not conversely.

The relativity of position was known from classical mechanics. The relativity of time, by contrast, came as a surprise. In particular, it sounded paradoxical that age turned out to be frame-dependent – the famous twins paradox. Still, this novelty was as nothing by comparison with this further consequence of the relativity of time: that an oscillation in one reference frame amounts to ("appears as") a propagating wave in a moving frame. This was the germ of Louis de Broglie's early wave mechanics (1924), which Erwin Schrödinger (1926) expanded into what soon became the nucleus of quantum physics. The wavelike aspect of matter, predicted by wave mechanics, was soon confirmed by experiment.

Thus special relativity, sometimes regarded as a mere offshoot of classical electrodynamics, had an unpredictable offspring: wave mechanics. And of course neither of these theories had any technological motivations. Rather on the contrary, the disinterested study of matter found eventually momentous technological applications, from telecommunications to nuclear engineering. Which brings us to one of the most famous formulas in the entire history of physics.

The equation "$E = mc^2$" in relativistic mechanics tells us that the mass and energy of a particle or body are equivalent. This quantitative equivalence is usually said to mean that mass and energy are the same modulo c^2. This is mistaken, for whereas E measures the ability to change, m measures inertia, or disposition to resist changes in the state of motion. Another popular construal of the same formula is that matter is the same as energy. This too is mistaken, because energy is a property of material entities, as shown clearly by the standard expression "The energy of thing θ, relative to frame φ, and in unit ε, equals number e."

Besides, while only particles and bodies are endowed with mass, energy is the universal physical property, as will be argued in Section 2.4. In other words, although matter is not the same as mass, and mass is not the same as energy, it is true that the predicates "is material" and "has energy" are coextensive, that is, they have the same instances.

Interestingly, the mass of a system is slightly smaller than the sum of the masses of its components; this difference is called "mass defect". Thus, the total mass of a system composed of bodies 1 and 2 is $M = m_1 + m_2 - m_{12}$, where m_{12} stands for the mass defect of the system. The corresponding energy, $m_{12}c^2$, equals the binding energy of the system. The fission of uranium occurring in a nuclear bomb is the most famous example of this equivalence: the fragments escape with part of the huge binding energy of the nucleus, now transformed into kinetic energy.

Likewise, the energy of a physical system is different from the sum of the energies of its constituents, because it includes the interaction energy. (The total energy is greater or smaller than the sum of the energies of the components, according as the interaction energy is positive or negative.) Note the contrast with distances, periods, electric charges, and other additive magnitudes. (This difference has escaped the philosophers and psychologists working in what they call "measurement theory", because they have confused *measurement*, an empirical procedure, with *measure*, a set-theoretic concept: see Bunge 1974. Worse, they have ignored the subadditive magnitudes, such as mass, as well as the intensive ones, such as mass density.)

Notice also the occurrence of the word "system" in the above: although few philosophers use it, it is rampant in all the sciences. The reason is that systems are not just collections of individuals, but individuals on a higher level: a wall is not a pile of bricks, a cell is not a collection of molecules, a regiment is not an accidental crowd, and so on.

The case of photons, or light quanta, is quite different from that of "ponderable" matter (or matter endowed with mass). In fact, photons have energy and momentum, but no mass. The three properties mentioned in the previous sentences are related by the equation "$E^2 - c^2p^2 = m_0{}^2c^4$". This equation is an invariant: it holds in all inertial frames. (m_0 denotes the rest mass, that is, the mass of the body in question relative to a reference frame attached to it.) The preceding formula is the dynamical counterpart of the space-time invariant we met previously.

The second of the above formulas shows clearly that energy is a more universal property than mass, since for $m_0 = 0$ the above formula reduces to $E^2 - c^2p^2 = 0$. This confirms the wisdom of regarding electromagnetic fields, which are massless, as constituting a special kind of matter. And it suggests that, whereas in mechanics mass measures the quantity of matter, energy measures matter of all kinds. (More on this in Chapter 4.)

To return briefly to light, recall that its velocity in void is the greatest speed. (However, light is accelerated when going through a gravitational field.) This fact is so important, that it has been erected into a principle. This is the *locality principle*, according to which no action can propagate faster than light. This principle is used to evaluate physical theories: nonlocal theories, that is, theories that assume superluminal velocities, are deemed to be severely defective. Thus, one of Einstein's main objections to quantum mechanics was that he thought that it enshrines action at a distance, which he rightly regarded as spooky. We have learned since that the theory involves entanglement instead (see Section 3.2). However, let us resume our discussion of invariance.

The fact that special relativity dethroned old invariants while anointing new ones shows that the choice of name for the new theory was unfortunate. It misled people into emphasizing the relative (or frame-dependent) at the expense of the absolute (or frame-invariant), while actually both are equally important. "Special absoluteness" would have been just as appropriate – or inappropriate. (The qualifier "special" refers to the particular kind of reference frames referred to by this theory: they are the inertial frames, that is, those that move with constant speeds relative to one another.)

But the main philosophical mistake was the belief that the theory involves subjectivism, just because Einstein tried to popularize it with reference to observers traveling in trains and measuring distances and durations. This popular conflation of "relative" with "subjective" might have been avoided using photocells and automated measuring devices – which of course were not widely available at the time. And no physicist should have indulged in this confusion, for any magnitude is either relative or absolute in a bounded way, namely relative to some particular group of transformations.

2.6 Gravitation

The next field concept to be invented was that of a gravitational field. Einstein's general relativity is actually a theory of gravitational fields. This theory would not have had any impact on ontology were it not that it involved an even deeper change in the concept of spacetime than special relativity. Indeed, it showed that the distance between two points in spacetime depends upon the distribution of matter, and is therefore likely to be different in different regions of the world – shrinking wherever matter density increases, and expanding where it decreases. In other words, matter deforms spacetime, and in so doing it alters the trajectories of classons such as planets, and of quantons such as photons.

Einstein's theory of gravitation (1915) transformed the view of spacetime as the universal passive container of matter into the conception of spacetime as the partner of matter. Before 1915 it could be speculated that, if God ever got tired of the world, spacetime would still remain as a hollow container ready to receive new material things – or none. After that date one had to face the possibility that, if matter were to disappear, so would spacetime: No matter, no spacetime.

This thought ought to cool down the enthusiasm for the Big Bang conceived of as an explosion of nothingness in the void. And it should also convey an excitement that the qualifier "general" fails to convey, namely that the new theory not only allows for accelerated frames, such as our planet, but is about gravitational fields and their sources or partners. So much so, that the solutions of Einstein's equations describe gravitational fields. Philosophers should have helped to identify the proper referents of the theory. But their semantics, if any, did not include any theory of reference (or aboutness), such as the one advanced by the present writer (Bunge 1974a).

Einstein may have died believing that his theory of gravitation had been confirmed by just three "effects", in particular the most surprising of them – the bending of light by gravitation. Since that time more than two-dozen further "effects" were found, and the existence of black (or rather gray) holes has been amply confirmed. Besides, all the cosmological models are expected to agree with Einstein's theory of gravitation. But let us face it: although relativistic cosmology is nearly one century old, it still has not decided whether there really was an absolute beginning (Big Bang), and if so what if anything exploded, or even whether the universe is spatially finite or infinite.

Closing Remarks

The vulgar opinion about classical physics is that it is basically wrong and therefore obsolete. (This opinion was largely shaped by the half-truth of Gaston Bachelard and Thomas Kuhn's, that the history of science is a sequence of total *ruptures epistémologiques* or scientific revolutions.) It is also widely believed that classical physics equals classical mechanics, which in turn would be reducible to Newtonian particle mechanics. That the latter is but only a very special case of continuum mechanics, is known to all mechanical engineers. And most admirers of Hegel, even

Frederick Engels, have swallowed that fantasist's assertion that Kepler's special laws of motion entail Newton's, rather than the other way round.

Another popular misconception is that classical physics, unlike its successor, was intuitive. But it does not seem such to our freshmen, who find it hard to understand that a moving body left to itself keeps moving (principle of inertia); that planets circle the Sun pulled by a force that is perpendicular to their motion; that a magnetic field is not the same as the iron filings that indicate its presence; that equal volumes of different gases contain the same number of molecules (Avogadro's law); that electric charges are multiples of an elementary charge; or that the basic laws of physics (unlike some physical properties) are the same relative to all frames of reference moving with constant velocities relative to one another (Galileo's relativity principle).

Because of the above-mentioned popular misconceptions, most philosophers have overlooked the rich problematics posed by continuum mechanics, classical electromagnetism, statistical mechanics, and the early atomic and electron theories. All of these theories – which are still being developed – have considerably transformed the idea of matter bequeathed us by Newton.

Even Euler, Newton's heir and the father of continuum mechanics, had a richer concept of matter than those influential twentieth philosophers of science who equated classical physics with Newtonian classical mechanics. And of course Faraday, Maxwell, Clausius, Boltzmann, J. J. Thomson, Arrhenius, and their contemporaries, knew about fields, fluids, thermodynamic systems, electrons, stars, and galaxies, neither of which is reducible to point particles. Add to these the accomplishments of the nineteenth century chemists, such as Berzelius and Avogadro, and biologists such as Bernard, Darwin, and Ramón y Cajal, and it becomes clear that the nineteenth century ideas of matter were far more elaborate than those of most philosophers of the past century.

By 1900, when the quantum revolution began, physicists, chemists and biologists had learned much about matter that was unknown one century earlier: that there are fields in addition to bodies; that most processes are irreversible; that atoms and randomness are for real; that vitalism was wrong, and biochemistry was the clue to life; that present-day organisms descend from very different remote ancestors; that material systems of a certain kind can think, and so on.

The new knowledge about matter gained in the course of the nineteenth century should have elicited a vigorous renewal of philosophical materialism. Paradoxically, the exact opposite happened: it triggered a strong idealist backlash, known in France as *la critique de la science*. In fact, a large number of authors "interpreted" physics as a refutation of materialism. Ironically, this attack on materialism in the name of science was perpetrated by distinguished scientists, while most professional philosophers went on writing on Kant and Hegel.

Indeed, at the turn of the nineteenth century Mach, Kirchhoff, Duhem, Ostwald, Poincaré, Pearson, and other practicing scientists reinvented positivism. They criticized both materialism and realism, reheated Kant's phenomenalism, and went as far as to repeat Comte's early condemnation of atomism and astrophysics. The great Ludwig Boltzmann was the only eminent dissenter.

Ironically, the most vehement critic of the new idealism was neither a scientist nor a philosopher, but a professional politician: Lenin (1908). His book on this subject, *Materialism and Empiriocriticism*, was a clever onslaught on positivism and conventionalism. Lacking a scientific background, Lenin confined his attention to secondhand sources. Yet, he was basically right, though for the wrong reason: because the opinions he criticized clashed with Engels's.

However, the revival of idealism in the name of science did not prevent the birth of experimental atomic and nuclear physics around 1900. But the quantum revolution (1925–1935) was interpreted in subjectivist terms, again by physicists rather than philosophers: as usual, the latter were out of sync with science. However, this deep transformation deserves a new chapter.

Chapter 3
Quantum Matter: Weird But Real

It is well known that the quantum theories introduced radical changes in the conception of matter. However, there is no consensus on which these changes are, or on what their impact on philosophy is or should be. Whereas some scholars have stressed the changes in physical theory, others have claimed that the quantum theory has forced us to give up the realist postulate that there are things in themselves, that is, things existing independently of the knowing subject. Having dealt elsewhere with the latter (or Copenhagen) interpretation (Bunge 1959b, 1967b, 1973a, 1985), I will concentrate here on the new ontology necessitated by the quantum theories.

All material things are either elementary, such electrons and quarks, or systems of such. In other words, things do not come in arbitrary amounts, and they cannot be divided into arbitrary parts. Thus, the ancient atomists were basically right. Moreover, some properties too are quantized. For example, the energy of an atom in a stationary state cannot take arbitrary values: it can only be in one in an infinite denumerable set. Hence, an increase or decrease in the energy of an atom is a discontinuous transition or quantum jump – an expression that has enriched ordinary language. Such discontinuities motivated the introduction of the neologisms *quantum theory*, *quantization*, and *quanton*.

However, it should be borne in mind that classical physics too contains quantized properties, such the quantity of electricity and the frequencies of a vibrating string or a membrane. And, contrary to the popular belief that the quantum theory introduced counterintuitive notions into physics, the history of the discipline teaches that the classical concepts of mass, energy, light polarization, stress, field, entropy, and even force were once regarded as puzzling.

True, Bohr claimed that the quantum theory forces us to revise the very concept of understanding; and Feynman famously stated that nobody really understands quantum physics. In my opinion this theory is hard to penetrate not only because it deals with unfamiliar events, such as the transmutation of a pair of electrons into *two or more* photons, but also because it is surrounded by a fog that emanates from the attempt to make it compatible with operationism. This is the positivist doctrine according to which the meaning of a construct consists in the operations performed to check it.

M. Bunge, *Matter and Mind*, Boston Studies in the Philosophy of Science 287,
DOI 10.1007/978-90-481-9225-0_3,

This philosophy, the so-called Copenhagen interpretation, demands that we perform the impossible feat of translating observer-free if initially unfamiliar sentences into the language of phenomena, in particular appearances to laboratory workers. The ancient atomists knew that we should do the exact opposite, that is, to account the perceptible in terms of the imperceptible. Chronometers do not make time: they only measure it. Likewise, photocells do not generate electrons out of nothing: they only detect electrons knocked off by photons from selenium atoms in the cells.

This is not to deny that electrons, photons, and the like are extremely sensitive to macrophysical systems such as measuring instruments, some of which alter the values of the property being measured, while others generate new properties in pre-existing things. The point is that all the things in question exist in the real world instead of being figments of the experimenter's imagination. This, the realist thesis, has not been damaged by quantum physics. Rather on the contrary, for, when assembling an experimental set-up, or when reading a measuring instrument, as well as when posing a theoretical problem, we assume that we are dealing with real things possessing physical properties, rather than mental ones. This is why no one in his right mind expects to find answers to physical questions through introspection. Psychologists, not physicists, are competent to investigate human experience.

Moreover, physicists take it for granted that the things they study are natural objects that pre-exist them, even if invasive measurements can alter some of their properties. If they did not make this realist philosophical assumption, physicists would study themselves, in particular their own experience: they would regard their science as a tool "for ordering and surveying human experience," as Bohr once said with the approval of Mermin (2009). That is, physicists would describe only their own experiences, and would do so in terms of secondary (subject-dependent) properties, such as color and taste, not primary (subject-free) ones, such as wavelength and acidity. But they do not do that. Hence the instrumentalist thesis, which the Inquisition brandished against Galileo's realism, is false.

The fact that in the course of their investigation of the external world physicists simplify ("stylize") matters and employ abstractions, and that sometimes they wrongly reify them, refutes naïve realism but makes no dent on scientific realism. Indeed, scientific realism stresses that our theories about real things are symbolic and roundabout rather than figurative and literal. But it insists that they refer to real material things, as shown by an analysis of expressions such as "the rest mass of the electron" and "the energy levels of the hydrogen atom". If physical theories did not have such intended reference, they could be neither be subjected to reality checks nor be used in technology. However, let us get on with the quantum/classical contrast.

The most important feature of quantum physics is not discontinuity, not even irreducible chance, but its contribution to the old question: What are the ultimate constituents of matter? Anaximander, the father of Western cosmology, conjectured that the various kinds of material were but variations on a primary matter (*apeiron*).

Robert Boyle, the founder of modern chemistry, revived this hypothesis two millennia later. And William William Prout (1815) updated it in the light of the recently revived atomic theory: He identified the primary matter with hydrogen, because he had found that the atomic weights are multiples of hydrogen's. Astonishingly, Prout came close enough: Around 1930 it was thought that all atoms are composed of protons (the hydrogen nucleus), neutrons, and electrons. Thus, diversity branches out from unity.

At the time of this writing we know more than 200 particle species, plus massless entities like the photon and the putative graviton. We also know that the proton and the neutron, far from being elementary, are composites of quarks and gluons. So, instead of a single primary matter we now count a hundred or so. Thus, the ancient idea, that the universe has ultimate constituents, has been kept. What neither the ancients nor Boyle nor Prout could anticipate, is that these basic constituents are utterly different from the tiny marbles that the earlier atomists, from Democritus to Dalton, had imagined. Reality is increasingly proving to be far more complex and less visualizable than anything that the most sophisticated theologians and science fiction writers could invent. Besides, many of the inventions of the quantum physicists have turned out to be amazingly realistic, thus contradicting the antirealist philosophy inherent in the orthodox or Copenhagen school. Let us take a look at some of the quantum-theoretical complexities.

3.1 Meet the Quanton

The quantum revolution, started in 1900 and still vigorously underway, dealt a heavy blow to the corpuscularian worldview and, by the same token, it reinforced the field-theoretic conception. Indeed, although the things described by the quantum theory are neither particles nor fields, they are more similar to fields than to particles, since they have no shape of their own, spread over space, diffract, and interfere. Thus, the position probability of an electron released into an empty room will spread out until occupying it in its entirety. In fact, its state function, the famous ψ, will expand, to vanish only at the room boundaries, as well as at special places in between – just like any other field. (And yet string theorists boast that, unlike quantum physicists, they conceive of the blocks of the universe as strings or sheets rather than point particles – as if quantum mechanics had retained the classical fiction of a point particle.)

Correspondingly, the dynamical properties of the quanton, in the first place its position, will be spread over the whole room – or indeed over the whole galaxy in the case of a neutrino that has been traveling unmolested for billions of years. However, a detector will localize the quanton at the regions where the state function ψ, or rather the square of its amplitude, peaks. In general, interaction with a macrophysical thing, whether natural or artificial, causes significant changes in the quanton, in particular a radical shrinkage of its position distribution.

Philosophical idealists and their unwitting followers in quantum physics have interpreted this result as confirming the idea that the observer can "conjure up" at will microphysical things and their properties. They forget that they themselves are composed of atoms and molecules, all of which exist most of the time outside their laboratories and on their own rather than thanks to scientific observers. When physicists do their job they study the external world, not themselves. And they take it for granted that most of the physical world lies far beyond their reach – e.g., in the center of our planet, or in another galaxy.

What happens is that a microphysical thing is likely to be overpowered by a macrophysical thing such as an invasive measuring instrument. This is why an experimenter intent on measuring a dynamical variable P will employ a device that may bring out P, thereby smudging its "conjugate" or partner Q. Two of the many pairs of conjugate variables are position and momentum in mechanics, and photon number and phase in electrodynamics. We shall return to this point in Section 3.4. Suffice it to note for now that the typical quantum-theoretical variables come in pairs: the only bachelors in the theory are the parameters inherited from classical physics, notably time, mass, and electric charge.

In any event, the field-like behavior of microphysical entities is the root of their remaining properties, and the reason given for calling them *quantons* (Bunge 1967c). Further, I propose to call *classons* the things that are correctly described by classical physics; and *semiquantons* (or semiclassons) the mesoscopic objects, such as DNA molecules and the objects of nanotechnology, which require semiclassical theories. Quantons are in evidence in any process where Planck's constant h plays a role. Contrary to folklore, a very small size is sufficient but not necessary for quantons to be in evidence: A superconducting lead ring, a glassful of liquid helium, the black-body radiation contained in a microwave oven, a neutron star, and other macrophysical things are quantons.

The best known trait of quantons is of course their very existence: the fact that there are such minimal units of matter as electrons and photons, just as there are units of money (such as the cent), and a unit of information (the bit). Shorter: Ancient atomistics has been vindicated, even though what we now call "atoms" have turned out to be composite, and some of their constituents, such as the protons and neutrons, are assumed to be composed of even more basic quantons, namely quarks and gluons. Not only bodies and corpuscles, but also fields are assumed to be quantized. (Actually, so far only electromagnetic radiation fields have been shown to be quantized, namely photons. The electrostatic and magnetostatic fields are not quantized, and the gravitational field is still to be correctly quantized.)

These findings force a radical change in the classical conception of the universe as made up of self-contained individual particles. Indeed, the basic ingredients of the universe are interpenetrable fields, not impenetrable particles. In particular, electrons and positrons are the quanta of the electron-positron field, which extends over space and accounts for some of the counterintuitive features of the quantum theory, such as the interference occurring behind a two-slit screen. If wave optics had emerged before mechanics, we might have been spared many a "paradox".

3.2 Loss of Individuality

One of the most puzzling features of quantons is that they do not have the individuality that allows us to tag classons. Take, for instance, the two electrons in a helium atom – or, for that matter, the zillion electrons in a copper wire. In principle these electrons are countable, but an exchange of such quantons makes no difference to the probability distributions of the system. (The state of the system is invariant under permutation of the components if these are bosons, or integral-spin quantons; and it only changes the sign or phase if they are fermions, or have half-integral spins.)

In other words, such "particles" are equivalent: They lose their identity or individuality when incorporated into a system. They are usually said to be *identical*, hence indistinguishable. But, since they are countable, the truth is that they are *equivalent*, hence exchangeable – but not indistinguishable. See Fig. 3.1.

An even more drastic case of absence of individuality is that of the quarks and gluons that constitute the protons and neutrons. According to the principle of confinement, they are inseparable from one another, and do not occur in isolation. But only a few years ago quarks (and antiquarks) were shown experimentally to be real: they were produced in the Large Electron-Positron Collider. The reaction schema was electron + anti-electron → ultra-high energy photon → quark + anti-quark.

(The last stage does not occur if the photon hits an atom, as happens every time a PET scan is performed.) Notice the chain of qualitative changes: "annihilation" (submergence of mass) and "creation" (emergence of mass). Mass was first lost, then regained, but energy and charge (as well as momentum and spin) were conserved throughout. Lucretius' principle of conservation of matter was vindicated, even while Lavoisier's principle of mass conservation was shown to fail for high energies.

Another remarkable peculiarity of quantons is entanglement or non-separabilty. Several experiments performed since 1981 have shown that, when a microsystem dismantles, its fragments appear to continue to be joined: their properties exhibit correlations at a distance ("nonlocal") even in the absence of couplings (attractive forces). Such entanglement, or loss of separability, is "paradoxical" (unfamiliar,

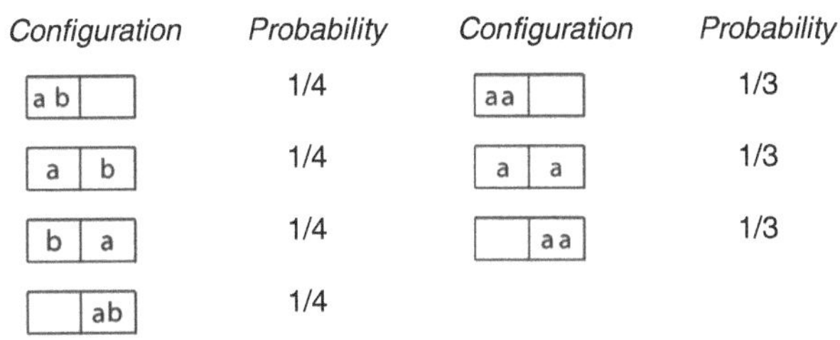

Fig. 3.1 Equivalent or interchangeable ("indistinguishable") components of a system. The cells are regions in the system's state space. The configuration in this space will differ or not according as the states "object a in cell 1 and object b in cell 2", and "object b in cell 1 and object a in cell 2" are assumed to be the same or different. From Bunge 1985, part 1, p. 218

unexpected, counterintuitive). This is because all the classical forces, except for the elastic one, decrease with distance; and because microphysical things are not expected to "remember" their past. (Nuclear forces have a very short range, and the forces among quarks are short-range as well, but they increase with distance. And all systems with memory known until recently, such as magnets and plastics, are macrophysical.)

Entanglement (non-separability) has been interpreted in several non-physical ways, including recourse to telekinesis; it has also been claimed that it refutes realism and confirms holism. In my view, all entanglement does is to confirm the thesis *Once a system, always a system*. However, this is not an independent postulate, but a consequence of conservation laws. Consider, for example, a system composed of two particles with anti-parallel spins: $\uparrow$ and $\downarrow$. If the system splits into two parts, and one of them is found with spin up, the other ex-component will be found with spin down, because the total spin, in this case 0, is conserved.

In other words, the original components of a system continue to be correlated (associated) no matter how far apart they move – until captured by another system. Thus, the past matters even at the microphysical level. And physical separation entails spatial separation, but not conversely. In sum, entanglement does not harm realism. It just confirms systemism, the thesis that very real thing is either a system or whole, or a system's component (Section 1.3). If realism had really been refuted, no one could perform reality checks, such as measurements.

Measurements on quantons are best seen as cases of entanglement by design. Indeed, to measure a quantum property one entangles it with a property of a meter, typically an indicator variable such as the angle of a pointer on a dial. Such entanglement produces a quanton-meter supersystem, whose state is a function of the variables in question. Thus, in the simplest case of a quanton with two possible states, a and b, entangled with a meter with only two possible positions, α and β, the combined state can be written in the form $\Psi = \Phi_1(a,\alpha) + \Phi_2(b,\beta)$. Clearly, this formula cannot be rewritten as a sum or a product of functions containing only the quanton or the meter states. Note also that the preceding formula is just a skeleton to be fleshed out with particulars about both the quanton and the meter. Incidentally, it is a reminder that there is no precision measurement without theory: the idea that measurements exude theories is just a philosophical myth.

Entanglement and its dual, decoherence, are particularly in evidence in the case of measurement, but they not rare laboratory contrivances. As Schrödinger (1935) noted, entanglement is even more characteristic of the quantum theory than quantization. Therefore, the sticky bur might be a better classical metaphor for the quanton than the smooth marble, and quantons may have to be rechristened *tanglons*. This is not to belittle quantization. After all, only the quantum theory settled the age-old question whether matter might be infinitely divisible, as Aristotle and classical physics supposed, or not, as the ancient Greek and Indian atomists held.

The Pauli exclusion principle is another case of quantum systemicity. Indeed, it states that at most two electrons of a system can be in the same energy state. (Actually this principle holds for all fermions, or spin one-half quantons, such as electrons and neutrons.) A classically minded physicist might wish to explain this principle by an ad hoc repulsive force, but the energy operator for a two-electron system contains no potential for such force.

A similar situation occurred a century ago with regard to special relativity, when some people tried to explain the Lorentz length "contractions" and time "dilations" in terms of pushes and pulls, instead of abstaining from using the words in quotation marks, and admitting that spatiotemporal features are relational, not intrinsic. All forces cause changes, and thus differences, but not all differences stem from changes. Moreover, some changes do not alter anything. For example, if all the positions and velocities in the whole world were to increase by given amounts, the physical laws would stay the same.

3.3 Loss of Vacuum and Stability

Another heavy loss is that of the vacuum, whose discovery had been an important piece of the Scientific Revolution. This finding refuted Aristotle's doctrine of the plenum and, by the same token, rendered atomism plausible. However, plenism was ultimately vindicated. Indeed, quantum electrodynamics discovered that, even when a container is thoroughly vacuumed, and all the electromagnetic fields are switched off, something remains in addition to the gravitational field: namely, the fluctuating quantum vacuum. This field may be regarded as the residual field that remains after all the electric charges have been removed. The intensity of this weird field fluctuates around zero, but occasionally it is strong enough to pull an atomic electron to a lower energy level. Thus radiative decay, which was initially assumed to be a spontaneous process, turned out to be caused – though occurring randomly rather than on a regular schedule. Moral: Never be in a hurry to bury once-powerful ideas.

The amazing fact that the vacuum has physical properties, so that it can exert a force on a bit of matter, might not have surprised Aristotle, Descartes, or the founders of the wave theory of light. But field-theoretic thinking is not for everyone. For example, chemists and mechanical engineers do not have much use for it; even quantum theorists call "virtual particles" the quantum fluctuations. Corpuscularian thinking is so much more intuitive than the field-theoretic one, that a few eminent physicists have tried to understand everything in corpuscular terms. For example, Feynman described the electrostatic repulsion of two electrons as mediated by a virtual photon – that is, one that violates energy conservation.

The Feynman diagrams became very popular despite containing unphysical ideas such as those of virtual particle, and positrons as identical with electrons moving toward the past. Physicists specialized in computation love those diagrams because they are mnemonic aids; and they are popular because they look intuitive, as they replace fields with corpuscles. (For a detailed criticism see Bunge 1959b.) See Fig. 3.2.

Newton's principle of inertia had challenged the alleged passivity of matter: it showed that, once in motion, bodies keep moving without having to be pushed or pulled. Moreover, the greater the mass (and the velocity), the greater the force

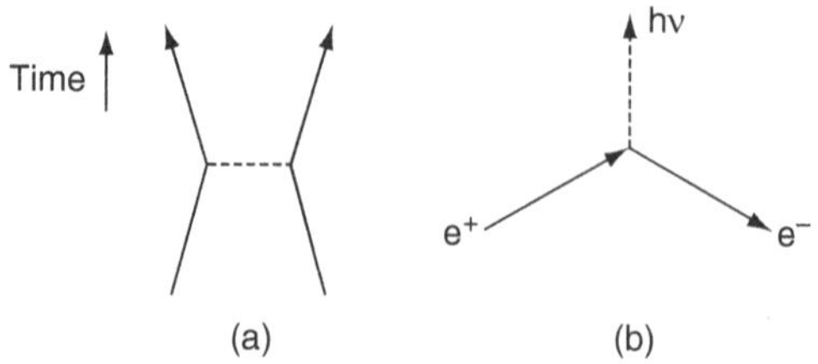

Fig. 3.2 Two Feynman (corpuscularian) diagrams. (**a**) Interaction between two electrons (*full lines*) mediated by a virtual photon (*dotted line*). (**b**) "Annihilation" of an electron-positron pair. The electron moves forward, to the future, whereas the positron is conceived of as an electron moving backward in time. The photon is imagined, à la Newton, as a tiny marble. Computation aid or gullibility test?

needed to stop them – hence the dual role of mass, as a measure of inertia as well as of quantity of matter. Inertia was so counterintuitive that Kant, writing a whole century after Newton, failed to understand it. This is why he invented a repulsive force that would balance the Sun's attraction. And yet many a historian of philosophy has regarded Kant and Hume as the philosophers of the Scientific Revolution, which they actually opposed – Hume explicitly and Kant unwittingly (Bunge 2006a).

Two centuries later came another shock: the discovery of radioactivity, whereby lumps of matter of certain kinds spontaneously emit electrons, alpha rays (the nuclei of helium atoms), and high-intensity electromagnetic radiation (gamma rays). And, of course, half a century after this discovery the nuclear bomb was invented, manufactured, and used. In between, nuclear physics had come to the rescue of astrophysics, by unveiling the nuclear reactions that generate the photons, electrons and other quantons emitted by stars.

The discovery of radioactivity also falsified the classical idea that matter is stable. Indeed, radioactivity involves the transmutation, natural or induced, of elements: the transformation of atomic species, earlier thought to be immutable. Nor is transmutation limited to the heaviest elements: even neutrons are radioactive. Indeed, after only about 15 min, a neutron decays spontaneously into a proton, an electron, and an antineutrino. (But when united to a proton forming a deuteron, a neutron does not disintegrate: It survives as such by surrendering its independence.)

However, such profound transformations in kind are accompanied by the conservation of energy, electric charge, and spin (or intrinsic angular momentum). Ironically, it took quantum mechanics to account for the existence and stability of the light atoms. Indeed, the very existence of the hydrogen atom, a system composed of a proton and an electron, is impossible according to classical physics, which predicted that it would implode because of the electrostatic attraction between its components. Bohr's semi-classical theory of 1912 postulated the said stability but did not explain it. This theory was the earliest version of quantum mechanics, and some of its features, such as the quantization of energy, have been kept. Most of the others, however, have not.

The neat orbits with precise velocities of Bohr's semiclassical model proved to be nonexistent. For example, whereas according to Bohr's model the electron in the first excited state of the hydrogen moves at $c/137$ cm s^{-1}, it does not move at all according to quantum mechanics. The latter, unlike Bohr's theory, does not contain a kinematics – as I discovered after completing my dissertation on the kinematics of the relativistic electron (Bunge 1960b). David Bohm's brave attempt to "classicize" quantum mechanics did not suggest new experiments. Einstein was wrong in denouncing quantum mechanics as being basically incomplete, although he was right in criticizing its subjectivist interpretation. Hence the famous Einstein-Bohr match ended up in a draw (Bunge 1979b).

The loss of kinematics at the quantum level is not due to the so-called wave-particle duality, for there is no such thing: Quantons are *sui generis* entities, neither particles nor waves (Bunge 1967c; Heisenberg 1930; Lévy-Leblond and Balibar 1990). But, since in certain environments quantons do behave *like* particles, and in others they act *like* waves, they may be said to be both *potential* particles as well as *potential* fields.

Let us now tackle another striking feature of quantons: the impermanence of some of them. Consider the following qualitative changes involving an electron (e^-), a positron (e^+), a gamma-ray photon (γ), and an unspecified nearby massive thing X:

"Creation" $\gamma + X \rightarrow e^- + e^+ + X$
"Annihilation" $e^- + e^+ \rightarrow \gamma + \gamma$

Notice that the electron and the positron did not preexist in the photon, just as the latter did not preexist in its precursors: both are cases of emergence of radically new things. These events are not laboratory curios: they happen all the time in stars and other "celestial" objects. In particular, our planet receives an intense gamma-ray flux originating in electron-positron collisions at the center of our galaxy. In both above reactions the total electric charge is the same, namely nil, before and after the transformation. The total energy too is conserved. (The massive thing called X in the first reaction schema absorbs or delivers the momentum that ensures the conservation of the total linear momentum.)

It will be noticed that mass emerges in the first reaction, while it submerges in the second. The words "creation" and "annihilation" have been put in quotation marks because they are absurd: they were given on the popular mistaken assumption that matter is the same as mass. Another misleading misnomer is that of "antimatter" for positive electrons, negative protons, and the like: antiquantons are just as material as quantons. The quantity of matter, measured by energy, is conserved even if mass emerges or submerges, as in the above reactions. Likewise, protons and neutrons are constituted by u and d quarks, which have tiny masses, as well as by gluons, which are massless: Mass increases, but energy remains constant.

3.4 Neatness Lost

Another surprising peculiarity of quantons is that they are blurry or fuzzy rather than neat or sharp. Whereas in classical physics all properties are sharp, in quantum physics only a few are: most are blunt or smudged. In fact, the dynamical variables, such as position, linear and angular momentum, and spin, are attached to probability distributions. They have sharp values only exceptionally; for example, the energy of an atom in its ground state, and the spin of an electron in a magnetic field. (Such fuzziness is often called "indeterminacy" or "uncertainty". This is wrong, because the distributions are subject to law, and uncertainty is a state of mind, not of physical things.) Only time intervals, masses, and electric charges are assumed to have definite values.

For example, an atom has either this or that mass, but it may be at the same time *here* with probability p, and *there* with probability $1 - p$. It is only when interacting with its environment that the atom "decides" to be either here or there. (The environment may or may not include a measuring instrument.) The loss of entanglement is called *decoherence*. All other correlations decay exponentially, hence cease only asymptotically; by contrast, decoherence may happen suddenly and totally ("sudden death"). See Fig. 3.3.

Thus, in general quantons have position, momentum (both linear and angular), spin, and energy *distributions* rather than *sharp values*. The latter emerge when the quanton interacts with its environment to become a classon or close to such. Consequently, in general, a dynamical variable A, such as the linear momentum, comes with a dispersion or standard deviation ΔA. The reason for this fuzziness is that ordinarily an isolated quanton is in a "coherent" state, that is, the combination or superposition (weighted sum) of two or more basic states (or eigenfunctions). The superposition or "entanglement" of states is a hallmark of quantum mechanics. And yet it is never observed.

However, when a quanton interacts with a macrophysical thing, whether natural such as a neuron, or artificial such as measuring apparatus, the said superposition collapses onto one of its elementary components, the way a vector projects onto a coordinate axis. It has been estimated that in a dense medium, such as the brain, coherence is so short-lived (on the order of 10^{-13} s), that quantum mechanics is unlikely to be relevant to neuroscience, hence to psychology. Therefore quantum neuroscience is a stillborn. However, let us get on with the superposition story.

According to the Copenhagen school, the superposition (or coherent state) collapses suddenly onto a sharp state (eigenstate), and it does so only as a result of a measurement or experiment – a sort of miracle, since the said interpretation has never proposed any mechanism for that event. But an analysis of the basic assumptions (axioms) of the theory fails to reveal the occurrence of any observer's coordinates in them (Bunge 1967b).

The emerging consensus is that the collapse or reduction does occur, though not as a result of observation but as an effect of the interaction of the quanton with

$$p \text{ here and } (1-p) \text{ there} \xrightarrow{\text{decoherence}} \text{Either here or there}$$

Fig. 3.3 The quanton → classon transition: Interaction with the environment localizes an initially delocalized quanton, which is here with probability p, and over there with probability $1 - p$

its macrophysical environment, which in particular might be an automated meter (see Schlosshauer 2007). Furthermore, the observed decoherence or collapse is continuous rather than instantaneous, and its rate can be experimentally controlled, by varying the distance between the state components: The boundary between the quantum and classical worlds can thus be shifted at will (Brune et al. 1996).

This late if partial vindication of causality might have pleased Einstein, de Broglie, Bohm, and Bell. However, they might end up being disappointed, for the realist view retains the hypothesis that decoherence is a random process: each sharp state emerges only with a certain probability. See Fig. 3.4.

Let us go back to the isolated quanton. A typical distribution of a dynamical variable A is bell-shaped, with width ΔA, called mean standard deviation (or square root of the variance). The widths of a position and momentum distribution are usually written Δx and Δp respectively. Each of these dispersions may take any value, but their product has a fixed lower bound, namely $h/4\pi$, where h is the tiny Planck constant. That is, $\Delta x . \Delta p \geq h/4\pi$: The sharper x, the more blunt p, and conversely. This is one of Heisenberg's famous inequalities, usually called "uncertainty relations". This misnomer originated in the belief that the position and momentum of a quantum are sharp but unknown, the reason for this uncertainty being that, in attempting to measure them, the experimenter disturbs the thing measured. Thus, ironically, the quantum "indeterminacies" were initially explained in classical and causal terms and as the effect of certain human interventions: they were seen as imperfections, and attributed to the so-called "observer effect." Moreover, a theorem in theoretical physics was regarded as a revolutionary epistemological principle: one about what a subject can know.

However, this is not exactly the interpretation that prevailed after the famous encounter of Bohr and Einstein in 1935, when most physicists adopted the official or Copenhagen interpretation of the mathematical formalism of the quantum theory. According to this interpretation, measurements do not *reveal* definite (sharp) property values, as Einstein thought, but *generate* them. Actually, as even the orthodox Pauli (1958) admitted, this only holds for *invasive* measurements such as those of position, spin, and polarization. There are also *non-invasive* measurements, such as those of time, mass, and wavelength, which only reveal pre-existing values.

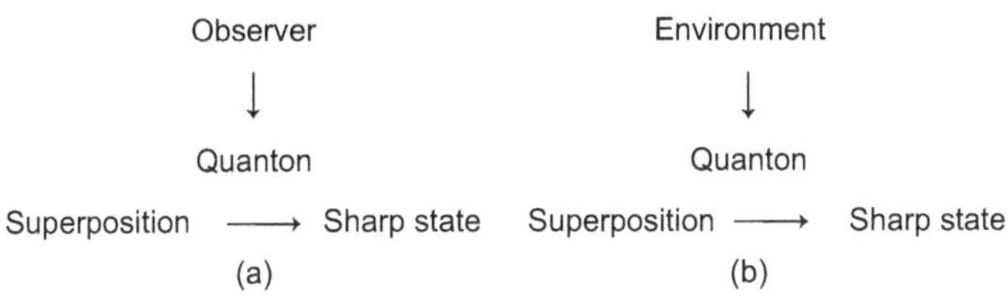

Fig. 3.4 Two conflicting views of the "collapse" or decoherence of the coherent state ("wave") function. (**a**) The subjectivist or Copenhagen view: the observer is the prime mover; (**b**) Interaction with the environment, whether free or controlled, causes the "collapse". Ironically, either interpretation partially vindicates causality

This is the place to distinguish two kinds of realism: classical or naïve, and new or scientific. A classical realist, such as Einstein, de Broglie, Bohm, and Bell, assumes that all properties have sharp values under all conditions and at all times. In other words, the classicist assumes that all properties are represented by "hidden" variables, that is, dispersion-free (or sharp rather than blunt) variables. And, since the quantum theory does not contain such variables, to implement this hypothesis, the classicist must enrich the standard quantum theory with such new variables and the corresponding hypotheses and definitions.

David Bohm did exactly this in his famous 1952 papers: he enlarged quantum mechanics by introducing two hidden variables, one for the position and the other for the momentum. He therewith restored kinematics. Bohm's papers angered the keepers of the Copenhagen flame, who were obviously reluctant to rethink the foundations of the theory. It was the old story of the revolutionaries who turn conservative the moment they attain power. But of course, pace the social constructivists, the conservatism in question had no political overtones: it was just intellectual inertia.

A decade later, Joseph Bell proved an empirically testable theorem, the famous Bell inequality, concerning a whole class of hidden variable theories. And in 1981 Alain Aspect and collaborators refuted this theorem in the laboratory. At the time, the journal *Science* reported that the experiment had "refuted realism." This is still the consensus – a sad comment on the loose use of a key philosophical concept. (More precisely, the claim is that the victim was "local realism". But this oxymoron has nothing to do with philosophical realism: it means the conjunction of the Einstein-Podolsky-Rosen classicism with the negation of the hypothesis of action at a distance. See Norsen 2007.) In any event, scientists happen to resort to experiment to check whether a hypothesis fits reality. Those who claim not to believe in the reality of the external world engage in navel contemplation.

I submit that actually the only casualty of the experiment in question was classical or naïve realism. The standard quantum formalism can be interpreted in an objective (or observer-free) way, that is, in terms of physical objects that exist on their own, though with blunt rather than sharp dynamical properties (Bunge 1967b, 1973a, 1985; Pérez-Bergliaffa et al. 1993, 1996). Let us call this the *new* or *non-classical* realism. Seen in this light, the famous Bohr-Einstein match of 1935 ended up in a tie: Bohr was right in upholding quantum mechanics, though not subjectivism, whereas Einstein was right in sticking to realism, though not to classicism – as was argued in Section 3.3.

In the new realist interpretation, the dispersions, "indeterminacies," or "uncertainties" in question are just as objective as the underlying probabilities: They are properties of the quantons in themselves, whether or not they are being observed (Bunge 1967b, 1973a, 1985; Gottfried and Yan 2003; Lévy-Leblond and Balibar 1990; Phillips 1949). In other words, quantons are neither point-masses nor marbles: instead, they are extended fuzzy blobs lacking a well-defined shape of their own. The density of any such blob equals $|\psi|^2$, a function that is space and time dependent. This quantum density is quite different from the classical, mass, energy, and charge densities. Indeed, the product of $|\psi|^2$ by the volume element ΔV at a given point in space equals the *probability* that the quanton be *present* in ΔV, whether or not it is being observed. This is de Broglie's realist version of Born's postulate, according to which that quantity would be the probability of *finding* the "particle" in question in ΔV upon performing a position measurement. Obviously, this second probability depends not only upon the probability of occurrence at the given place but also upon the measuring technique, whence it cannot be calculated using only quantum mechanics.

How about the so-called fourth Heisenberg inequality, "$\Delta E.\Delta t \geq h/4\pi$", between energy E and time t? Bohr proposed this formula on the basis of a thought experiment designed to refute Einstein's objections against quantum mechanics. But, although it occurs in nearly al textbooks, the formula does not follow from the principles of quantum mechanics (Bunge 1970). Nor could it be part of the theory, because it contradicts the law of energy conservation, as well as the presupposition that time is a classical variable, that is, one with nil dispersion Δt. Nevertheless, that ghostly inequality has been exploited in quantum electrodynamics, where it is used to justify the introduction of virtual particles, entities that would violate the conservation of energy. Indeed, it is argued that a particle can borrow the energy ΔE required to do something unlawful during a time Δt so short that no one will notice – like the thief who steals before the alarm sounds.

Thus, imaginary particles and processes are smuggled into the theory under the protection of a bogus formula. There would be nothing wrong if it were admitted that the argument is only a heuristic aid to facilitate extremely complicated calculations: that virtual particles are not real, and that energy is actually conserved. But usually all the bull about virtual particles and processes is delivered with a straight face.

Yet another warning: it is often said that spin has only two possible values, up and down. This is incorrect: What is true is that, when a spinning particle enters a magnetic field, its spin projects onto either of the said values, as shown by the Stern-Gerlach experiment. Otherwise the spin is in a superposition of its up and down sharp states, as shown by this hackneyed formula for an arbitrary spin state:

$$|\psi> = (a\,|\uparrow> + b\,|\downarrow>), \text{ were the weights } a \text{ and } b \text{ are such that } a^2 + b^2 = 1.$$

If the two sharp states, up and down, contribute equally to the arbitrary spin state, then $a = b = 2^{-1/2}$. This is the mark of a qubit, a device whose state is the superposition of the 0 and 1 states in equal measure: $|\psi> = 2^{-1/2}(|0> + |1>)$.

A classical analog is the flipped coin, which during its flight may be said to be in a superposition of the "sharp" head and tail states. Most physicists missed the centrality of superposition during the early days of quantum mechanics, when the emphasis was on the calculation of sharp energy levels, to account for the huge pile of data accumulated by spectroscopists since mid-nineteenth century. Schrödinger (1935), who was not interested in such calculations, duly emphasized the centrality of superposition and entanglement – two hallmarks of the quantum theory. Actually these two traits are one: indeed, superposition is nothing but self-entanglement. That is, saying that two or more sharp states of a single quanton superpose amounts to saying that they are entangled. Thus, the basic property is entanglement, whether of different basic (sharp) states of a single quanton, or of different quantons.

3.5 Irreducible Chance

It is generally admitted that the probability calculus elucidates the concept of chance or randomness. This opinion is false for the following reason. Consider the simplest, albeit idealized, example of randomness: that of a binary sequence of 0s and 1s (or heads and tails respectively) produced by a coin flipper. There are n^2 such sequences of length n. All of them, whether perfectly regular like 01010101, or irregular like 01100010, have the same probability, namely $(1/2)^n$. Such strings of equiprobable events are not well-defined mathematical objects, for no formula can define them. (In other words, such sequences have no precise general term, such as $x^n/n!$ in the case of the exponential function.) By contrast, well-defined sequences or series, whether finite or infinite, may be defined by writing down the first few terms, and then "...", which means "and so on." But in a random binary string there is no "so on," because it involves no "so": there is no overall order or regularity. In short, Randomness = Disorder.

Since in a binary random sequence 0 is just as probable as 1, not even a omniscient being could predict the term following any of the 0s or 1s. In short, the concept of chance, unlike that of probability, is non-mathematical. But of course it belongs in all of the factual sciences and technologies. In these, talk of chance is legitimate only if accompanied by a model of a randomization mechanism – which of course is a concrete thing that can produce only finite sequences of random events (see Bunge 2006a; Volchan 2002).

The situation in quantum physics is similar, only worse. An omniscient being would be incapable of deriving the state function of a quanton in any way other than by solving the corresponding state equation. But knowing the state function only allows us to calculate probabilities. That is, in dealing with quantons, God would be just as challenged as any of us. For example, He could only calculate the probability of a quanton going to the left or to the right of a sharp wedge facing an incoming "particle" beam. To take another example, two classons starting at the same initial state will end up in the same final state. By contrast, two quantons in the same initial state will most likely end up in two different states, each of which is just one member of a whole set of possibilities, every one of which has an objective probability. These possibilities are real, not imaginary, for they are the ones allowed by the quantum-mechanical laws and boundary conditions. (Contrast these real or lawful possibilities with the purely conceptual ones fantasized about by the possible-worlds metaphysicians such as Saul Kripke and David Lewis, according to whom anything goes.)

In the quantum theory, the really possible states a thing can be in are represented by the values of a state (or "wave") function. A state function ψ is basic, in the sense that it cannot be derived from any other functions. (What can be deduced is the future value of ψ from its initial value.) Hence, in the quantum theory, probability is just as fundamental a property as energy. (In fact, both occur on the same footing in a state or "wave" equation, which is of the form "$D\psi = 0$", where D designates an operator containing the energy operator for the quanton in question, and ψ is the corresponding state or wave function.)

In other words, according to the quantum theory, randomness is a basic trait of reality, whereas in classical physics it is a derivative property, though an equally objective one. Note, however, that this conclusion follows only under the realist interpretation of probability as the measure of possibility. If, by contrast, one adopts the subjectivist or Bayesian conception of probability as the measure of subjective uncertainty, then randomness is only in the eye of the beholder. Moreover, in this case, since quantum probabilities are calculated from states, nothing is in any objective state. In particular, whether an atom is in an excited state or in its ground state would depend on the observer, and the experimenters's only task would be to contemplate their own navels.

But causation is just as fundamental and objective as chance, as shown by the fact that the state function depends critically upon the forces and constraints on the quanton. For example, the state functions for a hydrogen atom in a void, in an electric field, or in a magnetic field, are different from one another; they also different if the atom is confined rather than in empty space. Moreover, as we saw earlier, a quanton can be overwhelmed by its macrophysical environment to the point of being transmuted into a classon: recall Fig. 3.1. So, the quantum theory has retained causation, though in a diminished capacity and combined with chance. By contrast, antecedence ("Input precedes output") has remained unscathed: There is no time inversion except as the conceptual operation of changing the sign of the time variable. True, Feynman and others claimed that positrons are nothing but electrons traveling to the past, just because the simultaneous inversion of time and the electric charge sign leaves the basic equations invariant. But this is science fiction (Bunge 1959b).

In collisions among classons, such a bullets and their targets, the result depends crucially on the target's geometric cross-section, which is constant – in particular, velocity-independent. By contrast, when quantons collide with one another, the geometric cross-section plays no role, because quantons have no shape of their own. For quantons, the characteristic area is the so-called effective cross section. Far from being constant, this quantity is inversely proportional to the square of the velocity of the incident quantons.

In other cases, the scattering will be inelastic: that is, the projectiles will be absorbed by the target or cause the ejection of qualitatively new quantons. For example, a beam of protons colliding with nuclei of hydrogen atoms will generate a beam of positive and negative π mesons according to the reaction schema: $p + p \rightarrow p + p + \pi^+ + \pi^-$. (Some of the protons are scattered elastically, while the rest combine with the targets to produce qualitatively new things, mesons.) This and similar reactions occur not only in high-energy colliders but also in stars and in the upper terrestrial atmosphere (as effects of cosmic ray showers). The total energy, electric charge and spin are conserved in these processes, but the number of things and their masses can increase. Bread loaves and fish cannot be multiplied at will, but quantons can – though with energy expenditure, not incantations, of course.

3.6 Paradoxes

It as been said that, where the quantum theory is concerned, no question is too silly, and no answer too absurd. In fact, this theory has given rise to plenty of "paradoxes" – that is, counterintuitive results. However, it is all too often forgotten that classical mechanics and field theory too seemed paradoxical at birth: How could a body, once set in motion, keep moving without being pushed or pulled; how could the moon cause tides at a distance; how could electromagnetic waves exist other than riding on a substratum, the ether; and how could a system left to itself become increasingly disorderly? These and other "paradoxes" were eventually shown to be just ideas inconsistent with intuition or common sense, that is, fossil knowledge.

Many of the so-called quantum paradoxes turned out to be of the same kind, namely perplexities due to unfamiliarity. But others can be shown to derive from the subjectivist interpretation put forward in the 1930s by the Copenhagen school led by Bohr and Heisenberg. They all fall under the heading of "observer effect." The best known of them are the "wave" function collapse (Section 3.4), Schrödinger's cat, and the quantum Zeno paradox (see Bunge and Kálnay 1983a, b).

I submit that all of the quantum paradoxes are variations on two familiar games involving randomizers. These games are either classical, like coin flipping, or quantum-mechanical, such as radioactivity effects. Let us briefly analyze the two cases. We may distinguish three stages in the coin game: coin flipping (that can be mechanized to avoid bias), coin flight, and landing. We all agree that, during flight, the final states, head and tail, have the same probability, namely ½. Disagreement arises only when asking the question "What is the probability of *seeing* a head once the coin has landed?" A follower of the Copenhagen school will respond that observation caused the coin to land either with the head or the tail uppermost, and that either event has probability ½. By contrast, a realist will state that observation, unlike experiment, has no causal effect. He will also assert that, once the coin has landed, the real or physical probability has vanished: *alea jacta est*. The uncertainty has passed from coin to mind, which lies beyond the ken of physics. Only the superstitious tend to believe that the player, by sheer mental power, can force the coin to show either side once it landed.

The quantum counterpart is this: A device is activated by one of the products of the disintegration of a small sample of radioactive material. Call E the observable effect of this process, such as a sound signal or the switching on of an electric circuit. The Copenhagen fundamentalist will claim that, since no observation is complete while the final stage has not been recorded, the occurrence of E must be attributed to the observer or even to her consciousness. But the realist will note that E cannot be recorded unless it has occurred to begin with, and that this event is the final link of a chain of strictly physical events. This is so to the point that the probability of E occurring within the next time unit is calculated exclusively on the strength of the law of radioactive disintegration. And the realist may add that the epistemic situation would be the same in the case of a physical experiment, where the scientist designs and triggers a purely physical chain of events, such as a collision of a beam of high-energy protons with the nuclei of hydrogen atoms, which reproduces the initial particles and produces positive and negative π mesons, as we saw in the previous section. Even in psychological experiments, the experimenter is careful to remain at arms-length from his experimental subjects once the latter have been prepared.

If quantons are paradoxical (counterintuitive), dark matter is even more so, despite being thought to constitute 95% of the total matter in the universe. It is called "dark" because it neither emits nor reflects light, but it is known to exist because visible bodies orbit around it. At present the favorite conjecture is that it consists mainly of WIMPs, or weakly interacting material particles. But of course no one doubts that eventually something more will be discovered about dark matter thanks to the usual interplay between observation and theory.

For the time being the only philosophical lessons to be derived from the discovery of dark matter are that matter comes in a large variety of kinds, and that we must be prepared to find new and amazing kinds of matter if we stick to materialism and scientism. The medieval king Alfonso X el Sabio had an inkling of such complexity: He said that, had God consulted him when He was about to create the world, he would have advised something simpler.

3.7 Materialism vs. Idealism

Occasionally one reads that the power of mathematics in describing external reality is such, that the latter must be mathematical (e.g., Tegmark 2008). This argument omits a crucial ingredient of every physical theory: the interpretation of some of its symbols in terms of physical entities and their properties, as when one assumes that "E_n denotes the energy of the atom in its nth state". Such symbol-fact bridges are absent from pure mathematics, for the simple reason that mathematical objects have no physical properties. This is why it would make no sense to try and measure, say, the energy of a number, or the lifetime of a vector space. That is why a correct axiomatization of a physical theory must include one interpretation postulate along every mathematical postulate (Bunge 1967b). And that is why there are alternative interpretations of quantum mechanics (see, e.g., Bunge 1956b; Neuenberg 2007).

At other times it is held that the universe is mental, for "nothing exists but observations" (Henry 2005). But this extraordinary sentence violates the grammar of the verb "to observe". Indeed, this verb stands for a very special relation between an animal endowed with sense organs and its object of observation – a thing or event. There can be no observation if either term of the relation is missing. Yet, we are assured, in ungrammatical terms and on no proper evidence, that the whole of reality – present, past and future – is an "observer effect". Berkeley and Fichte would have been delighted. Ditto the science administrators strapped for funds, since mentalism has no need for laboratories or observatories.

Would it be possible to find any empirical support for the subjectivist interpretation of quantum mechanics? Let us see. As suggested above, and as Campbell (1920) argued persuasively nearly a century (Campbell 1920), a physical theory consists of a mathematical formalism together with an interpretation. The latter is a set of semantic assumptions (aka "correspondence rules" and "operational definitions"). These are statements of the types "Set *S* is a set of things of kind *K*", and "Attribute (e.g., function) *A* represents the property *P* of *K*s". Calling *M* the mathematical formalism and *N* its interpretation, a physical theory *T* may be conceived of as the logical sum of *M* and *N*, i.e., $T = M \cup N$. Hence, altering *M* or *N* results in a different theory. Since the formalism *M* is philosophically neutral, the theorist's philosophy will be included in the semantic assumptions *N*. No wonder, then, that if *N* agrees with philosophy *P*, the partisans of *P* will claim that *T* confirms *P*: actually, this is contained in the premises *N*.

Let us now confront two theories, T_1 and T_2, that share the same mathematical formalism *M* but differ in its interpretation. That is, $T_1 = M \cup N_1$, and $T_2 = M \cup N_2$. Next, let us enrich T_1 and T_2 with indicators I_1 and I_2 respectively, bridging the theoretical concepts in the rival theories with their empirical or observable counterparts. Now let us derive testable consequences C_1 and C_2 from the union of the theories with their corresponding batteries of indicators: $T_1 \cup I_1 \vdash C_1, T_2 \cup I_2 \vdash C_2$, where $\vdash$ designates the entailment (or logical implication) relation. Finally, design and perform a crucial experiment capable of deciding between C_1 and C_2.

Obviously, if the semantic assumptions *N* are materialist and realist, the corresponding indicators will be relations between physical variables, whereas if the *N* are idealist and subjectivist, they will be paired off with relations between psychological variables. In the former case the tests will be physical observations or experiments, such as measurements of spectral lines or of intensities of particle beams. In the alternative case, where the semantic assumptions *N* are idealist and subjectivist, we will perform psychological tests, or just contemplate our navels.

Which kind of empirical procedure, physical or psychological, will count as yielding empirical evidence relevant to the theory *T* in question? The answer is up to the reader. She should be able to decide by herself whether the physicists who opt for either of the interpretations at play do execute the empirical tests that confirms or falsifies it. In this way she will be able to check whether the physicists in question practice the philosophy they preach. Thus, the final decision about whether the universe is physical or mental will depend on the size of the experimenter's budget: if adequate, he may afford to perform a physical experiment, and conclude that physics studies matter; whereas, if the physicist's budget is small, he may only think about his own thoughts, and conclude that the universe is mental, as the Copenhagen school has always held. Thus, subjectivism comes far cheaper than objectivism.

Closing Remarks

In conclusion, quantum physics has retained some classical ideas, in particular those of time and antecedence; it has radically altered others, in particular those of causation and chance; and it has given up many of the properties formerly attributed to matter, such as those of matter and mass conservation, stability, separability, individuality (or independence), and the sharp values of every property. There is no question, then, that quantum physics has necessitated radical changes in metaphysics, even if most metaphysicians have not noticed them. (Bunge 1977a, 1985 and Maudlin 2003 are exceptions.)

Epistemology, by contrast, has not been altered by the quantum revolution, except that the latter has given the coup de grâce to both nave realism and positivism, in confirming the principle of ancient atomism, that the perceptible is to be explained in terms of the imperceptible. True, the dynamical variables, such as position, momentum, and spin, are often called "observables". But this convention is just a curtsy to positivism, for the corresponding properties of quantons are anything but directly observable. The observation of all quantum "effects" involves amplifiers.

All experimental results, on all scales – micro, meso, and macro – are explained with the help of theories that do not contain phenomenal attributes. The design, calibration, and improvement of even the simplest of laboratory devices – such as scales, ammeters, photographic plates, and Geiger counters – calls for several theories. Thus in the laboratory, by contradistinction to ordinary life, empirical data are produced with the help of theories rather than given (Bunge 1967a, 1973a). Empiricism is only suitable for hunters and gatherers.

True, quantons are so sensitive to experiment, that it used to be thought that their very existence depends on it. Thus, Heisenberg (1969, 171) held that atoms are not things-in-themselves but "parts of observation situations," hence things-for-us, to use Kant's terms. But the use of quantum theory to explain events occurring far beyond any laboratories, such as the nuclear reactions in the sun, should have dispelled any doubts concerning the autonomous existence of quantons. The denial of such autonomy is either a relic of the Berkeley-Kant-Comte-Mach philosophy, or a product of the confusion between "reality" and "separate (or independent) existence". (Analog: Just because infants are inseparable from their caregivers, it does not follow that they are not real.) By contrast, the centrality of the concept of energy has never been challenged. Let us therefore take a closer look at this concept, which is just as central to ontology as to physics.

Chapter 4
General Concept of Matter: To Be Is To Become

One may peruse an encyclopedia of physics without ever encountering the word "matter", even though the entire work deals with nothing else. For example, solid state physicists study material things in the solid state, not the latter in itself. A major reason for the omission of the noun "matter" and the adjective "material" is that the concept of matter is a very general ontological category: in physics there is no general theory of matter, just as there is no general theory of processes.

Steven Weinberg (1992, 3), an eminent student of matter on the microphysical level, believes that the concept of matter has lost its central role in physics. This is because he has kept the obsolete definition of "material" as an object endowed with mass. And some philosophers, such as Daniel Stoljar (2006), have interpreted the absence of "matter" from the vocabulary of contemporary physics as evidence that scientists ignore what "the physical" is. Chomsky (2009) concurs, and offers as evidence for such alleged ignorance the idiosyncratic opinions of the contemporary physicists John A. Wheeler and Henry Stapp. Wheeler held that the ultimate constituents of the universe are bits of information [about what?]; and Stapp claimed that quantum mechanics treats events as "experienced increases in knowledge [of what?]". Do not ask for evidence supporting either obiter dictum. And note that both are nutty and barren, rather than promising unorthodoxies; and they are motivated by the old idealist dream of destroying materialism.

Getting hold of a handful of attention-grabbing antirealist and antimaterialist eccentricities is far easier than studying physics or chemistry, the sciences of matter *par excellence*. Besides, such new-fangled pearls of wisdom are not worse than the much earlier phenomenalist pronouncements of Locke, Berkeley, Hume, Kant, Comte, and Mach, that knowledge is limited to describing appearances: that things in themselves are bound to remain "mired in mystery," or might not even exist. (See Bunge 2006a for a criticism of phenomenalism.) I submit that mysterianism is not just barren: it I also an obstacle to the advancement of knowledge and an alarming indicator of intellectual decadence.

The closest thing to a general physical theory of matter is classical thermostatics, which deals with isolated macrophysical systems of arbitrary composition, hence regardless of any particular properties, such as specific heat or viscosity. But, as Clifford Truesdell (1984) rightly noted, thermostatics is excessively simplistic even for macrophysical systems, because it ignores processes, such as flows, and assumes

M. Bunge, *Matter and Mind*, Boston Studies in the Philosophy of Science 287,
DOI 10.1007/978-90-481-9225-0_4,

isolation. Indeed, like neoclassical microeconomics, thermostatics is concerned only with closed systems, and it treats every change as a sequence of slow timeless transitions between equilibrium states. Hence, it is neither as general nor as and accurate as Planck and Einstein believed it to be.

So, if we wish to craft a general concept of matter, we must turn elsewhere. I suggest we look at the most abstract formulations of the most general physical theories: those based on variational or extremal principles, such as Hamilton's (see Lanczos 1949). These principles are all about the one universal physical property: energy.

4.1 Energy

Is there a property common to all material objects, and such that no immaterial (ideal, imaginary, abstract) objects can posses? In other words: Which, if any, is the universal physical property, hence the one that individuates matter? Answer: Energy. But there are several kinds or species of energy: kinetic and potential, elastic and thermal, electric and magnetic, nuclear and chemical, and so on. In fact, there are as many kinds of energy as kinds of process. Correspondingly, there are as many special concepts of energy as chapters of the physical and chemical sciences. (But there are no biological, psychological, or social concepts of energy.)

Unlike other species, the various kinds of energy are mutually equivalent, in that they can be transformed into one another. There is one restriction, though: Thermal energy cannot be totally transformed into work, or macromechanical energy, for some of it remains bound to the thermodynamic system. This partial unidirectionality is expressed by the second law of thermodynamics. But this restriction does not affect the microphysical processes, that is, those where Planck's constant h plays a role. And, of course, the average total energy involved in any process occurring in a closed system remains constant. (The rider "average" is requires because, according to the quantum theory, in general a thing has an energy distribution rather than a single or sharp value.) Such quantitative conservation is the reason that all the kinds of energy may be regarded as being mutually equivalent. And this in turn justifies talk of energy in general.

Yet both the general concept of energy and the general principle of conservation of energy, though rooted in physics, overflow it. Indeed, every branch of physics defines its own special concept(s) of energy, and assumes or proves its own law of conservation of energy, but none defines the general concept. To add energies of two or more types we need to join the corresponding disciplines. For example, the total energy of a jet of an electrically charged fluid or gas can only be calculated in the interscience of electro-magneto-thermo-hydrodynamics. Maybe this is why the celebrated *Feynman Lectures on Physics* states that present-day physics does not know what energy is. Might philosophy help? Let us see.

Here is a clue: All the fundamental physical theories, from classical mechanics and classical thermodynamics to Einstein's theory of gravitation and quantum electrodynamics, can be cast in the Hamiltonian formalism or its cognate, the

Lagrangian one. The centerpiece of any such theory is a function or operator H representing the total energy of the referent. H depends on time and certain basic (undefined) variables called generalized coordinates and generalized momenta. The equations of motion (or of field propagation) of the thing in question, whether particle, body, field, or quanton, constitute a system of equations for the rates of change of H with respect to those variables. (In particular, the gradient of H equals minus the force.) All change involves a variation of energy with regard to the generalized coordinates. In other words, for something to happen there must be an energy inhomogeneity or gradient. Pierre Curie said it in his pithy formula: *L'asymétrie crée le phénomène*.

So, we stipulate the following

Definition Energy = Changeability.

A pedant, such as this author, would rewrite this statement as follows:
For all x: (x has energy $=_{\mathrm{df}}$ x is changeable).
We now put this definition to work. We start off by assuming

Postulate 1 All and only material (concrete) objects are changeable

That is:
For all x: x is concrete if and only if x is changeable.

Remark 1 We have equated "concrete" with "material". This convention is more common in philosophy than in physics. According to our stipulation, physical fields are just as material as extended bodies. (Recall Section 2.2.)

Remark 2 Our definition of the general or ontological concept of energy in terms of changeability shows that it is defined rather than primitive (undefinable, defining). By contrast, the concepts of energy occurring in the theories cast in the Hamiltonian format are primitive in them.

Remark 3 As Aristotle noted, change of place is the simplest kind of place, in that it is merely quantitative. Moreover, unlike qualitative change, which is absolute, change of place is relative to the frame of reference: what is at rest relative to one frame is moving relative to other frames. Hence rest, which Aristotle took to be the "natural" state of things, is a very special case of motion. The concept of stable equilibrium generalizes that of rest. Correspondingly, the statement that systems in equilibrium react to perturbations in such a manner as to restore equilibrium (Le Chatelier' "principle") would seem to vindicate Aristotle's basically static worldview. But this is not so, because the "principle" in question is not a law of nature but a criterion of stable equilibrium: a rule for identifying this particular state. Likewise, the postulate of standard economics, that all markets are in equilibrium, or return to it if disturbed, is only a piece of wishful thinking. In sum, nature is not particularly fond of rest, and society does not worship equilibrium. As Heraclitus said, every thing is in a process of change: reality and changeability are coterminous.

The above Definition and Postulate 1 jointly entail

Theorem For all x: If x is a material object, then x has energy, and vice versa.

This theorem has two immediate consequences. The first is

Corollary 1 The abstract (ideal, imaginary, non-concrete) objects lack energy.

For example the concepts of energy have no energy. (Warning: P-less, where P stands for a quantitative property, is not the same as $P = 0$. For example, it is not that the balance in a cow's bank account is zero: cows just cannot have bank accounts – except of course in some of the possible worlds imagined by world-weary metaphysicians.) When we say that concepts, hypotheses or theories change, we mean that the brains that used to think of them are now thinking different ideas. For example, one may redefine to death the concept of energy. But every such concept is timeless. If preferred, one's successive conceptual creations do not change by themselves. The same holds for the concept of matter: regarded as a concept, matter is immaterial, just as motion is immobile and life lifeless.

The second immediate consequence of the above theorem is

Corollary 2 Energy is a property, not a thing, state, or process. (Warning: some authors treat "energy" and "radiation" as synonyms, which they are not. Radiation is a thing and, as such, it possesses energy.)

Remark 3 Because energy is a property, it can be represented by either a function or an operator. In classical physics, one may say that $E(c, x, t, f, u)$ is an arbitrary value of the energy of kind E of thing c situated at point x, and time t, relative to reference frame f, and reckoned or measured in the energy unit u. The function in question has then the general form $E: C \times E^3 \times T \times F \times U \rightarrow \mathbb{R}$, where C is the set of all possible concrete things, E^3 stands for the Euclidean 3-space, T for the set of instants, F for the set of reference frames, U for that of energy units, and $\mathbb{R}$ for the real line. In the case of an interaction energy, such as the gravitational or the electric one, C will be replaced with the set $C \times C$ of pairs of concrete entities. In quantum physics, energy is represented by a Hamiltonian operator. (For example, the kinetic energy operator of a quanton of mass m is $(ih\nabla/2\pi)^2/2m$.The corresponding property is the energy density $\psi \cdot H\psi$, which depends on place and time. Incidentally, all the densities are intensive or non-additive quantities, unlike lengths and time intervals, which are extensive.)

Remark 4 All energy values are frame-dependent. Hence one and the same thing has as many energies as reference frames moving at different speeds relatively one another. For instance, the kinetic energy of a particle relative to a frame attached to it is nil, whereas it is nonzero relative to any moving frame. Similarly, the total energy of a thing embedded in a field becomes zero when its kinetic energy equals its potential energy. However, zero energy is not the same as lack of energy, just as zero temperature (on some scale) is not the same as lack of temperature. In these cases, unlike the case of the (nonexistent) photon mass, zero is just a special numerical value; moreover, it often depends upon a conventional scale.

Remark 5 Admittedly, there are some dubious cases. For example, are books, bank bills, contracts, blueprints, circuit diagrams, musical scores, and maps, material things? Yes, but they are also something more: they convey meanings. They are *semiotic* objects, whence they belong to culture. This is why they are useless unless accompanied by codes allowing their users to read or interpret them. And this is why they can be realized in multiple ways, that is, by multiple physical carriers: Think of the many "incarnations" of money: coin, bill, credit card, money order, promissory note, etc. (More in Bunge 2003a.)

Remark 6 Corollary 2 entails that the concept of a material thing cannot be replaced with that of energy. There is no such thing as energy in itself: Every energy value is the energy of some thing. This is why energetism, which a century ago was advanced as the alternative to both materialism and idealism, is logically untenable. However, the energetists, particularly the great physical chemist Wilhelm Ostwald (1902), were right in holding that energy is universal – a sort of cross-disciplinary currency. They would have been even more right had they proposed the following

Postulate 2 Energy is the universal physical property: the only trait common to all material things.

Remark 7 One might think that position in spacetime is another universal physical property. It is but, according to any relational (as opposed to absolute) theory of spacetime (e.g., Bunge 1977a; Bunge and García Máynez 1977; Pérez-Bergliaffa et al. 1998), the latter is not basic but derived: it is the basic structure of the collection of all things, every one of which possesses energy. Roughly, space is rooted to the spacing of things, and time to their change. No space without things, and no time without change. Aristotle nods.

Remark 8 Postulate 2 does not state that every thing has a precise energy value at any given time and relative to any given reference frame. It does not, because sharp energy values are the exception rather than the rule. Indeed, according to the quantum theory, typically a quanton is in a superposition of infinitely many energy eigenfunctions, the corresponding eigenvalues of which scatter around a central value, such as an atomic energy level or the spatial average of an energy distribution.

Our final assumption is

Postulate 3 The total energy of an isolated material object does not change in the course of time.

Remark 9 This is of course the general principle of conservation of energy. It is so extremely general that it belongs in philosophy rather than in physics.

Remark 10 It has been held that energy is not conserved in an expanding universe. But, since the universe has no walls, it is not a closed system, so that the principle in question does not apply to it.

Remark 11 Any violation of the above principle is suspect. Example 1: The steady-state cosmology, popular in mid-twentieth century, postulated the continuous creation of matter, hence of energy too, out of nothing. This hypothesis contributed to the discredit and ultimate downfall of the theory (Bunge 1962). Example 2: Telekinesis, or the ability to move things by sheer mental power, involves the violation of energy conservation. This did not deter a number of philosophers, among them C. D. Broad and K. R. Popper.

Remark 12. According to quantum electrodynamics, the vacuum energy is not zero but fluctuates irregularly around zero. This result does not invalidate the characterization of energy as the universal property of all things. All it does is to restrict the domain of validity of the classical definition of the vacuum as the absence of material entities endowed with mass. The electromagnetic field that remains in a region of space after all the electric charges have been neutralized, and all the electric currents have been switched off, is a concrete though tenuous thing. It is so concrete, that it exerts the measurable Casimir force on the outer sides of two parallel conducting plates, thus sticking them together. The Lamb shift of atomic energy levels is another of several measurable properties of the void. In short, there is no total void: The universe is a *plenum*. Aristotle and Descartes might feel vindicated. But, whereas their ether was sheer fantasy, the quantum vacuum is described in exact terms, and its existence has been confirmed by many experiments.

Remark 13 There has been much talk of "dark energy" in recent cosmology. This expression is unnecessarily mysterious, for it is admitted that "dark energy" is the name given to the unknown entity or process that counteracts gravity. In turn, such counteraction seems required to account for the apparent accelerated expansion of the universe. But such acceleration may turn out to be illusory: that is, it may be explained by ordinary physics.

Remark 14 The concept of energy may be used to define that of causation, and to distinguish the latter from correlation. Indeed, causation may be defined as energy transfer, as in the cases of the light beam that burns a dry leaf or activates a photocell. (In both cases the cause is light absorption, not light; likewise, the effects are processes: combustion in the first case, and electron emission in the second. To generalize, the relata of causal relations are events or processes.) By contrast, no energy transfer need be involved in a correlation between two "factors" or variables. If only for this reason, it is wrong to define causation as a particular case of probability, namely when the latter equals unity – the way Suppes (1970) did in his probabilistic metaphysics.

This completes our minitheory of energy. The upshot is that energy is the universal physical property; that the various special laws of conservation are so many examples of the general principle of conservation; and that "matter" is definable is "having energy" (or "having the ability to change"). This philosophical result has been reached through combining ontology with physics

4.2 Information

Let us briefly explore the relation between the concepts of matter and information.

There are several concepts of information, in particular the semantic and the technical ones. The former is roughly equivalent to that of knowledge bearer, such as a text or a diagram. By contrast, the technical concepts of information refer to concrete things such as communication systems, nervous systems, and systems composed of DNA molecules and the proteins they code for. No material bearer, no information.

Claude Shannon published his theory of information the very same year of 1948 that Norbert Wiener launched cybernetics, or the general science of control. Both disciplines became instantly famous and, because of their generality, they were promptly applied and misapplied. Wiener (1948, 155) famously wrote: "information is information, not matter or energy. No materialism which does not admit this can survive at the present day." However, all the technical definitions of "information" show clearly that, far from being self-existing like matter, information is a property of very special material systems, such as nervous systems and TV networks, and as such inseparable from stuff and energy.

The vast majority of things do not carry information, whereas all information rides on energy fluxes, all of which are material. Thus, in the end, information is just as material as energy, though far less ubiquitous because it involves coding, and codes are conventional as well as artificial. For example, traffic lights work only for people who know the convention that green light denotes "go" and red light stands for "stop". By contrast, the traffic in neural networks "obeys" only certain natural laws, such as those that "govern" the corresponding electrical and chemical signals.

In sum, the concept of information is derivative, nor primary; in particular, it depends upon that of matter. Indeed, all information is transmitted by some physical process, just as every bit of energy is the energy of some material entity, and every energy transfer is a physical process that connects two or more physical entities. (What is true is that the information *theories* are so general, that they do not specify the composition of the communication systems: they are stuff neutral.) An example of such a thing is a switch, since it can be in either of the two states "on" and "off," and thus encodes one bit of information. To invert John A. Wheeler's famous slogan, we always get *bits from its*, never the other way round.

And yet the idea that information is more basic than either matter or energy became rather popular in all research fields during the second half of the past century, from physics to biochemistry to molecular biology to psychology to sociology. Suffice it to remember the attempts to understand DNA as constituted by the "letters" A, C, G, and T; Hilary Putnam's proposal to view the mind as a set of computer programs; and Niklas Luhmann's assertion that society is a communication system where people do not matter – a graph without nodes. These fantasies hang on the coattails of the information revolution, but they have not contributed anything to our understanding of reality. This is because they focus on a single aspect of things and processes, and propose shallow views of the mechanisms that make material things tick. For instance, while it is true that all social transactions are accompanied by information exchanges, it is equally true that we only understand them when studying their specific features, such as what distinguishes manufacturing from politicking, or courting from trading.

4.3 Digital Metaphysics

Digital metaphysics is the doctrine that bits are the building blocks of the universe (and that God is a computer programmer). This is a new version of the Pythagorean myth – as Gregory Chaitin (2006), a prominent member of the new esoteric brotherhood, has admitted. This idea seems to have several sources. A first root, obvious in Wilczek (2008) and other distinguished scientists, is the confusion between real things and our models of them. This confusion is a feature of the magical thinking involved in sticking pins on dolls as a cheap and safe murder method. Another source of digitalism is the computer cult and the concomitant worship of programmers as omniscient beings, and of computer models as infallible and as having a life of their own.

A third, independent source of "digital philosophy," is the old positivist confusion of evidence (How do you know?) with reference (What is it about?). Indeed, we know that an observation or a measurement has not been completed until its outcome has been recorded – that is, translated into a piece of information. From this, Wheeler inferred that information creates reality. As Paul Davies (2004, 10) put it, Wheeler sought "to turn the conventional explanatory relationship

matter → information → observers

on its head, and place observership [sic] at the base of the explanatory chain

observers → information → matter

thus arriving at his famous "it from bit" dictum, the "it" referring to a physical object such as an atom, and the "bit" being the information that relates to it. In it from bit, the universe is fundamentally an information-processing system from which the appearance of matter emerges at a higher level of reality."

There is a missing step in the preceding argument, namely Berkeley's formula "To be is to perceive or to be perceived." Indeed, Wheeler's reasoning looks correct if that link is explicitly interpolated. But then it becomes clear that Wheeler's thesis is just a rehash of Berkeley's subjectivism – the original source of all the anti-realist interpretations of science (Bunge 1955, 2006a). Worse, the existence of the universe prior to the emergence of atomic physics becomes then highly problematic.

A further gap in the it-from-bit story is the mechanism whereby material things emerge from symbols. How is it possible for energy to be created from abstract objects that, like 0 and 1, are devoid of energy and obey man-made rules rather than physical laws? How is it possible to get an electric shock from touching a wiring diagram? And how is it possible for physicists to concoct one more version of Pythagorean mysticism, with no concern at all for empirical evidence?

A fourth and rather unexpected source of digitalism is empiricism (or positivism), the view that all knowledge is a bunch of data, and in particular that theories are just data condensates. This view involves the thesis that we only need computable numbers or, to put it negatively, that we do not need real numbers other than fractions, since every measurement outcome is a fraction. A tolerant positivist will allow mathematicians to play with transcendental or noncomputable numbers, such as π, definable as the infinite series

$$\pi/4 = 1 - 1/3 + 1/5 - 1/7 + \cdots$$

This number can be computed to any desired approximation, by truncating the above series and adding up the resulting polynomial. But a radical positivist will seek to persuade mathematicians that they should give up real numbers, and reconceptualize lines as strings of fractions. That is, the radical empiricist would attempt to digitalize all the sciences. This would involve giving up the vast majority of numbers, which are noncomputable. Consequently, it would also force us to bury the infinitesimal calculus and replace differential equations with finite-difference equations. Fortunately, the vast majority of mathematicians and scientists have not even heard the digital siren.

What about virtual reality? How real are the virtual worlds that computer buffs like to navigate? Admittedly, those worlds do not exist outside computer screens and the brains that look at them, yet virtual realities are every bit as real as those brains, since these conjure them up – just as dreams. Location in someone's external world is sufficient but not necessary for real existence: her internal world is in your external world. Let us dwell in it for a while.

4.4 What's Out There

"Why is there something rather than nothing?" This question, often credited to Heidegger, was actually asked by Leibniz (1956, II:1038), and it has always been debated in Islamic theology The question makes sense to theologians wondering what reason could God have had to create the universe. But it has no place in a secular ontology, which takes the existence of the world for granted and attempts to explain it. Only the existence of particular things, by contrast to that of the totality of things, might interest some of us. For instance, biologists would like to know why do organisms exist, particularly in view of Richard Dawkins' popular myth that they are just funnels for gene transmission, hence dispensable in principle, and thus "paradoxical." Economists must wonder why emotion occurs, since it can only distort or even block the clever calculations of a rational actor. And political philosophers must ask themselves whether democratic socialism could exist, given the force of individual interests.

In his main work, *Being and Time*, the famous existentialist writer Martin Heidegger only dealt with human existence, or rather a handful of traits of it. This is all he had to say about being one generation later: "The essence of Being is IT itself" (Heidegger 1954). And Jean-Paul Sartre, his erstwhile imitator, stated that "there is no existence except in action." Neither sentence helps us understand questions such as "Do gravitons exist?", "Do atoms exist while not being manipulated?", or "Are there free markets in equilibrium?"

Logicians, who occupy the other end of the philosophical spectrum, assure us that the "existential" quantifier $\exists$ exactifies the concept of existence. Does it? Consider these two sentences: "There are superstrings," and "There are irrational numbers." According to logical imperialists, both statements are formalized in exactly the same form, namely "$\exists xPx$," or "There are *P*s." And yet the first sentence makes a claim to real or material existence, whereas the second asserts the ideal existence of certain mathematical objects, not found in the real world.

Logic is just not equipped to distinguish, let alone check, the two main kinds of existence: real and ideal. This is not a problem in mathematics, which deals exclusively with ideal objects; but it places the problem of real existence in the hands of factual scientists and sane philosophers. The ambiguity of the symbol $\exists$ suggests interpreting "$\exists xPx$" as "*Some* individuals are *P*s," without making any existential commitment, the way an atheist might say that some (imaginary) deities are (imagined to be) benevolent if bribed. We shall return to this problem in Chapter 5.

How is real existence related to materiality? It won't do to *define* materiality as real existence, the way dialectical materialists do, because an objective idealist – such as Plato, Hegel, Dilthey, or a possible-worlds metaphysician – might claim that ideas in themselves are just as real as, or indeed even more real than, pieces of matter. Hence the materiality-existence relation must be asserted or denied explicitly and separately from the definition of materiality. This is why we posit

Postulate 4 All and only material objects exist objectively (really).

Shorter: Reality = Materiality. Notice that this is not a definition (identity) but an assumption. An immediate logical consequence of it is

Corollary 3 No ideal (or imaginary) objects exist objectively (really).

In turn, this statement entails

Corollary 4 No mathematical structures exist out there.

The last two corollaries are intended to discard objective idealism, which is still going strong among the philosophers, mathematicians, and even theoretical physicists who need reminding that it makes no sense to ask, for instance, which numbers run the fastest, or whether Boolean algebras could cure common colds. If in doubt about the presence of Platonists among us, consult the abundant literature on parallel universes (e.g., Everett 1957, Lewis 1986, Tegmark 2004).

Our preceding assumptions are hoped to elucidate what is *meant* by "real existence", but they do not help us *recognize* an object as existing out there. To accomplish this second task we need to bridge ontology to normative epistemology, that is, methodology. That is, we need explicit real (or material) existence criteria. Here are two in two parts each:

Criterion 1 (necessity and sufficiency). An object

(i) is likely to exist really (materially) if, and only if, the hypothesis of its existence is plausible, i.e., it belongs to a theory that has been

confirmed empirically, or to one that coheres with other well-corroborated theories;
(ii) exists really (materially) if, and only if, it has been detected directly (by the senses) or indirectly (with the help of scientific instruments).

Part (i) of this criterion was used when judging the likelihood of the existence of electromagnetic waves (Maxwell), positrons (Dirac), and gravitational waves (Einstein) before acquiring experimental evidence. As for part (ii), direct detection is of course perceptual, whereas indirect detection is achieved with the help of observation instruments, whether simple like magnifying glasses or complex like ammeters or radio telescopes. In addition, we need

Criterion 2 (sufficiency). An object exists really (materially) if

(i) it reacts (kicks back) when acted upon;
(ii) it has been reproduced or made.

Reactivity to a stimulus, involved in part (i), is sufficient but not necessary, because our action may not be energetic enough to provoke a testable reaction – as when aiming a flashlight at the Sun. And part (ii) applies in the laboratory, the bench, or the field. A recent example is the experiment performed at CERN to produce and detect the Higgs boson that had been postulated theoretically.

Physicists will note that our reality criteria differ substantially from those advanced by Einstein, Podolsky, and Rosen (1935) in their famous paper. With hindsight we realize that they had conflated "real" with "classical", since they presupposed that what they called "elements of reality" are classical entities or properties, such as separability and simultaneous sharp position and velocity values (Bunge 1979b). They had also subordinated the ontological problem of reality to the epistemological question of predictability: they had stipulated that X exists if X can be predicted with certainty. Bohr and his followers used this unfortunate confusion of realism with classicism to defend their own confusion: the contraband of subjectivist positivism under the guise of quantum mechanics. With hindsight one realizes that realism can be kept along with the quantum theory (Bunge 1967b).

However, epistemological realism without materialism is vulnerable and useless to science, because one may adopt Plato's or Hegel's objective realism, that is, the view that ideas exist by themselves. To a Platonist the existence (or reality) criterion is non-contradiction: an idea exists provided it is not self-contradictory. Obviously, this criterion does not apply to material entities, for contradiction and its dual, namely consistency, are peculiar to constructs, and checking for them does not require any empirical procedures.

The best way to shore up epistemological realism is to combine it with materialism, producing what may be called *hylorealism* (Bunge 2006a). Thus a hylorealist will admit a scientific theory that posits the existence of previously unknown things, however weird or "paradoxical," provided they are well-confirmed and do not contradict the bulk of the background knowledge. By contrast, a hylorealist will reject the so-called digital metaphysics ("its from bits") by saying that the notion of information, and in particular the concept of bit or information unit, makes sense only with reference to information systems, all of which are concrete artifacts.

Closing Remarks

The concepts of matter and energy have undergone huge changes along the history of science – just like other key ontological concepts, such as those of space, time, causation, chance, life, and mind. In our view what is common to all those concepts, that is, the most general concept of a material thing, is this:

For any x: (x is material = x is changeable).

Since in turn

Changeability = Having energy,

it follows that

For any x: (x is material = x has energy).

Further, since energy is conserved, so is matter – not mass, though. And since matter cannot be destroyed, philosophers had better pay close attention to the contemporary concept of matter – before physicists alter it once again. This concept is so broad, that it applies to living beings and social systems.

The broad concept of matter suggests broadening physicalism to what may be called *inclusive materialism*, which can be compressed into the following postulates:

Materiality = *Changeability*
Reality = *Materiality*.

We shall expand this view in Chapter 7. But before doing so we had better take a closer look at some of the supraphysical (yet not nonphysical) kinds of matter.

Chapter 5
Emergence and Levels

The worldview that informed the vanguard of the Western scientific community between ca. 1600 and ca. 1850 conceived of the universe as the maximal self-winding mechanism: recall Chapter 2. By the end of that period a number of discoveries and inventions contributed to the decline of this, the earliest scientific world view: the marvels of field physics, chemical synthesis, embryology, and biological and social evolution. These and more showed that matter, far from being the passive substance imagined by the traditional metaphysicians, was capable of spontaneous self-organization – the transmutations and metamorphoses dreamed up by the ancient alchemists and naturalists. Some of these processes turned out to be real, and they were not admired with "natural piety," in the manner of the holists: they were now analyzed into their elementary components. In some cases, it was attempted to replicate them in the laboratory, and even to better nature – for instance, by manufacturing artificial materials such as paper and plastics, as well as transuranians, GMOs, and, of course, by organizing utterly unnatural social systems such as schools, churches, businesses, armies, and governments.

By mid-twentieth century it was known that the elementary constituents of lead and DNA are the same, namely electrons and nucleons, as well as the gluons that keep the latter together; platinum has only one electron more than iridium; we share with chimpanzees about 99% of our genes; both New York and a primitive Amazonian village are inhabited by members of the same biological species; sublime poems and vile slanders can be composed of the same words – and so on. Such basic commonalities do not entail fundamental identities: platinum is not iridium 77 parts in 78; the reader is not 99% chimp, the US Senate is not basically a primitive tribe's assembly of elders, and so on.

The philosophical moral is clear. Composition, though essential, is not everything: structure and mechanism are equally important (see Bunge 2003a for the composition-environment-structure-mechanism model of a system.) Hence compositionism, or naïve reductionism, or methodological individualism, is simplistic. In other words, analysis or decomposition, is always necessary but never sufficient to understand wholes. We must also find out the bonds that keep the parts together and explain the emergence of global properties – for instance, the hydrogen bonds among the molecules of a water body, and the psychological and economic bonds among the members of a business firm. Besides, we must place the thing of interest

M. Bunge, *Matter and Mind*, Boston Studies in the Philosophy of Science 287,
DOI 10.1007/978-90-481-9225-0_5, © Springer Science+Business Media B.V. 2010

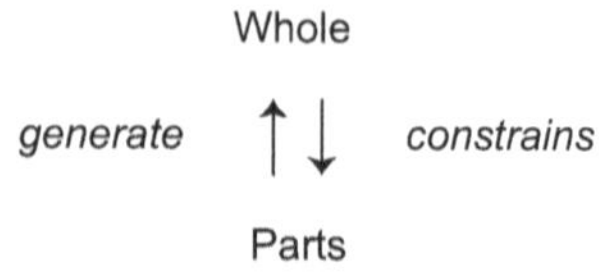

Fig. 5.1 Top-down and bottom-up analyses complement each other

in its context instead of treating it as a solitary individual. For instance, atoms in the center of the Earth lose their outer electrons; neurons behave differently in different neuronal networks; and tyrannical bosses may behave tamely and lovingly at home.

The methodological lesson is that we should supplement every bottom-up analysis with a top-down analysis, because the whole constrains the parts: just think of the strains in a component of a metallic structure, or the stress in a member of a social system, by virtue of their interactions with other constituents of the same system. See Fig. 5.1.

This conclusion contradicts the principle of the physicalist (or vulgar-materialist) worldview, according to which the universe is laid out on a single level, that of physical things, whence physicists may eventually craft the theory of everything. (In the nineteenth century physicalism, and later on physico-chemicalism, succeeded the mechanism that had prevailed in the two previous centuries.)

The methodological counterpart is that the atomistic program of explaining the whole by its parts fails every time the behavior of the part is strongly influenced by its position in the whole. For example, the entire family of rational-choice theories, so popular in social studies, has failed because individual behavior is strongly influenced by macrosocial circumstances, such as the political and economic situation.

In general, it is more fruitful to interrelate levels than to attempt to reduce them either downward (micro-reduction) or downward (micro-reduction). Boudon-Coleman diagrams (Bunge 1996; Coleman 1990) are therefore likely to appear in all the disciplines that involve two or more levels of organization: see Fig. 5.2.

In the course of the last two centuries, chemistry, biology, and social science have found systems of millions of different kinds, thus confirming the impossibility of crafting an all-fitting theory. Indeed, we know about 200 different kinds of "elementary particles"; more than 100 different kinds of atoms; over two million kinds of molecules; hundreds of millions of biological species; and hundreds of kinds of social system. In other words, some collections of things, whether atoms, molecules, specks, fibers, Lego building blocks, people, or what have you, may combine

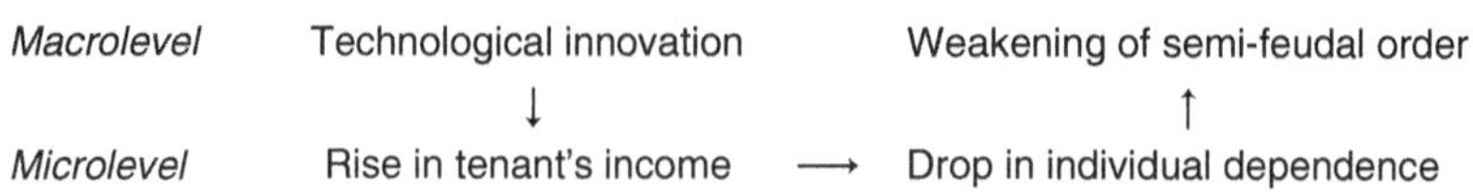

Fig. 5.2 The Boudon-Coleman diagram explaining the political impact of advances in agricultural technology in a semi-feudal society. From Bunge 1996. Other collective processes, such as boiling, growth through cell division, inflammation, and epidemics, start at a microlevel

into qualitatively different systems. This, the set of processes involving qualitative jumps, is what emergence in the ontological sense is all about (see, e.g., Alexander 1920; Bedau and Humphreys 2007; Blitz 1992; Bunge 1969, 1979a, 2003a; Lewes 1874; Luisi 2006; Morgan 1933; Needham 1943; Sellars 1970; Wimsatt 2007).

Emergence and levels are found everywhere except in classical mechanics, which deals only with bodies small and large. This science overlooks the fact that a collection of particles, such as water molecules, can organize itself in different ways: as snowflakes, ice pellets, liquid drops, lakes, rivers, seas, or clouds. Each of these things has properties that the others lack. Other familiar examples of emergence and levels are pile of bricks-wall-house-town, population-gang-firm-conglomerate, and letter-word-phrase-sentence-text. The higher levels are dependent upon the lower ones but they are not reducible to them in an ontological sense even if the complex things (systems) are shown to emerge from the interaction of their constituents. Analysis should explain emergence, instead or trying to eliminate it. (More on reduction in Section 9.8.)

Although all that has been known for quite a while, the explicit and general notion of an integrative level, or level of organization, is rather recent (Bunge 1960a, 1969, 1980b; Hartmann 1949; Novikoff 1945). And neither this concept nor that of a system has yet reached mainstream metaphysics. Worse yet, because more expensive and time-consuming, is the oversight of the level structure of the universe, for it has facilitated the unchecked proliferation of basement-level theories such as the promised "theory of everything," quantum theories of the universe, and fantasies on the physical (not biosocial) nature of the mind, as well as on the reducibility of social science to biology. It is therefore necessary to recall some of the salient points of the levels doctrine.

5.1 Physical Matter

The ancient atomists, and two millennia later on statistical mechanics as well, contradicted the one-layer doctrine, asserting instead that everyday-life things are composed of imperceptible entities. They also held that macrophysical things possess emergent properties, such as temperature and entropy, that their microphysical components lack. Furthermore, the micro-macro bridges involve the concept of objective chance, which occurs neither in classical mechanics nor in classical electrodynamics.

The epistemological consequence of the micro/macro split was obvious: We need two sets of physical theories, microphysical and macrophysical, to account for physical reality. Quantum physics confirmed this conclusion, although the finding of quantum macrophysical entities, such as superconducting rings, shifted the boundary between the two strata, from micro/macro to quanton/classon, or quantum/classical. In any event, the hope for a single theory of everything was dashed before birth, by both materialist philosophy and modern physics.

The initial program of atomic physics since antiquity was clear: to account for perceptible macrothings in terms of imperceptible microthings. In our terminology, the goal was to construct classons from quantons. Much of this program has been successfully carried out. For example, a century ago photocells were explained in terms of the electrons ejected by incident photons; magnets were explained in terms of the spins – and the associated magnetic moments – of the atoms composing the magnets; and the macroproperties of pieces of condensed matter of certain kinds, particularly the semiconductors used in electronic devices, are explained in terms of the properties of their atomic constituents. Of all these achievements, the latter had the most important impact on industry, for the digital electronic computer is the child of the marriage of quantum solid-state physics, and the mathematical theory of computation.

However, the standard (or Copenhagen) interpretation of the quantum theory tells a different story: According to it, this theory only calculates the possible outcomes of observations. For instance, it is postulated that the eigenvalues of an operator representing an "observable" (dynamical variable) are the possible results of measurements of the said observable, as if there were universal meters. Furthermore, it is assumed that all measurements and experiments involve exclusively classically describable devices. But this is obviously false, since explaining the ionization occurring in the cloud chamber and in the Geiger counter, as well as the photochemical reactions occurring in a photographic plate, require quantum physics. In any event, the Copenhagen orthodoxy erects a strict duality between quantons and classons, where the latter are to be regarded as irreducible. Actually the dogma states that the observer, his apparatus, and the object of observation, constitute an unanalyzable whole. Astrophysics, which relies heavily on the quantum theory, belies this thesis, for no measuring instruments, let alone observers, could get even near a star.

This micro-macro duality, which is in stark contrast to the atomistic program, is not a scientific result but a philosophical interpolation. In fact, it derives from the subjectivist (or positivist) principle, that there is no such thing as objective reality: there would be only phenomena, that is, appearances to some observer. In particular, the quantum facts would arise only from observation. Actually, as Bohr and Heisenberg once declared, even macrofacts, such as the passing of a streetcar, are results of repeated observations. The naive reader might ask: "Observation of what, if not atoms, streetcars, or what have you?" But he would be told – as my teacher, Guido Beck, was told when working for Heisenberg in Leipzig – that this and similar questions should not be asked. Mussolini's slogan, "Believe, obey, fight," was adapted to read "Believe, obey, calculate."

Until recently the Copenhagen orthodoxy could only be countered by conceptual analysis: by showing that neither the axioms of the theory nor experimental data refer to observers. This situation changed in the 1990s with the development of decoherence theory and experiment (see Schlosshauer 2007). Recall, for instance, the diffraction of bucky balls, such as C_{70} molecules, which are about 1 nm across. At temperatures below 1,000 K degrees, these things diffract when passing through double slits: they are quantons. But as the molecules are heated to 5,000° by a laser beam, the interference pattern gradually vanishes: the quantons gradually become classons as a result of decoherence. A result of these experiments is that the

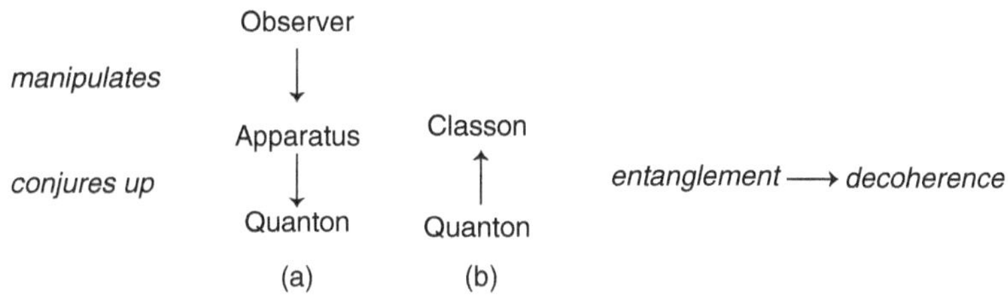

Fig. 5.3 The quantum/classical duality. (**a**) Copenhagen orthodoxy: the omnipotent observer manipulates the apparatus, which in turn conjures up quantum facts. (**b**) Decoherence program: Classons emerge form quantons, particularly when the latter get entangled with the environment (natural or artificial)

quanton-classon boundary can be shifted toward the macrophysical environment, which may, in particular, be the apparatus. See Fig. 5.3.

These recent developments have vindicated the original goal of the quantum program, namely the derivation of classical physics from the quantum theory. Does this entail that we will eventually be able to dispense with such classical concepts as those of friction, heat, temperature, viscosity, vorticity, elasticity, magnetization, surface tension, or wetting? These concepts will continue to be needed because they stand for objective bulk properties and processes that emerge from myriads of quantum facts. Likewise, the neuroscientific explanation of cognitive and affective processes does not allow us to dispense with such words as "fear", "imagination" and "love". Explained emergence is still emergence.

In other words, epistemology cannot erase ontology because the universe is objectively stratified. The qualitative differences among the levels of reality impose a plurality of levels of description, even if the higher ones can in principle be explained in terms of the lower ones. True, the proponents of the thesis that the quantum theory is universal write symbols said to designate state functions for cats, observers, measuring instruments, and even the universe. But I submit that these symbols are fake, for they are not solutions of any equations containing Hamiltonians: they are just squiggles. See Fig. 5.4.

In short, classons emerge from quantons, whence it is legitimate to try and derive classical physics from the quantum theory instead of regarding the former as an absolute given and starting point. But the ultimate goal of the "decoherence program," namely to dispense altogether with classical ideas, seems sheer fantasy, because the classical properties, such as shape, viscosity, and temperature, are just as real as the quantum ones, such as spin and non-separability. Shorter: the distinction between the quantum and classical levels is objective, not just a matter of levels of description and analysis.

If in doubt, recall the chain that starts at the center of our star and ends as sunshine, wind, rain, ocean current, chemical reaction, or metabolism. All the energy we get from the Sun comes ultimately from the thermonuclear reactions that occur in the center of stars. Most of these reactions involve transmutations (the alchemist's

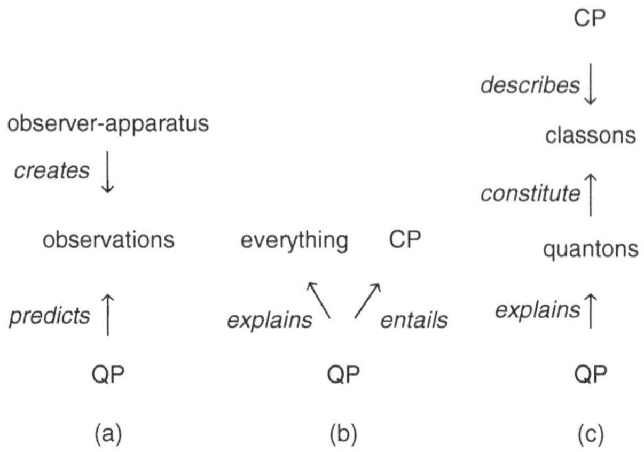

Fig. 5.4 Levels of description. CP = classical physics. QM = quantum physics. (**a**) Copenhagen doctrine. (**b**) Decoherence Program. (**c**) Realistic view

dream), such as Deuterium + Tritium → Helium + Neutrons. In turn, the sunlight produced by nuclear reactions in the Sun is critical for photosynthesis, a quantum process without which there would be no plants, hence no animals either. Note the successive levels: Quantons-Light molecules-Macromolecules-Organelles-Cells-Multicellular organisms.

5.2 Chemical Matter

The inputs (reagents) and outputs (products) of chemical reactions are physical things; more precisely, they are quantons, classons, or semiclassons. But the matter that emerges the instant a chemical reaction starts, and subsists till it stops, is qualitatively different from the inputs and outputs, even if it lasts, as it often does, as little as a femtosecond, or 10^{-15} s. This stuff is, indeed, a piece of matter undergoing relentless radical (qualitative) changes. Familiar pieces of chemical matter are flames, rusting iron pipes, fermentation vats, the contents of cooking pots, and decaying living matter. (Chemists may be tempted to regard living beings as very complex chemical reactor. But biologists know that the thousands of chemical reactions going on in an organisms at any time are scheduled and coordinated in such a way that three non-chemical traits – homeostasis, self-repair, and survival– are assured.)

It is generally thought that molecules are bits of matter composed of atoms. However, their mode of composition is far less well known. For example, the hydrogen molecule, H_2, the simplest of all, does not emerge from the mere juxtaposition of two H atoms. In fact, these atoms are the *precursors* of the H_2 molecule rather its *components* since, when combined, the original protons and electrons are redistributed in a counterintuitive fashion. Indeed, despite their mutual electrostatic repulsion, the two atomic electrons get together and interpose between the two protons; and this pair constitutes the chemical (or covalent) bond. This kind of bond, discovered in 1916, is quite different from the ionic bond, such as that which keeps the chlorine and sodium ions together in the salt molecule. Whereas the ionic bond exemplifies electrostatic attraction, known since antiquity, the explanation of the covalent bond involves the concept of an electron, which was invented only in 1898. Incidentally, when *Nature* commemorated the 50th anniversary of this feat, it included a discussion on whether it was a discovery or an invention. To a realist the answer is clear: the electron was discovered, but the various theories of the electron were invented. So much so, that electron beams can be generated and manipulated in the lab, whereas electron theories can only be expanded, corrected, or replaced with paper and pencil.

Like all emergents, H_2 has properties that its precursors lack, such as a characteristic dissociation energy, and a band spectrum rather than a line one. A far more complex molecule is the buckyball C_{60}, a major resource of nanotechnology. This unexpected molecule, discovered only in 1985, is a mesophysical thing, for it has a definite shape (similar to a soccer ball), but it diffracts through a slit system, just like quantons: it is then a semiclasson (or semiquanton). Note that the key ontological concepts in the preceding description – those of matter, energy, system, precursor, and emergence – are conspicuously absent from mainstream metaphysics, and even from the standard philosophical dictionaries.

Consider next the molecules that emerge from non-elastic collisions between simpler molecules. One of the simplest is the water molecule, usually symbolized as H_2O. This formula exhibits the atomic components – or rather precursors – of the system, but does not suggest its structure, which is partially shown by the structural formula H—O···H, together with the data that this system is V-shaped, the angle being about 104°, and the inter-atomic distances about 1 Å.

If water molecules get close together in large numbers, a body of liquid water emerges – one that belongs in the fourth integrative level after those of elementary particles, atoms, and molecules. We all know something about liquid water, but only quantum chemists know why it exists at all. Indeed, since H_2O is rather light, water would be a gas at ordinary temperature were it not for the strong hydrogen bond, a non-classical type of inter-molecular coupling. Paradoxically, this and other types of chemical bonds involve two classical properties, inter-atomic distance and angle, neither of which plays an important role in atomic physics, where dynamic properties, such as momentum and spin, take precedence over geometric and kinematical properties. Shape emerged along with molecules: the components of the so-called early universe were shapeless.

Water has unique properties that make it essential to life. One of them is that it is gregarious: the H_2O molecules tend to clump into snowflakes and ice pellets, droplets and lakes, glaciers and seas. The bonds that bring such systems into being are of two kinds: covalent and hydrogen bonds. As we saw above, a covalent bond is a pair of electrons shared between two atoms, as in the case of the hydrogen molecule H_2 mentioned above. By contrast, a hydrogen bond is an interaction

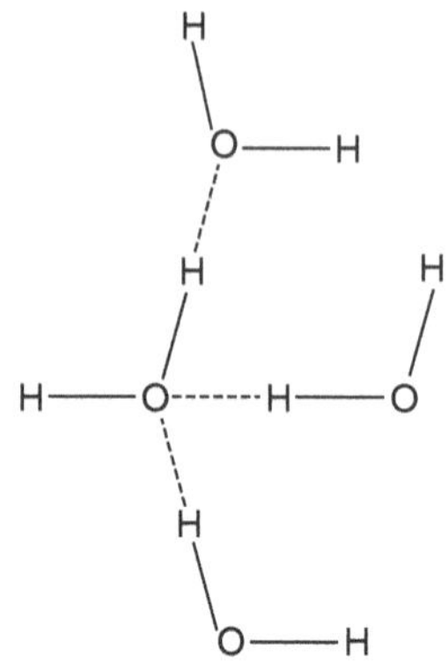

Fig. 5.5 Imaginary snapshot of a system composed of water molecules. The solid lines represent covalent bonds, whereas the dotted lines represent hydrogen bonds

between a hydrogen atom and an electronegative atom such as oxygen. Both bonds occur in a water body, and they exchange places in rapid succession – a dynamical feature that is absent from the familiar ball-and-rod model of molecules. Water is unique in that the total number of bonds of a water molecule is up to four – which explains the high boiling point of that liquid. See Fig. 5.5.

Suppose for a moment that quantum chemistry, which is still in its infancy, were to explain satisfactorily the emergence of some of the typical biomolecules, such as DNA, as well as some of the most complex chemical reactions, such as the synthesis of RNA. If this were to happen, chemical theory would be reduced to (deduced from) quantum theory together with some subsidiary assumptions. Most scientists would claim that such reduction would disprove the ontological thesis that there is such thing as a distinct chemical level. I beg to differ: I submit that there is such a supra-physical level, defined as the set of all systems wherein chemical bonds change. Besides, quantum mechanics has to be enriched with subsidiary hypotheses in order to yield quantum chemistry (see Bunge 1982a).

Chemical matter should draw a closer attention of metaphysicians, if only because of its remarkable ability to spontaneously synthesize molecules of several million kinds. Such spontaneous morphogenetic power, particularly in the case of carbon compounds, is far greater than that of physical matter, which is limited to crystal growth and the formation of rocks and large liquid bodies. Synthetic chemistry, the synthesis of molecules in the laboratory, harnesses the spontaneous self-organizing forces. And synthetic life, the production of living cells from chemical matter, is presumably a doable if distant goal of biochemistry.

In sum, we should acknowledge the existence of pieces of matter, namely molecules, belonging to a structural level of their own, and originating in equally supra-physical processes, namely chemical reactions, as exemplified by fire, oxidation, reduction, hydrolysis, electrolysis, and molecular synthesis. The level in question is, of course, the chemical level. It is generally accepted that the components and the precursors of the members of this level are physical. All of this suggests that chemistry, though dependent on physics, is not a chapter of it.

5.3 Living Matter

Living things are so very different from non-living ones that, until a century ago, they were widely believed to be characterized by goal-seeking behavior, or even by an immaterial entity, variously called *entelechy, Bildungskraft,* or *élan vital.* This doctrine, vitalism, was destroyed by the so-called mechanistic conception of life. This bold research project was sparked off by the biomechanics and medical physics inaugurated by Giovanni Borelli ca. 1650. Three centuries later this mechanist view of life developed into physico-chemicalism, according to which organisms are nothing but very complex chemical reactors (Loeb 1912).

However, living things have a number of inter-related properties that chemical systems lack. Here are a few of them: cellularity, metabolism, homeostasis, cell division, heredity, mutation, morphogenesis, self-repair, evolution, value, sickness, and death. Cellularity is the biological counterpart of atomicity: it consists in the fact that the units of living matter are cells, which are systems endowed with a semi-permeable outer envelope or membrane. In other words, all organisms are either cellular or multicellular: there is no subcellular life. Metabolism is of course the process of chemical self-renewal, in particular through protein synthesis and the incorporation and transformation of environmental items (nutrients). Homeostasis is the ability to maintain a fairly constant internal milieu (in particular, temperature and acidity) by virtue of control mechanisms (feedback pathways) built into the organism. Cell division, which involves RNA self-duplication and the splitting of DNA strands by enzymes, is the ultimate fate of all cells except for most neurons. Morphogenesis is a process involving cell differentiation and the construction of organs with specific functions, that is, processes that no other components of the organism can undergo.

Heredity is the ability of reproducing organisms to transmit some of their traits to their offspring. Such transmission is usually rather faithful ("true to type") because the heredity units, DNA molecules, are rather robust and inert. But of course these molecules are sensitive to strong environmental stimuli, such as cosmic rays and certain enzymes. For example, two identical twins may possess the schizophrenia gene but, whereas one of them also possesses the enzyme that switches that gene on, his twin may lack it, so that he is spared that terrible mental disorder. This fact should suffice to put the selfish-gene myth to rest.

As for evolution, most biologists agree that it has resulted from the concurrent operation of mutation, environmental selection, niche construction, hybridization (mainly in plants), and a few other mechanisms present on all levels, from molecule to whole organism to population to community. A popular view is that evolution is nothing but a change in gene frequencies, whence DNA is the evolutionary unit. This idea is wrong, because what drives evolution is natural selection, and what gets selected (to reproduce) is the organism as a whole, not its genome. Besides, not only genes are subject to mutation: also proteins can mutate, hence evolve, as first suggested by the resistance to new drugs that some proteins can acquire.

Unsurprisingly, the concept of emergence was first suggested by evolutionary biology, which showed that the history of life is one of speciations and extinctions.

Interestingly, evolution suggested gradualness to some people (Darwin the first of them), and saltation to others. It is now generally agreed that denial of saltation is just as incompatible with evolution as denial of continuity: that evolution at all levels is gradual in some respect and discontinuous in others (Blitz and Bunge 1989).

Finally, it may be surmised that value emerged along with life, since the questions of the form "What is the value (or worth) of X for Y?" only make sense with reference to organisms and their creations, such as burrows, nests, dams, machines, and organizations. For example, compost heaps may be said to metabolize; but they do so to their owner's benefit, not their own. A question such as "What is the value of this thing or process for that molecule (or stone, or light beam, or star)?" deserves no answer. By contrast, a question such as "What is the value of substance, organ, or process X for organism Y?" is in principle answerable: We can find out what role X has in keeping organism Y in good health, by experimentally inhibiting or even removing X, and checking the effect of this alteration on the organism's state of health. Value is coextensive and coeval with life, but this does not entail that all we do is adaptive, that is, life-preserving. For example, neither music nor philosophy is adaptive; and unlimited procreation and war are definitely maladaptive.

In short, life is not characterized by one or more special properties. Rather– as Alexandr Oparin (1953) emphasized nearly one century ago – it is characterized by a whole system of special inter-related properties. However, this is also the way physicists characterize physical things. For example, even the humble electrons and positrons are characterized by a dozen basic properties, related to each other by a peculiar state (or wave) equation. So far, biologists have found some of the basic biotic variables, but not the counterpart of a physical state equation. The most they deliver is lists of key properties, such as the one proposed above.

The materialist conception of life, suggested by philosophy as well as by biochemistry and evolutionary biology, inspired the ambitious research project of synthesizing cells in the laboratory. This experimental project, started by Oparin in 1924, is currently under way in several laboratories around the world (see, e.g., Lazcano 2007a; Luisi 2006). Experts agree that life on Earth emerged about 3.5 billion years ago, but they still disagree on the precise sequence from molecule to earliest cell. In fact, there are at least 10 competing hypotheses about biogenesis.

However, they can be grouped into two main schools. Most workers (the genetic-first or heterotrophic school) conjecture that genetic material (in particular RNA) and Darwinian evolution were essential. But a strong minority (the metabolism-first or autotrophic school) speculates that biogenesis was a purely physico-chemical process that led straight from non-genetic catalytic networks of macromolecules to the first cells. Since both schools employ the scientific method, their rivalry has stimulated their productivity instead of degenerating into barren controversy. Besides, the rivalry vanishes when the controversy is placed in a wider context, since the genetic school admits that the early genetic material self-assembled from abiotic material. See Fig. 5.6.

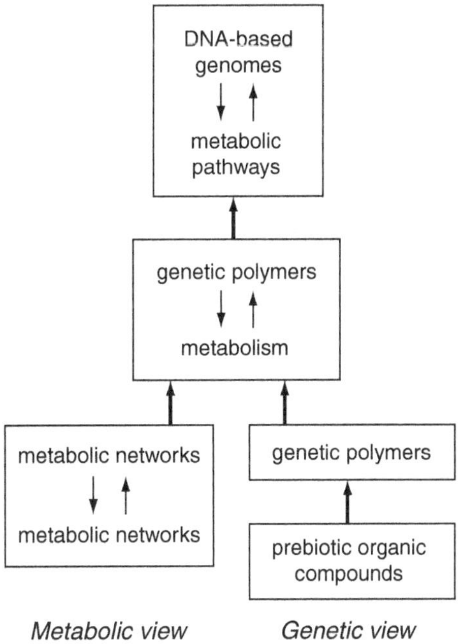

Fig. 5.6 The two main current scientific hypotheses about the origin of life. From Lazcano (2007b, 60)

In 1953 two sensational advances reinforced the hypothesis that living matter is just as material as nonliving matter: the molecular-biology revolution, and the first successful synthesis of amino acids by Miller and Urey, followed by Khorana's synthesis of long chains of nucleic acids.

On the down side, molecular biology led to the reductionist exaggeration that genes rule life, whence the successful sequencing of the human genome would reveal the secret of life. Another negative aspect of the molecular-biology revolution was the careless way in which the ambiguous expression "genetic information" has been used. Indeed, it has often been interpreted anthropomorphically as the set of instructions for assembling an organism as if it were a Lego toy. ("Genetic information" is actually short for the order of the nucleotides in nuclei acids. This concept is unrelated to both the physical concept of information as entropy or disorder, and the statistical information concept used in communications engineering.) However, we should not complain too loudly, because there is a price to be paid for any conceptual novelty, namely vagueness.

In short, biology has given philosophy at least three gifts in the course of one and a half centuries: the concept of an integrative level, and the theses of the materiality of life, and of bioevolution through modification and natural selection. By the same token, biology got rid of creationism, vitalism, and radical reductionism – the view that only the composition of a system matters. The reader is invited to compare this rich crop to the contributions made by metaphysicians to both their own discipline and to science during the same period.

5.4 Thinking Matter

We adopt the psychoneural identity hypothesis. That is, we assume that the *res cogitans* is not the mythical immaterial soul but the brain of highly evolved animals. This philosophical assumption, which originated in ancient Greek medicine, has been the guiding principle of cognitive and affective neuroscience – the recent fusion of neuroscience and psychology. This thesis will be developed in Chapters 9–11. It is stated here only so as not to miss a step in the levels ladder to be introduced later on.

It might be thought that psychology becomes dispensable if mental processes are regarded as brain processes. But such elimination – actually proposed by the self-styled eliminative materialists – would not be advisable, because psychologists study mental processes in highly gregarious animals, whose brains are bombarded by social stimuli coming from other brains, and in turn affect the affective and cognitive processes undergone by other people. This is why we need social psychology, the interscience that investigates the way thinking matter interacts with social matter. We shall come back to this subject in Chapter 9.

5.5 Social Matter

Social interactions, from sexual pairings to political fights, generate, maintain or alter social systems, from families and firms to nations and beyond. Such systems are composed of animals, but they have no biological properties – they do not metabolize, for one thing. But they are just as concrete or material as their components: they can be said to constitute social matter, just as organisms constitute living stuff.

Though obvious to a materialist, the preceding assertions will be rejected not only by idealists, who claim that everything social is spiritual (*geistig, moral*) and therefore the subject of the *Geisteswissenschaften* (spiritual or cultural sciences).

The claim that there is such thing as social matter will also be rejected by the so-called methodological individualists, for they deny the very existence of social wholes – and consequently cannot make any original contributions to their study. But such denial is tantamount to denying the existence of solid bodies on the ground that they are composed of atoms or molecules, or to denying the existence of sentences on the ground that they are composed of letters.

Anyone who wishes to study individuals rather than social wholes will have to engage in human biology or psychology; and even so his research will miss no less than the social matrix of human existence. And if the social environment of the person is ignored, then her behavior in love, learning, parenting, work, stress, rebellion, and much more will remain unknown. In general, a system is not just the set of its components. Every concrete system must be analyzed into its composition, structure (set of bonds among its constituents), environment, and mechanism (the processes that make it what it is). This is why the understanding of systems requires approaching them in a systemic rather than in either an individualist or a holistic fashion. Mathematicians, physicists, chemists, and biologists have always known this truism (see, e.g., Bunge 2003a).

5.6 Artificial Matter

Artifacts, or made things, can be said to embody or materialize ideas or feelings, because they are designed, often with the help of high-level technical knowledge. This is how the young Marx distinguished between a house and a beehive – a distinction that escaped Popper when he put houses in the same bag as nests, anthills, rabbit warrens, and beaver dams. No doubt, animal artifacts can be very complex, but there is no indication that they are built according to plan: The ability to build them is inborn, and the building process is an automatic outcome of the combination of genetic programming with environmental materials and circumstances.

Because artifacts embody ideas or feelings, they may be said to constitute matter of a special kind, namely artificial or made, by contrast to natural or found.

To realize the differences between natural and artificial matter, let us recall the salient traits of the manufacture of any good, such as a loaf of bread or a book. See Fig. 5.7.

Artificial matter is of either of two kinds: meaningless, such as a bridge, a house, a machine, or a formal organization; and meaningful, such as a text or a diagram. True, made social systems are often said to have meanings, but this use of the word "meaning" is metaphorical: what is actually intended is, that those systems are purposeful or, better, that they are used as means to goals.

Another useful classing of artificial matter is this:

artistic: music, painting, sculpture;
semiotic: poetry, fiction, non-fiction literature;
technological: technical diagram, machine, pipeline, transportation means, TV network
social: formal social organization (business firm, school, market, state, church).

All formal social organizations, such as schools, corporations, and governments, are artifacts, for they are designed, if often only sketchily, before being

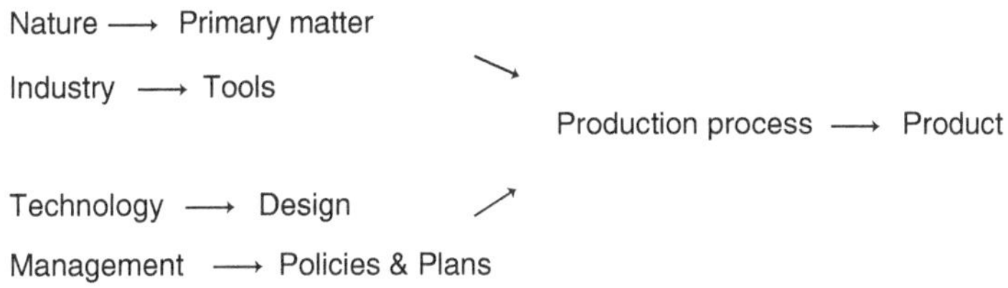

Fig. 5.7 Artificial matter is natural matter transformed by actions guided by engineering and management ideas

put together. Thus they contrast with informal organizations, such as families and gangs, which emerge, develop and dismantle rather spontaneously. The artificiality of formal organizations bears emphasis because of the comparative popularity, in Conservative circles, of Hayek's thesis, that everything social is spontaneous – while at the same time not existing as social since, according to the same author, there are only individuals.

5.7 Emergence

The preceding pages should have set the stage for the formal introduction of three key concepts of our ontology: those of emergence, level, and levels ladder (or "hierarchy"). The polymath George Lewes introduced in 1877 the now fashionable word "emergence", but the concept had been invented earlier by John Stuart Mill (see Blitz 1992). Indeed, Mill noted that the synthesis of water out of hydrogen and oxygen involved the emergence of properties, those of water, which its precursors lack.

A Platonist would attempt to account for emerging properties without their bearers. Aristotelians and scientists know better. For example, the above synthesis of water is a particular case of a chemical reaction of the form: $A + B \rightarrow C$. Chemical kinetics describes this process with the following equation for the rate of change of the concentration of the product of kind C in terms of the concentrations of the reagents of kinds A and B: $d[C]/dt = k\,[A]\cdot[B]$. No properties without bearers. Speaking logically and grammatically: Every predicate is attributed to at least one subject, as in Px and Rxy.

The concept of emergence was originally understood as an ontological category, synonymous with the occurrence of qualitative novelty in the course of a process and, therefore, clearly exemplified by crystallization, chemical synthesis, morphogenesis, biological development and evolution, learning of new skills, social organization, and history (see, e.g., Bunge 2003; Coleman 1990; Luisi 2006; Moessinger 2008; Piaget 1965).

However, because some irrationalist philosophers and psychologists stressed the occurrence of emergence but denied the possibility of explaining it in terms of lower-level processes, the word was eventually taken to stand for an epistemological category. (The obscurantist opinion that the new and complex cannot be explained in terms of the old and simpler is sometimes misleadingly called "strong emergence".) We shall keep the original ontological concept, or rather concepts: "emergence" will be equated with the rise of qualitatively new objects, properties, or processes.

We distinguish two ontological concepts of emergence: synchronic or emergence$_1$, and diachronic or emergence$_2$. The former coincides with that of bulk property, or Gestalt quality, whereas the second has to do with process. Indeed, a bulk (or emergent$_1$) property of an object is one that no part or component of the object in question possesses. For example, liquidity is a property of large collections of water molecules, and meaning that of concatenations of phonemes or letters. Christian von Ehrenfels introduced this concept in 1890, and it became the buzzword of Gestalt psychology, which trumpeted it against associationism, according to which all novelty is combinatorial. Regrettably the Gestalt movement denied that emergence_1 is an analyzable concept: it joined intuitionism.

On the other hand, an emergent_2 property is a qualitatively new property of a thing that none of its precursors possesses. For example, adaptation is an emergent trait of most organisms, both in that it is a bulk property, and in that it arose in the course of evolution along with the synthesis of the first cells from their prebiotic precursors. The dual or complement of emergence_2, as in speciation, is of course submergence, as in species extinction. (For more precision see Chapter 14.)

The following figure illustrates the idea that new properties, far from arising *ex nihilo*, emerge from pre-existing ones. If a thing α with properties *A* and *B* generates a thing β with properties *A, B* and *C*, then thing β and the new property *C* are emergent relative to α, *A*, and *B*. Furthermore, if *A* and *B* are linked by law L_{AB}, then *C* is related to its precursors *A* and *B* through the additional emergent laws L_{AC} and L_{BC}: see Fig. 5.8.

The idea that something radically new may ever arise has been resisted since the Ecclesiastes asserted that "there is nothing new under the Sun." (For a history of this idea and its dual see Merton and Barber 2004.) The most popular idea about novelty is that whatever appears to be new actually existed previously in a latent form: that all things and all facts are "pregnant" with whatever may arise from them. An early example of such neophobia is the conception of causes as containing their effects, as expressed by the scholastic formula "There is nothing in the effect that had not been in the cause".

Preformationism, the ancient biological doctrine that all the stages of a morphogenetic process are contained in the zygote, involves yet another denial of newness. Its contemporary successor is the doctrine that whatever new forms arise are contained in the inherited "coded instructions": Genome would be destiny. Genetic information is thus the inheritor of theological fate and secular preformationism. (See Mahner and Bunge 1997, 280–294). By contrast, modern chemists have avoided preformationism. In particular, no chemist has speculated that hydrogen and oxygen are "pregnant" with water, just as no sane person has claimed that girls are born pregnant. Of course, potentiality precedes actuality, but the latter is not contained in the former, but emerges from the conjunction of possibility and circumstance.

There is no universal emergence mechanism: Since all mechanisms are stuff-specific, things of different kinds change differently. However, when near their critical points, many things belonging to different kinds have similar macrophysical properties – something referred to as *universality*. For example, water and carbon

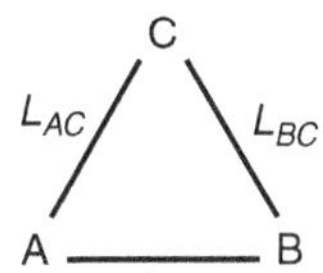

Fig. 5.8 Laws relying three properties: **a**, **b**, and **c**

dioxide behave thermodynamically in the same way when their liquid and gaseous phases separate. Besides, all evolutionary (or historical) processes share certain common traits. They involve the formation of new kinds from old in accordance to laws and under environmental circumstances, some of which favor while others oppose the emergence of certain novelties. In particular, nurture will either favor or frustrate nature, which in turn will react on the former.

Although there is no universal emergence mechanism, it can safely be conjectured that nothing emerges *de novo*: that every new thing develops or evolves from pre-existing things, so that there is continuity or conservation along with discontinuity or novelty. In particular, biological evolution is tree-like, as Darwin surmised, so that complex organisms of very different genera have usually simpler common ancestors. (The same holds for social evolution.) For example, the eyes of all modern organisms, whether spiders or people, descend from the photoreceptors of single-celled animals instead of having emerged independently. Something similar holds for the limbs of tetrapods. The clue is the commonality of certain blocks of genes (homeboxes) and genetic regulatory mechanisms (Shubin 2009). Thus, idiosyncrasy in some respects blends with universality in others.

These and related findings suggest that there may be evolutionary laws after all. Two possible candidates are these: (1) Similar environments favor the emergence of similar phenotypes (*convergent evolution*). (2) Different environments favor the emergence of different genotypes (*divergent evolution*). These are not wild fantasies, but experimentally testable hypotheses. For example, MacLean and Bell (2003), working on a common bacterium for ca. 1100 generations in different environments, found convergent phenotypic evolution along with radically different sets of beneficial mutations (divergent genotypic evolution).

In other words, evolution involves both conservation and therefore repetition in some regards, together with qualitative change (emergence), and therefore uniqueness, in others. Graham Bell (2008) proposes this suggestive musical metaphor: Evolution involves a few themes – the major genes where beneficial mutations can occur – and countless variations on the themes, some due to mutations, others to environmental circumstances, and all of them subject to selection. And once in a while variations accumulate to the point that new species emerge – or old ones become extinct. Emergence (speciation) and its dual, submergence (extinction), constitute the main plot of evolution: the details are commentaries and footnotes

Furthermore, evolution starts at the molecular level. For example, recent experiments have shown that, as a result of mutations, certain proteins acquire new functions in an irreversible manner (Bridgham et al. 2009). This and other findings vindicate Gould's (2002) emphasis on historical uniqueness and contingency (circumstance). Evolution is irreversible (Dollo's law) because every phase of it builds on its predecessors. A systems engineer might put it this way: If one important trait of a system changes, then other traits of it must alter too for the system to keep working. If they do not, the system crashes.

Irreversibility holds not only for complex molecules, organisms, and societies, but also for materials with memory. For example, magnets "remember" previous magnetizations, and swords "remember" the tempering process. In all of these cases, exact reversibility (or recovery of all ancestral traits) is impossible. The ancient myth of eternal recurrence, popularized by Nietzsche, is just that: a myth.

Whereas some thinkers have emphasized law and uniformity, others have stressed circumstance (or contingency). However, there is no opposition here: One may assume that present circumstances (or initial conditions and constraints) are but the final state of a previous process following different laws. Thus, whether a seed swept by an air current will fall on fertile land and germinate, or on a stone and be eaten by a bird, depends not only on the laws of aerodynamics, but also on the distribution of soil, rocks, and birds in the region concerned. And there is nothing haphazard about such distributions: they involve accidents but not chance or randomness in the technical (statistical) sense of disorder.

To conclude this section, notice that the concepts of emergence analyzed above are only faintly related to that of supervenience, familiar to philosophers. This idea was introduced by G. E. Moore, the philosopher of "naturalistic fallacy" fame, when arguing against the possibility of defining axiological concepts, such as "good" and "beautiful," in terms of natural-science concepts, such as those of symmetry and efficiency. Jaegwon Kim (1993) attempted to elucidate the concept of supervenience. In my opinion he failed to achieve this worthy goal because he dealt with properties in themselves, detached from their bearers, and adopted a static viewpoint, that overlooks the fact that all emergence occurs in the course of a process. In short, Kim unwittingly adopted Plato's Doctrine of Forms. Furthermore, he identified properties with their conceptualizations, namely predicates. Indeed, he postulated that the set of all properties constitutes a Boolean algebra. But this assumption has the absurd consequences that for every property there is a negative property, and that for any two properties there are both a disjunctive property and a conjunctive property. Thus, a thing might possess traits that it lacks: it could have the properties of not having a spin, of being a fish or an accountant (inclusive "or"), and of being at once an elephant and light. Acquaintance with scientists' ideas of properties and their emergence might have avoided these mistakes. (More in Bunge 1977a, 2003a; Mahner and Bunge 1997)

5.8 Levels

Another key concept of our ontology is that of a level or, more precisely, structural, organizational, or integrative level – by contrast to level of analysis (see, e.g., Bunge 1959b, 1960a; Needham 1943; Novikoff 1945; Simon 1962). A structural level is a set of objects each of which possesses properties peculiar to them – such as, e.g., being able to metabolize, or to form social systems. However, this definition is defective because it is roughly the same as that of a species of things, such as a chemical or biological species. In both cases the concept in question is defined by a system of properties, that is, a set of interrelated properties. Let us aim for greater precision.

To uniquely characterize the notion of a level of reality we also need that of relation between two adjoining levels. We stipulate that level L_{n-1} *precedes* level L_n if every object in L_n is composed of objects in L_{n-1}. Thus, the part-whole relation defines the level precedence relation. We posit that the set of all levels is ordered by the relation of level precedence. It may be called the *levels ladder* or, if preferred, the *great chain of being* – a grand ancient idea (Lovejoy 1953).

The levels ladder should not be likened to an onion's layers, and this for three reasons. First, because the peels of an onion are all qualitatively alike, whereas the point of using the concept of a level is to mark qualitative distinctions. Second, whereas onion peels and other layers are real, the structural levels are sets, and therefore conceptual objects: they are not independently real, although we hope that they are realistic. (Note that, whereas "real" is an ontological category, "realistic," or "true," is a semantic and epistemological one. For example, all theories are unreal, since they lack independent existence, but some are more realistic, i.e., truer, than others.) Our levels differ then from Nicolai Hartmann's (1949) layers.

Level ladders are also called "hierarchies", which is all right as long as this word is not understood in the traditional sense as involving a power or dominance relation, as in formal organizations and the celestial hierarchies imagined by Plotinus, Dionysius the Pseudo-Areopagite, and Dante – not to mention uncounted modern philosophers. Belonging in a rung of a natural levels ladder confers no value or power on members of another rung, whether higher or lower, for dependence may be reciprocal. For example, survival depends on water; but in turn the ability to explore the environment facilitates access to water.

We propose the levers ladder shown in Fig. 5.9. However, each of the levels indicated may be split into several sublevels; as a matter of fact we have already distinguished two physical levels, those of classons and quantons; chemists draw a line between small molecules and macromolecules such as polymers; biologists, between unicellular and multicelllar organisms; and social scientists between the microsocial, the mesocial, and the macrosocial.

The concept of a level allows us to refine that of emergence, namely thus. Instead of saying that property P is an emergent, we should say that P emerges on level L_n, which suggests that P was not possessed by any entities on the preceding level L_{n-1}. For example, presumably the property of being valuable emerged long with the life level; and that of acting purposefully emerged along with the level composed of animals endowed with a prefrontal cortex – roughly mammals and birds. Notice the virtuous circle: the concept of emergence defines that of level, which in turn refines the emergence concept.

Fig. 5.9 A levels ladder: Matter genera. The reader is invited to propose the splitting of every level into sublevels

↑ *Emergence*

Artificial
Social
Thinking
Living
Chemical
Physical

Fig. 5.10 The ladder of epistemic levels. Caution: It is not claimed that the Indicated reductions have been accomplished

↓ *Reduction*

Technology
Social Science
Psychology
Biology
Chemistry
Physics

5.9 Epistemological Partner

The levels examined in the previous Section are ontic, not epistemic: their constituents are material things, not ideas. The epistemic counterpart of the ontic ladder 5.9 is shown in Fig. 5.10.

Whereas the ontic levels are ordered by the ontic relation of precedence, the epistemic levels are ordered by the relation of reducibility, as when physicists explain magnets in terms of the spin and the associated magnetism of their constituent atoms, and social scientists claim that all social facts emerge from individual actions.

The radical reductionist thesis is that all the higher-level constructs are deducible from (reducible to) lower-level ones, so that eventually all of the factual sciences will be derivable from physics. The most famous case of reduction is that of thermodynamics, which indeed is deducible from laws satisfied by the microphysical components of thermodynamic systems. But there are not many examples like this, not even within physics (see Bunge 1973a). The limited success of the reductionist program is rooted in two features of the real world. One of them is that emergence and the accompanying level structure of reality are pervasive (Bunge 2003a).

The second feature that limits the feasibility of reduction is that, although wholes emerge from their constituents, once in place they constrain the latter (recall Fig. 5.1). For example, the strong bonds among the molecules of a water body, shown in Fig. 5.5, explain the comparatively high boiling point of this liquid: a physical-level property is here explained in chemical terms. Second example: Half a century ago Ronald Melzack discovered that dogs raised in isolation are insensitive to stimuli that cause pain in normal dogs: pain (like love and shame) would be largely learned. Or take the familiar case of a person who is a tyrant at home and a slave at work: his personality changes with his environment. Again, macroeconomic constraints and circumstances explain some of what a business firm (a microeconomic thing) does. In short, while in some cases our knowledge of things in L_n is derivable from pieces of knowledge regarding L_{n-1}, in other cases the converse is the case. In short, the epistemic levels are not isomorphic to the levels of being – which suffices to ruin the reflection theory of knowledge (Bunge 1959b).

The noted differences between the higher and the lower should not lead us to the holist and idealist thesis that the higher owes nothing to the lower. And yet, all idealists have held that the spirit or mind (the secular successor of the theologians' soul) is free from material constraints, in particular biological and economic ones. Kant, the neo-Kantians, hermeneuticists, intuitionists, phenomenologists, and existentialists have claimed that their ontological thesis that the spiritual is way above the material entails the methodological principle that the study of the spirit and its works calls for a method of its own, such as intuition or "interpretation" (*Verstehen*), that owes nothing to the scientific method. In short, the ontological dualists of all stripes are wedded to methodological dualism.

At first sight, the levels doctrine gives aid and comfort to the dualist party since it, too, seems to violate monism, or the continuity principle upheld by all naturalists and most materialists. The levels doctrine appears to conflict also with scientific realism, which is scientistic, in that it affirms the universal validity of the scientific method. But here, as elsewhere, appearances can deceive. Indeed, the levels doctrine is compatible with substance monism, though not with property monism: in fact, substance monism can be combined with property pluralism. Let us recall the meanings of these terms.

Thus emergentist materialists agree with Spinoza (one substance, infinitely many properties), while disagreeing with Descartes (two substances, each with a single property – extension in the case of matter, and thought in that of mind). As for method, the levels doctrine only confirms the widespread opinion in the scientific community that, while the general *method* is the same across all fields of serious inquiry, every level calls for its distinctive set of special *techniques*. This is why the mystic's navel contemplation, Hegel's dialectical "method," Dilthey's sympathetic understanding, Bergson's global and sudden intuition, and Husserl's intuition of essences, have never produced any new truths. These alternatives to the scientific method are barren because they are inward looking and dogmatic, and because they reject science but accept hunches without checking them. This failure justifies the scientistic slogan: Go scientific or bust.

Closing Remarks

Reality is layered rather than flat, because things tend to assemble into systems, and every system has (emergent) properties that its components lack. Consequently materialism, to be consonant with science, must be at once systemic (rather than individualist) and emergentist (rather than physicalist). If this injunction is taken seriously, naturalism is seen to be insufficient. However, naturalism is such an important worldview that it deserves a chapter of its own – the next one.

Chapter 6
Naturalism

Things may be split into natural and non-natural. In turn, the natural things may exist either in nature or in society; and the unnatural things may be artificial like books or supernatural like gods. Naturalism is the philosophical view, first advanced by Thales of Miletus, that the universe and nature are the same, so that there is nothing supernatural and nothing irreducibly social. For example, according to Aristotle (1941: 1253a), "it is evident that the state is a creation of nature, and that man is by nature a political animal."

We may distinguish two kinds of naturalism: naïve and sophisticated. The naïve naturalists hold that everything desirable "comes naturally," that is, is part of human nature, innate, or hard-wired in the brain. Thus, both selfishness and aggressiveness – or else fairness – would be "in our genes," and rationality and science would be just extensions of common sense. Moreover, everything natural would be superior to anything artificial – hence the preference for "natural" medication (quackery) and the abhorrence of genetically modified organisms. The Romantic slogans, "Back to nature!," and "Feeling trumps reason," are examples of naïve naturalism. But so is animism, the precursor of religion, that postulates that all things are "ensouled."

Although naïve naturalism itself may "come naturally," it is false and self-destructive. It is false because artificiality is known to be just as distinctively human as versatility: even the most primitive hunter-gatherers make or use tools of various kinds, invent social conventions, and organize themselves into social systems – all of them being quite unnatural since they are constructed and eventually reformed or dismantled. And naïve naturalism is also self-defeating because it may be used to justify the law of the land, religion, and other social constructions: remember the claim that slavery, or else capitalism, is the natural social order.

Sophisticated naturalism is something else: it is just a reminder that, for all our sophistication, humans are animals, and consequently should attend to their biological needs, and anchor the various human sciences to biology. Sophisticated naturalists demand, in particular, that social science, ethics, and epistemology be "naturalized." In turn, sophisticated naturalism comes in three main varieties: humanistic (like Spinoza's), vitalist (such as Nietzsche's), and pragmatist (such as Peirce's, James's, Dewey's, and Hugo Dingler's). Let us peek at them.

Spinoza was a humanist naturalist, in that he held at once the identity of reality and nature, the further identity of nature and God, and the sacredness of our fellow

M. Bunge, *Matter and Mind*, Boston Studies in the Philosophy of Science 287,
DOI 10.1007/978-90-481-9225-0_6,

human beings. Kant rephrased the latter as the norm that all persons should be regarded and treated as ends rather than as tools – a principle that, if taken seriously, would be seen as endorsing egalitarianism.

Vitalism is the doctrine that all our ideas and actions should serve the individual's survival, whereas pragmatism is action-oriented. Though different, vitalism and pragmatism are rather close to each other. Thus Goethe, the great Romantic poet and amateur naturalist, stated in his masterpiece, *Faust,* that "the green tree of life" is far superior to "gray theory," and that "in the beginning was action," not word.

The vitalist and the pragmatist versions of naturalism are utterly different from Spinoza's naturalism: they are anthropocentric and rather egocentric. Whereas Spinoza wrote always in the third-person mode, vitalists and pragmatists are centrally interested in what is convenient for them to believe and do. Otherwise pragmatists are very different from vitalists.

Whereas vitalists reject reason and science altogether, pragmatists have no use for disinterested inquiry: they wish to make full use of science and technology to improve the human condition. This explains the well-known connection between vitalism and fascism (through the Blood and Soil doctrine); but it also explains the pragmatism-democracy connection. A major trouble with pragmatism is that, in holding that disinterested inquiry is a waste of time, it underrates the humanities and it neglects or even discards disinterested basic research, which not only enriches our lives but is the ultimate source of technology.

A recent example of pragmatism is the science-policy reluctant to fund basic research. A related example is the contention that all research is conducted with utilitarian motives, so that one should talk about "technoscience" rather than about science and technology as distinct though intertwined endeavors. Hence doing cosmology, evolutionary biology, history, and philosophy would be a waste of time. However, let us go back to naturalism in general.

Sophisticated naturalism promotes the scientific exploration of nature but neglects mathematics and attempts to force social inquiry into the naturalist box – witness human sociobiology and its offspring, Evolutionary Psychology. I hasten to note that naturalism is limited not because it excludes the cultural, which it does not, but because it denies its specificity and irreducibility. In short, naturalism promotes inquiry but only up to a point.

And naturalism discourages research when blended with radical skepticism, as in the cases of Sextus Empiricus, Hume, and "transcendental naturalism" or "mysterianism". According to the latter view, "[r]eality itself is everywhere flatly natural," but "[o]ur epistemic architecture obstructs knowledge of the real nature of the objective world" (McGinn 1993, 2) How can epistemically challenged beings know that the universe is both "flatly natural" and basically unknowable? Mystery. And how can scientists realize when they have reached the end of the road? Another mystery. And so, what scientists took to be the boldest and most rewarding of explorations turns out to be but a secular version of a Christian mass.

Whether crude or refined, naturalism opposes spiritualism in all its guises, from religion to idealist philosophy to the social studies that underrate or even overlook the so-called material basis of human existence. Standard economic theory is a case in point, for it postulates, as Adam Smith did in his foundational work of 1776, that only labor is the source of value, in particular wealth. In thus dethroning nature, that theory condones the industrial practices that have been destroying all the natural resources for the past two centuries. The nature/culture dichotomy, and the corresponding schism of science between natural and social, is still an open philosophical and practical problem.

Hence the plan of this chapter: a quick review of spiritualism followed by a glimpse at the realizations of naturalism, ending with the suggestion that scientific materialism includes whatever is valuable in naturalism while transcending its limitations.

6.1 Spiritualism

The traditional worldviews were spiritualist, in that they revolved around imaginary beings: They attempted to understand facts in terms of fictions, and the reputedly lower in terms of the allegedly higher. Recall the following outstanding examples. Plato claimed that concrete things are but the shadows of ideas (or "forms," or universals). John the Evangelist thought that "[i]n the beginning was the Word, and the Word was with God, and the Word was God" (John: 1); Aristotle held that God was the prime mover or first cause of all changes; and Hegel, that the magnet is the materialization of the argument (*Schluss*).

Every philosophical school or religion is characterized by the stand it takes with regard to the nature of reality and its knowledge. Thus, the logical positivists prescribed that we should always employ the "formal mode" rather than the "material mode": we must say "the word X means Y," not "X is Y" or "X does Y"; Heidegger maintained that "the word is the house of being," which Derrida translated as "there is nothing outside the text"; and Charles Taylor famously wrote that social facts "are texts or like texts." All religions have claimed to explain the worldly in otherworldly terms. And the dominant philosophies of history assert that society is moved by either individual minds or the "World Spirit." Thus, the UNESCO initial manifesto stated that "[w]ars are made in the minds of men." In short, according to magical thinking, "the higher" trumps "the lower" both in time and in power, hence in explanatory power as well. Theology would explain metaphysics, which in turn would explain physics. Likewise, philosophy would explain psychology, which would in turn explain social science.

By contrast, scientific thinking takes the real existence of the external world for granted: it is realist or objectivist. And the sociologies of science, technology and religion attempt to explain in social terms the emergence and fortunes of their subjects. Moreover, the science-oriented worldviews are free from disembodied souls, world spirits, and other fictions: they are also tacitly naturalist or materialist in some sense – without however necessarily denying the existence and power of the mental.

Nor do naturalists and materialists reject spirituality, although they redefine it. Indeed, whereas in traditional cultures spirituality was identified with religiosity, in modern cultures spirituality consists in cultivating or appreciating the basic sciences, the humanities, and the arts. This includes learning theorems, reading philosophy, history, or poetry, admiring Notre Dame or the Sydney Opera House, and being overwhelmed by Beethoven.

6.2 Naturalism

Naturalism is the worldview according to which all existents are natural, and none are spiritual or supernatural. Put it negatively: there is nothing outside nature (see, e.g., Bechtel 2008; Kanitscheider 1996, 2007; Krikorian 1944; Papineau 1993; Ryder 1994; Sellars 1969; Shimony 1993; Vollmer 1986). Many materialists call themselves "naturalists", some to avoid being mistaken for greedy hunters after material goods, and others to dodge the accusation of being "crass materialists" or, worse, "soft on Marxism." Thus, naturalism is often only timid materialism, just as many agnostics are bashful atheists. But just as often naturalism is a form of nothing-but-ism or radical reductionism. This view focuses on the ultimate components of things, with disregard for their structure and modus operandi or mechanism; or, when it does propose mechanisms, it denies that they are lawful – a view that invites miracles.

Naturalism is particularly obvious in the cases of physicalism, geneticism – according to which genome is destiny – and sociobiology, which holds that everything social is *au fond* biological. It is also clear in the physiocrats, the economists who held that all wealth comes from land (rather than from labor or trade), and that there is a natural social order, that governments should respect rather interfere with. It is debatable whether the Natural Law school is another instance of naturalism, because it was proposed and upheld not only by the ancient Stoics, who were radical naturalists, but also by such theologians as Thomas Aquinas. And Jeremy Bentham criticized Natural Law but attempted to explain all human behaviors in terms of pleasure and pain.

The terms "natural" and "naturalism" "come naturally" in all fields, but they are ambiguous, so that their use gives rise to equivocations. For example, neophobia, the dislike of anything unfamiliar, comes naturally because most novelties, not having been designed by us and for us, are likely to harm us, or to demand extra effort or even sacrifice. In short, conservatism looks "natural." But we are also naturally curious and eager to improve our condition, particularly when unsatisfactory, as is normally (naturally) the case for the vast majority of people. So, neophilia, curiosiy, and the concomitant reformism seems to be just as "natural" as neophobia and the conservatism it inspires. The "natural" (obvious) question is whether a "natural" (reasonable) balance between the two opposites is viable and desirable. Try this injunction: Preserve whatever is (objectively, naturally) good (beneficial to the many), and alter whatever is not. In any case, "natural" is just as ambiguous and potent a rhetorical tool as "useful", "noble", and "beautiful", and must therefore be handled with care.

As for naturalism, it stands for somewhat different though related doctrines in the philosophy of religion, value theory, ethics, ontology, logic, epistemology, and the philosophy of the social sciences. Let us recall briefly these different meanings, in order to clarify some recent philosophical controversies.

In relation to religion, "naturalism" usually involves the rejection of supernaturalism. The ancient Greek Stoics, Spinoza, and Einstein may be regarded as religious naturalists for having identified God with nature. But the religionists – whether Jewish, Catholic, Lutherans, or Calvinists – were never fooled by this tactic: they smelled disguised atheism. Their suspicion was justified, for, if every existent is natural, then nothing is supernatural. That is, for the naturalist there is no God in the common acceptation of the word, which is that of the being characterized by supernatural properties, such as immateriality, omnipotence, and omniscience. Hence Spinoza's equation "God = Nature" is self-contradictory, and must be seen as a subterfuge. After all, Spinoza's motto was *Caute*, i.e., "Be prudent." Consequently, the vulgar characterization of Spinoza as a pantheist is incorrect: He was a secret atheist because he was a naturalist. The same holds for Einstein, who once declared that his religion was Spinoza's – that is, none.

At other times, "naturalism" has stood for the attempt to derive religious beliefs from pure reason rather than revelation ("natural religion"). In still other cases it has characterized the search for, and interpretation of, the footprints allegedly left by God when creating the world ("natural theology"). The doctrine of "intelligent design," favored by most contemporary American Republicans, is the latest attempt in this direction. However, let us confine our discussion in this section to philosophical naturalism.

A number of components or branches of philosophical naturalism should be distinguished (see, e.g., Koppelberg 1999; Mahner 2007a). Let us sketch and briefly analyze the following varieties, starting with the root.

In *metaphysical naturalism*, the universe and nature coincide, so that the supernatural is just a fiction. This ontology comes in two strengths: radical and moderate. Radical naturalism denies the existence of the mental or spiritual and, a fortiori, that of consciousness and free will – which is why it is often called "eliminative". For example, the eminent neuroscientist Rodolfo Llinás (2001,128) has claimed that the self is a construct that "exists only as a calculated entity," namely "a complicated eigen (self) vector" – but he did not disclose the corresponding operator or matrix, as a consequence of which his assertion is puzzling. In any event, the reader is likely to resent the claim that she is just a mathematical object and, as such, incapable of doing anything of her own free will.

Many eliminative naturalists regard the brain basically as a computer and, as such, devoid of disinterested curiosity, self-knowledge (consciousness), initiative, and free will (e.g., Churchland and Sejnowski 1993; Paul Churchland 1984). Consequently they do not explain creativity – the ability to come up with radically new ideas. Moderate naturalists, by contrast, admit creative minds. And some of them – in particular Donald Hebb (1980), the father of contemporary cognitive neuroscience – have even acknowledged free will. But, unlike scientific materialists, the naturalists of both varieties tend to underrate the influence of the social context, and consequently they ignore developmental and social psychology, which account for the differences between the cognitive abilities of moderns and primitives (Mithen 1996), and the educated and the uneducated (Vygotsky 1978), as well as between identical twins reared apart.

Incidentally, the free will hypothesis started as a theological fantasy that was a convenient means to blame humans for what might be taken as God's perversity. (This is how Augustine used free will to combat Manichaeism.) But morally justified punishment presupposes self-consciousness, the ability to tell wrong from right, and free will. This is why in advanced nations nonhuman animals, children, and the mentally handicapped, and in recent years victims of injuries to the frontal lobe as well, are exempted from the rigors of criminal law. To my knowledge, eliminative naturalists have not tackled this problem. So, one does not know whether they should hold that no one should be blamed for his crimes, or that there can be no extenuating circumstances, in both cases because consciousness and free will are illusory.

Logical naturalism comes in two versions: strong and weak. The former says that logic is general ontology: that it contains the most general laws of all objects, real and imaginary – which is why the mathematician Ferdinand Gonseth (1937) called logic "the physics of the arbitrary object." This view, first hinted at by Aristotle, is false, for the laws of science are material-specific, whereas logic is topic-neutral (Nagel 1956). What is true is that logic *refers* to anything even though it *describes* nothing extra-logical in particular (see Bunge 1974b). Indeed, the predicates in predicate logic, just like the sets in set theory, and the functions in functional analysis, are arbitrary, which is why logic refers to anything – which in turn makes it eminently portable across research fields. But formal logic does not contain the concept of change, which defines that of matter (recall Chapter 4). Presumably, this is why Hegel (1926 [1812]) thought that formal logic is subjective, whereas his own "logic" (metaphysics) pivoted around the notion of becoming.

The weak version of logical naturalism holds, with George Boole, that the laws of logic are laws of thought, and as such psychological (or neuroscientific) laws. Dewey (1938) was even more of a logical naturalist: to him logic was a biological product and the summit of evolution. This view is false, as shown by the fact that most "natural" (spontaneous, untutored) thinking is logically invalid (see, e.g., Johnson-Laird and Wason 1977). These are some of the most common fallacies: (a) confusing "some" with "all," or "a" with "the"; (b) identifying necessity with necessity and sufficiency, or "$A \Rightarrow B$" with "$A \Leftrightarrow B$"; (c) concluding A from "If A then B" and "B." The laws of logic are actually rules, and as such artifacts. Moreover, there is no single logic: in addition to the standard or classical logic, there are many nonclassical logics (see, e.g., Haack 1996).

Much the same holds for the laws of pure mathematics: they are not natural or intuitive; if they were, they would not look mysterious to the vast majority of people. A few examples will bear this out. First, it is natural to think that there are half as many even numbers as integers – but, it is false, as Galileo proved four centuries ago. Second, most subjects, when asked to draw an arbitrary triangle, will choose an acute triangle, even though there are three times more obtuse than acute triangles.

Third, most people disbelieve the equality $0^0 = 1$, although it is just a special case of the identity "$x^0 = 1$ for all x." In short, mathematics does not come naturally to most of us. (See Kanitscheider 2006 for an analysis and defense of mathematical naturalism; and Bunge 2006b for a comment.)

The discrepancy between mathematics and intuition explains why logic and mathematics emerged less than three millennia ago: they had to be invented, just like bronze, money, writing, and the state. In short, logical naturalism is untenable. By contrast, materialists have something to say about logic and mathematics: namely, that, just like plows and poems, they are human creations rather than residing either in mines or in Plato's lofty realm of ideas.

Quine once proposed what he called "mathematical realism," which Putnam adopted in one of his many phases. This sounds like mathematical naturalism, and actually is a sort of generalized Platonism, for it states that the mathematical objects exist in the same way as atoms and stars. Moreover, it sounds obvious to anyone naïve enough to believe that there is single concept of existence, namely the one exactified by the ill-named "existential" quantifier $\exists$, as in the polytheistic principle "$\exists$x (x is a god)." But, just as we distinguish between abstract objects, with conceptual attributes, and concrete objects, with non-conceptual properties, we must distinguish between formal (or conceptual) existence, on the one hand, and real (or material) existence on the other.

If there were a single existence concept, there would be a single method for proving the existence of anything. But, whereas formal existence must be either postulated or proved in a strictly formal fashion, material or real existence is a given. Furthermore, it must be justified by empirical tests: by showing that, in fact, the thing whose existence has been assumed can be "kicked" and can "kick" back. This is so because material things, unlike abstract objects, possess energy and the ability to absorb it and release it: they are changeable (recall Chapter 1).

Like all stray opinions, mathematical realism is sketchy, fuzzy, and arbitrary. Because it is not accompanied by an exact definition of the concept of a physical object, mathematical realism is not even capable of distinguishing material existence from conceptual existence. Worse, Quine never formulated an exact and comprehensive ontology sketching the salient traits of the things that compose his world. This is why all of his philosophical opinions, like Wittgenstein's, have gained instant popularity: because they come in small doses and strikingly worded, on top of being extravagant, and thus original after all.

An even weaker version of logical naturalism is this: Some assumptions are more natural (familiar, obvious, verisimilar?) than others, and some proof methods (particularly Gentzen's and Beth's) are more natural (simpler, intuitive, didactic?) than others. But here again there is no consensus on the very meaning of "natural". And in any case, since formal logic is just over two millennia old, and quite hard to teach, it would seem unnatural to call it "natural".

Semantic naturalism says that meaning and truth, the two foci of philosophical semantics, should be accounted for in a naturalistic fashion. For instance, in the late nineteenth century Franz Brentano (1960) equated reference – a component of meaning – with "intentionality." In fact, he claimed that the peculiarity of mental phenomena is their reference to an object other than the said phenomena – a clear case of *obscurum per obscurius*. Searle (2007, 6) too conflated intention (or intentionality), a psychological category, with reference (or aboutness), a semantic one. Which is like claiming that a single theory can cover *race* in the two main acceptations of this term: competition and ethnic group.

John Dewey (1958) claimed that meaning is not a mental object but a property of behavior: words elicit dispositions to overt behavior. Quine (1968) adopted this behaviorist project, only to jump suddenly to the crypto-idealist thesis that "it makes no sense to say what the objects [referents] of a theory are, beyond saying how to interpret or reinterpret that theory in another" (op. cit.: 202). While this suggestion may be useful in discussing the reduction of numbers to sets, or some other subject in the foundations of mathematics, it is utterly useless to find out anything about the referents of a scientific theory, suh as quantum mechanics. This task is performed by analyzing the postulates of the theory and the way the theory is used to account for facts of a certain kind.

For example, no extra scientific theory is needed to discover that mechanics is about bodies: suffice it to note that all the specific or technical predicates in this theory represent properties of bodies. The "ontology" (universe of discourse or reference class) is the set of all bodies. Any "ontological [referential] relativity" would be out of place here. But I confess that I do not know how to "interpret or reinterpret" mechanics in another theory – and anyway Quine offers no hints. Because of his repeated mention of the Skolem theorem, I suspect that what he had in mind was models (examples) of abstract theories such as set theory. But the theories in factual science are utterly different: their universes of discourse or reference classes are fixed from the start by their postulates, such as "$X(p, f, t)$ represents the position of particle p, relative to reference frame f, at time t." To understand a factual theory one tries to get into it, not out of it.

In a later work, Quine (1973) attempted to flesh out his earlier endorsement of Dewey's behaviorist philosophy of language, and tried to explain reference by the way children and foreigners learn to master a word "attesting to the presence of an object." But learning how to use a word is irrelevant to finding out the object that the word names. For example, one does not learn economics by investigating the way children acquire the concepts of good, price, market, and the like. Besides, a psychological account of reference cannot help us find out, say, whether or not quantum mechanics refers to observers rather than to physical things. In short, naturalism does not supply a theory of meaning, any more than it can explain the origin of legal codes – which is why the expression "natural law" is a contradiction in terms.

There are two kinds of semantic proposition: linguistic conventions, such as "Let f designate an arbitrary reference frame," and semantic assumptions, like "Function f represents property P." Like all conventions, the former are arbitrary and therefore freely replaceable. By contrast, semantic assumptions are testable and therefore subject to error and eventual rectification. (This is why their usual names, "operational definition" and "correspondence rule" are wrong: definitions are conventional, and rules are more or less efficient.) For example, there are rival hypotheses with regard to the operator that correctly represents the velocity of a quantum-mechanical thing

(Bunge 2003c). In any case, whether conventional or hypothetical, semantic stipulations do not grow on trees: they are just as unnatural as legal codes. However, let us now move on from meaning to truth, the other focus of semantics along with meaning.

Nietzsche, a notorious naturalist of the vitalist and pragmatist variety, regarded truth as only a tool in the struggle for life: He neither *defined* the concept of truth nor proposed truth *criteria* or conditions. Worse, Nietzsche exalted lies, in particular Plato's "noble lie," as an even more potent weapon than truth in Superman's struggle for supremacy. More than just a naturalist, he was a brutalist. This why Nietzsche has been the hero of uncounted reactionaries, among them Hitler, Heidegger, and Leo Strauss.

However, the fact that the pragmatist conception of truth is false, does not disqualify all attempts to naturalize "truth." In fact, it is possible to read the concept of factual truth, or truth as correspondence, or adequacy of thought and external fact, in terms of a brain process that somehow "corresponds to" its referent (or "truthmaker," as it is fashionable to say). This is indeed what this author has proposed (Bunge 1983b, 119; Chapter 15 of the present work).

Epistemological naturalism holds that cognition is a natural process and, as such, the subject of scientific research: that Plato's Realm of Ideas is fictitious, and so is the idea of knowledge in itself, that is, without a knowing subject. The strong version of epistemological naturalism also holds that epistemology has lost its autonomy: that it is being replaced with cognitive science. But, since human development and evolution are biosocial rather than purely biological, it is doubtful that naturalized epistemology, aka cognitive science, can solve any important problems without the help of social epistemology, aka social cognitive neuroscience. This holds, in particular, for evolution.

Methodological naturalism may be understood in three very different ways. What may be called *tacit* methodological naturalism is the common practice of filtering out the supernatural and the paranormal in the design, construction, and operation of measuring instruments. In other words, these should exclude the possibility that they might be interfered with by unnatural entities, such as ghosts, Descartes' malicious imp, or alleged paranormal abilities, such as telekinesis. In other words, no spiritual entity shall interpose itself between the observer and the observed thing. The reason is that, if instrument readings could be accounted in non-physical ways, they would inform us about thing-spirit compounds rather than about "normal" (natural or social) things.

In addition to tacit methodological naturalism, there is what may be called weak, strong, and extra-strong methodological naturalism. The *weak* version says that philosophy should use methods and findings in the natural sciences. *Strong* methodological naturalism is identical with scientism, or the thesis that the scientific method is applicable in all research fields, the social sciences among them. This version of naturalism, often called *scientistic*, has been championed by the positivists, logical positivists, and Quine among many others. By contrast, Putnam, Davidson, Rorty and others have defended what may be called *nonscientistic* naturalism, which is little more than a perfunctory nod to ontological naturalism (see De Caro and Macarthur 2004).

Finally, ultrastrong methodological naturalism is the program of reducing the social sciences to the natural ones. Edward O. Wilson's sociobiology is the paradigm. This bold project was doomed from the start, because it overlooked the differences between the artificial or the conventional, on the one hand, and the natural on the other. The same holds for sociobiology's successor, Evolutionary Psychology, which centers on sex. While it is true that mating is natural, courting and marriage are social constructions that vary across societies. The failure of sociobiology should have been anticipated, because the very expression "social naturalism" is an oxymoron. We shall return to this matter in Section 6.5.

6.3 Phenomenalism

It is well known that lizards like to bask in the sun. But they do not know that they owe the heat to the sun, and the warmth they feel to their nervous system. This is because, as Lovejoy (1955, 401) put it, a knowing subject "will have the power of imputing otherness and beyondness to what it perceives, and thus need not be blind to the rest of the universe." She has this power because her nervous system, unlike a lizard's, can think in addition to being able to perceive. In other words, whereas the lizard's worldview is necessarily phenomenalist, hence lizard-centered, a person's can be naturalistic and objective.

Phenomenalism states that there are only phenomena, that is, appearances to someone. It is the ontology of empiricism and, in particular, logical positivism. The logical positivists – in particular the members of the Vienna Circle – were the heirs to Mach, who followed Comte, who in turn owed much to Kant's subjectivist idealism, which had been invented by Berkeley. In his first *Critique*, Kant (1787) had held that the world is a pile of phenomena (appearances) rather than a collection of noumena or things existing in themselves; that, although all things exist in space and time, these are in turn forms of intuition rather than traits of the real world, whence things are in the mind; and that "God is a mere idea." Thus Kant was a timid naturalist – a position that could satisfy neither materialists nor idealists.

The same applies to Hume, his precursor – although, unlike Kant, Hume granted the independent existence of the external world. But Hume denied the possibility of knowing anything other than phenomena (appearances) – and this at a time when physicists and chemists were studying non-phenomenal facts such as planetary orbits, imperceptible gases, and invisible chemical reactions. It is not that Hume was unaware of these novelties: He rejected them explicitly because they contradicted phenomenalism.

A century later Ernst Mach (1914), the great experimental physicist and physiological psychologist, embroidered Berkeley's, Hume's and Kant's phenomenalism. He stated unambiguously the thesis that the building blocks of the universe are sensations. Obviously, this claim is naturalistic but also unscientific, because it is anthropocentric. A possible root of this claim – apart from its historical origin in Berkeley and Hume – is the confusion of reference with evidence (Bunge 1967a). Thus, a piece of empirical evidence for the hypothesis that this lump of material is radioactive, is hearing the clicks of a Geiger counter placed in its vicinity. In my time one could buy for a dime a tiny sample of radioactive material mounted on a screen that would be seen to sparkle in the dark. But of course one did not identify radioactivity with the perception of such tiny brief flashes: this was just evidence for the presence of the said material.

In sum, phenomenalism is still rather popular because it is commonsensical: after all, it takes theory and experiment to unveil the thing-in-itself underneath the thing-for-us or phenomenon. And yet, when Bas van Fraasen (1980) stated that the aim of science of is "to save the phenomena", many philosophers felt this as a breath of fresh air, while actually it was nothing but Ptolemy's two-millennia old epistemology. H. G. Wells would have been pleased to learn that philosophers can drive his time machine.

6.4 Physicalism

Physicalism is the thesis that everything is physical, and that nothing is supraphysical, let alone supernatural. For example, a consistent naturalist will claim that flags are nothing but pieces of cloth; consequently she will not understand why so many people have died defending their flags. Like all *isms*, physicalism comes in at least two varieties: hard and soft. Hard physicalism holds that there are only physical entities and properties, whereas soft physicalism prescribes that we use only the language of physics and, in particular, that we describe everything in spatio-temporal terms. Thus hard physicalism is identical with what is ordinarily called "crass (or vulgar) materialism". By contrast, soft physicalism is often subjectivism in scientific disguise, namely phenomenalism.

An example of soft physicalism is the view advanced by Otto Neurath (1981), and at one time shared by Rudolf Carnap – two of the pillars of the Vienna Circle. This is the empiricist thesis that, in the last instance, all the scientific statements are reducible to "protocol sentences," which in turn are of the same kind as "Otto, at place X and time Y, perceived Z." By collecting all such empirically "basic statements" one obtains the most general physics, the unified science, or physicalism, which is but "a tissue of laws that express spatiotemporal connections" (op. cit., I: 414). Clearly Neurath, a sociologist trained as a mathematician, did not realize that the basic laws of physics, such as Maxwell's equations, did not contain any reference to test procedures, for the same reason that Reference $\neq$ Evidence.

Energetism is the now forgotten variety of physicalism according to which everything, from bodies to values, is made of energy. Its creator, the eminent physical chemist Wilhelm Ostwald (1902, 373), regarded matter as "a secondary appearance that occurs as the coexistence of certain kinds of energy." And he thought that, because energy is neither matter nor idea, energetism was the alternative to both materialism and idealism. He dedicated his book to Ernst Mach, who could not have approved of it because for Ostwald energy takes precedence over sensation.

But of course there is no such thing as energy in itself: energy is a property of concrete (material) things, indeed their peculiarity (Chapter 4). That this is so, is shown by inspecting any physical formula containing the concept of energy, such as the most famous of all: E(closed system) = constant, and E(body of mass m) = mc^2. No energy bearers, which are material things, no energy.

Finally, strong or authentic physicalism is the thesis that all existents are physical, if not manifestly so, at least when analyzed into their ultimate constituents. For example, the brain would be a physical system, and feeling and thinking would be physical processes. Physicalism may also be understood as radical naturalism, in the sense that it affirms that everything in the world is natural, and denies that there are qualitative differences between the physical, on the one hand, and the chemical, biological, social, semiotic, and technological, on the other. Understandably, physicists are the most enthusiastic of physicalists.

However, it is not easy to be a consistent physicalist philosopher. For example, Papineau (2003, 353), a self-styled physicalist, conceives of a mental state as "the instantiation of a mental property." But this locution is doubly Platonist rather than materialist. First, because it suggests that properties precede their "instances" (bearers), whereas for materialism properties come with their bearers. Second, because the phrase in question tacitly identifies the properties possessed by physical entities with the corresponding attributes or conceptualizations.

A sophisticated materialist will stress the property-attribute distinction, and this for two reasons. The first is that only general concepts and propositions can be instantiated (exemplified). Second, because, whereas natural properties are traits of natural objects, hence inseparable from the latter, attributes are components of our views, in particular theories, about them (more in Bunge 1977a). For example, bodies possess mass, but this property is conceptualized differently in classical and relativistic mechanics. In any event, for a materialist, like for any scientist, every state is a possible state of a material entity. (See Chapter 14.)

It is often stated that the phenomenal properties, such as redness and bitterness,cannot be accounted for in physical terms. For example, David Chalmers (1995) asserts that, although it is widely agreed that experience arises from a physical basis, we have no good explanation of why and how it so arises. This statement is true, but it does not refute naturalism, for physicalism is only the earliest and coarsest version of naturalism. A non-physicalist naturalist will argue that phenomena occur only in sentient organisms when interacting with their environment. Phenomenal facts, or experiences, are facts of animal life, and therefore they call for biology in addition to physics and chemistry. To hold that the occurrence of phenomena falsifies naturalism or materialism betrays ignorance of the many varieties of naturalism and materialism. A contemporary emergentist materialist, in particular, is bound to know that vision, hearing, olfaction, and the like, are tackled by neuroscientists, not by physicists.

6.5 Biologism

Let us now examine a far more restricted version of naturalism, namely biologism, or the program of explaining nonbiological facts in biological terms. Biologism comes in two versions: strong or radical, and weak or moderate. Strong biologism is the ancient animist view that everything in the world is alive (or "animated"). This worldview was common in ancient India, and it was partly shared by Plato, who in his *Timaeus* held that the Earth is an animal. However, this particular belief of Plato's did not contaminate his remaining works, which were thoroughly rationalist. By contrast, the consistent vitalists, such as Nietzsche and Bergson, were irrationalists, in particular intuitionists.

Weak biologism holds that everything human is understandable in purely biological terms, without recourse to psychological or sociological concepts. A number of modern doctrines fall under biologism: Nietzsche's vitalism, social Darwinism, the Nazi cult of race and instinct, human sociobiology, the Gaia hypothesis, and current Evolutionary Psychology. Ludwig Feuerbach's philosophical anthropology (1845), which influenced Marx and Engels, belongs in the same tradition. His famous apothegm *Man ist was man isst* ("Man is what he eats") inspired the Catalonian biologist and philosopher Ramón Turró, who in 1912 claimed that what generates knowledge is not curiosity, as Aristotle taught, but hunger – whence the name "trophology", the study of eating, for his doctrine. Trophology is alive and well: Hillard Kaplan and coworkers (2000) held that the distinctive feature of our species is the tendency to procure calorie-dense, large-package, and skill-intensive food resources. This is possibly true provided it is supplemented with cooperation and food sharing.

In his *Genealogy of Morals* (1887), as well as in some of the aphorisms collected in *Will to Power*, Nietzsche demanded the "naturalization" of all human knowledge, in particular that of ethics. He realized that standard morality thwarts our instincts, which he wished to free. Nietzsche rejected all the "unnatural" limitations to egoism, in particular the "will to power": let the stronger prevail, let the "law of the jungle" rule. Hence Nietzsche's contempt for compassion and cooperation, and his hatred for democracy, trade-unionism, and socialism. Unsurprisingly, Nietzsche was Hitler's favorite pop philosopher – an admiration Hitler shared with is comrade Professor Heidegger. True, Nietzsche was also admired by anarchists and other nonconformists, but only for his iconoclasm and fiery pamphleteering.

Nietzsche was radical, superficial, inconsistent, and vitriolic enough to win the admiration of half-educated rebels – as well as of supporters of "savage" capitalism, such as the pop philosopher Ayn Rand, a naturalist, realist, and vehement advocate of "rational egoism." Her star pupil, the super-banker Alan Greenspan, stands accused of being partly responsible for the 2008 financial meltdown. He admitted that this crisis took him by surprise because he was confident that "rational egoism" would lead bankers and financiers to make only wise decisions. However, let us move on.

Human sociobiology is the program of accounting for everything social in biological terms, thus "preempting social science," as Alexander Rosenberg (1980) put it. That program has had a strong impact on both social science and popular culture, and it has been ambivalent in both cases. Indeed, human sociobiology has insufflated some biology into the social sciences; by the same token, it has weakened both sociologism and idealism. Sociologism involves the thesis, championed by Karl Marx and Emile Durkheim, that social wholes determine the behavior of their components (top-down action); correspondingly it also holds that social science owes nothing to biology and psychology. And the idealist school in social studies, championed by Wilhelm Dilthey, spread by Heinrich Rickert, paid lip service to by Max Weber, and practiced by Pitirim Sorokin, Alfred Schütz, Clifford Geertz, and the ethnomethodologists among others, holds that the social scientist should focus on the inner or spiritual lives of people, whence the names *sciences morales* and *Geisteswissenschaften* – spiritual or cultural sciences.

The naturalist reaction to both sociologism and idealism was particularly pronounced and successful in pop anthropology, as evidenced by the bestseller *The Imperial Animal*, by Lionel Tiger and Robin Fox (1971). These pioneers of human sociobiology stressed our animality, strong sexuality, selfishness, competitiveness, and aggressiveness. They also proclaimed the superiority of the male, as well as the primacy of brute force and domination over work, intelligence, cooperation, compromise, and morality. According to them, hunting, fighting over females, and having sex has always been more important than sociality. The lack of empirical evidence from archaeology did not inhibit those fantasists. In particular, the absence of warring scenes from the realistic cave paintings, which covered 25 millennia, told them nothing about human nature. Hobbes and Nietzsche had finally made it in social science departments.

Contrary to the majority of their colleagues, Tiger and Fox adopted ontological and methodological individualism: they focused on individuals rather than social systems such as families, bands, tribes, business concerns, armed forces, or churches. And they were not interested in the manner in which people made their living and adapted the environment to their needs: they pictured men as obsessed with power for the sake of sex. Accordingly, Tiger and Fox had no patience with democracy, feminism, the welfare state, or pacifism. In their view, human nature was fixed about 100,000 years ago, and it is so robust that social reforms are doomed to fail. What men need, they suggested, was not democratic leaders, let alone workers in the "human betterment industry," but ruthless zookeepers.

Some of the behavioral geneticists and molecular anthropologists who followed used Richard Dawkins' folk genetics to try and "naturalize social inequalities," as Jonathan Marks (2002) put it in his devastating critique. The least inhibited of these attempts was *The Bell Curve* (1994), the bestseller by professor Richard Herrnstein and journalist Charles Murray. Their crude message was in tune with the virulent neoconservative ideology of the day: Since science has proved that social standing depends on intelligence, and the latter is innate, there is nothing that can be done to improve the lot of the poor.

Evolutionary biology had from the beginning a strong impact on social theory and politics, both on the left and on the right. Marxists saw in it indirect evidence for the view that institutions are historically changeable. And conservatives interpreted it as confirming the dogma that people are born commoners or upper-class, as well as racially inferior or superior. Biologism has also inspired purely academic pursuits, such as the attempt to explain the evolution of languages by analogy with biological evolution, with utter neglect of economic and political factors. Yet it seems obvious that nowadays English would not be the language of choice if Britain and America had remained culturally backward and had not conquered huge tracts of the planet.

Languages evolve along with the cultures that speak them. In particular, they prosper or decline together with the scientific, technological, artistic and humanistic communities they serve; and when people migrate, they carry their words along with their genes (see, e.g., Cavalli-Sforza et al. 1994).

Likewise, rock music would not dominate the music market today if it had emerged in Albania or Nepal rather than in Britain and the United States, and if it had not been linked to the records, electric guitar, drugs, show-business, and television industries; and if young people in the affluent nations had been kept on the small allowances that their ancestors had received. Music historians cannot ignore that the Wagner cult peaked together with the European colonial empires, any more than they cannot overlook the powerful business factor in the rise of large-scale commercial music around 1950. In short, cultural historians have much to learn from social, economic and political historians, but nothing from biologists, except that even the most sophisticated artists must eat. Could biologists explain why Mozart and Schubert died poor, whereas Elvin and the Beatles finished wealthy? They cannot even explain why death merchants and warmongers make far more money than good parents. After all, the measure of biological success is not body count but Darwinian fitness (progeny size).

Fast, radical social changes, such as the emergence of agriculture and the state, urbanization, industrialization, militarization, alphabetization, democratization, and secularization refute biologism, for those processes are not etched in the genome and they do not result from biological alterations. On the contrary, they cause changes in nutrition, metabolism, hygiene, and modes of thinking. For example, the emergence of cities favored the spread and lethality of plagues; the decline of the Maya civilization was followed by a decrease in stature; the "discovery" and looting of America decimated the indigenous populations because it spread Western germs among them; social inequality induces stress and the accompanying pathologies – and so on. On the positive side, suffice it to recall the health benefits of sanitation, the reduction in the length of the workday, female emancipation, and progressive social legislation. In short, social organization has a strong biological impact. This is why advancing the biosocial sciences and technologies is far more important and urgent than attempting to reduce the social sciences to biology (Bunge 2003a, 2009).

In conclusion, the zoological (or ethological) approach to the social was initially positive insofar as it reminded social scientists that we are animals. But its negative impact was far stronger and noxious: it diverted attention from the specifically social, and was used by the American New Right to justify its attack on the welfare state and its preference for military aggression over international cooperation. In short, biologism, from Nietzsche to Nazism to human sociobiology and Evolutionary Psychology, failed as a new foundation for social science, and ended up as a right-wing ideology.

Biologism is on the decline in academia. The rising view is that we are the products of both genome and culture, and that these two "factors" have coevolved (e.g., Cavalli-Sforza and Feldman 1981; Richardson and Boyd 2005). This could not be otherwise, because sociality is an essential property of humans, and nearly everything social is artificial, not natural. For instance, eating is natural, but eating junk food is not; sitting is natural, but sitting long hours in front of a screen is not. And the combination of bad eating habits with sedentariness have caused the current epidemics of obesity, diabetes, arthritis, and cardiovascular diseases, all of which are contributing to the decline of fertility.

In short, biologism is wrong because it ignores the obvious fact that humans are largely artificial. As Merlin Donald (1991, 382) put it, "the Darwinian universe is too small to contain humanity. We are a different order." But the failure of biologism is that of naturalism, which is a version of vulgar materialism, not that of emergentist materialism, a much broader version of materialism, to be defended in the next chapter.

6.6 Naturalism's Three Musketeers

The writings of the mechanist or "vulgar" materialists Ludwig Büchner, Jacob Moleschott, and Karl Vogt, were hugely popular in the second half of the nineteenth century. Büchner, a physician, was by far the most original and influential of the three. His *Force and Matter* (1855) circulated in several languages throughout Europe, and remained in print for half a century. True, the annexation of biology, psychology and social science remained at the programmatic stage; but the spiritualist cosmology was severely wounded in almost all domains. Frederick Engels' (1954) mockeries did nothing to discredit mechanistic materialism, particularly since he opposed to it the alleged depth and subtlety of Hegel's dialectics. It was only natural for the friends of science to side with Büchner, Moleschott, and Vogt, rather than with its most strident enemy.

One century later, another troika lit the enthusiasm of the naturalists: that composed of Richard Dawkins (e.g., 1976), Steven Pinker (e.g., 2003), and Daniel Dennett (e.g., 1995). These popular writers enjoy an advantage over their predecessors: they preach from the academic lectern. Regrettably, as will be argued anon, the naturalism of these Three Musketeers involves some bad science of their own making, as a consequence of which they have actually weakened the naturalist cause.

To begin with, all three writers in question have propagated nativism, whose central dogma is "Nature trumps nurture" – or "Genome is destiny." This doctrine is in turn based on an extravagant version of genetics, according to which the DNA molecule is "selfish," or intent on spreading its own kind as far and wide as possible; genes would also be self-sufficient, in particular self-duplicating, as well as the subjects and units of natural selection; by contrast, the very existence of the whole organism would be "paradoxical," since its only function would be to serve as the means for the transmission of genetic material from one generation (of organisms!) to the next.

Ironically, geneticism is at variance with genetics. Indeed, DNA is rather inert, and only enzymes can divide it. Furthermore, DNA does not "specify" the mode of folding of proteins, which is strongly influenced by the cellular milieu. Second, no one knows yet how to induce a bag containing DNA molecules and a mixture of water and biomolecules to metabolize, which is the very first condition for being alive. Consequently, synthetic biology, which intends to manufacture cells from their abiotic components, will have to encompass much more than genetic engineering.

Third, the whole organism, not the genome, is subject to natural selection, and therefore the unit of evolution. The reason is that natural selection is about survival and reproduction, which only organisms can perform. If one were to die before reproducing, one's precious genes would not be transmitted to the next generation. Furthermore, unlike the passive gene, the organism reacts back to its environment: it takes part in the construction of its own habitat – it adapts itself and its environment. (See Lewontin and Levins 2007; Odling-Smee et al. 2003)

Fourth, the writers in question share the widespread misconception that natural selection preserves all adaptations and eliminates all dysfunctional traits. Although natural selection is indeed a wonderful adaptation mechanism, it is far too slow and imperfect to completely eliminate such dysfunctional things as toenails (which nowadays benefit only podiatrists) and wisdom teeth (obviously designed to enrich orthodontists) and debilitating disorders such as depression, that only benefit psychiatrists – even though Nesse and Williams (1994) have seen it as adaptive for leading the patient to giving up hopeless struggles. (More against adaptationism in Gould 2002.)

Another original if wrong contribution of our self-appointed "brights" is the idea that evolution has been programmed by "evolutionary algorithms" (Dennett 1995). But of course a process cannot be at the same time natural and guided by algorithms, for these are artifacts. Moreover, every algorithm is designed to produce unerringly an outcome of a prescribed kind, whereas speciation is hardly predictable – except of course in the deliberate creation of hybrids such as the tangelo.

Moreover, as Stephen Jay Gould and Richard Lewontin have rightly stressed, evolution is characterized by contingency – by non-biological external circumstances, such as geological and climatic catastrophes. And, as François Jacob (1977) suggested, the evolutionary process looks more like the work of an opportunistic tinker who works with whatever odds and ends he finds in his toolbox, than that of an engineer who can start from scratch, knows in advance what he wants, and plans for it. That is, molecules and organs change functions as opportunity allows. For instance, sponges, which are extremely primitive organisms, contain the same neurotransmitters that in humans play eminent roles; and humans' suprarenal glands, which in a well-designed body would be inside the brain, straddle the lowly kidneys. Nature is a collagist, not a sculptress.

Fifth, the three self-appointed "bright" champions of Darwin's admire and propagate the fantasies of the so-called evolutionary psychologists (see, e.g., Buss 2004). In particular, these writers share the belief that the human mind was basically adapted to "the Pleistocene environment," a period that started about 1.6 million years ago. So, we would be basically living fossils: we would walk the mean streets of Miami, and some of us the safe Harvard Yard, or even the moon surface, endowed with prehistoric brains. Nothing, not even the Neolithic, Industrial, or French revolutions, could possibly have altered human nature: human evolution stopped about 100,000 years ago, so that Evolutionary Psychology could be invented. The reason, we are told, is that human nature resides in the genome, which is assumed to be unchanging, basically self-sufficient, and the first motor of everything human.

True, none of the writers in question denies the "influence" of the environment, in particular the social context. But the point that all three miss is that sociality and artificiality are part of human "nature" – a part that falls beyond the ken of biology. It is not just that humans engage in social relations and use artifacts, but that these are unnatural, and yet co-define humanness just as having 23 chromosomes, descending from hominids, and having a plastic brain capable of learning almost anything. True, thc latter ability seems to be rooted in a collection of RNA genes (the human accelerated regions 1 and 2) peculiar to humans. But individual potentialities cannot be actualized in the absence of favorable social circumstances.

Finally, both Pinker and Dennett, along with Hilary Putnam, Margaret Boden, Patricia Smith Churchland, and other philosophers, have championed the computer metaphor of the brain. At first sight this view looks naturalistic because it dispenses with the immaterial soul—although some of the above-mentioned have stated that what matters is function, not stuff. But computers are not exactly natural. Worse, unlike live human brains, they are limited to performing algorithmic operations. They lack spontaneity, creativity, insight (intuition), the ability to feel emotions, and sociality. Indeed, computers have to be programmed; there can be no programs for coming up with original ideas; an emotional computer would be unreliable and therefore unmarketable; and we would have no use for computers who were to freely associate among themselves. (More on this in Chapter 11.)

In sum, the cause of naturalism has not been well served by the Three Musketeers: Muskets are outmoded.

6.7 Psychologism

Psychologism is the attempt to account for social behavior and social traits in psychological terms, with disregard for irreducible social items like social structure, institution, and social movement. The social forays of Freud and Searle are so many examples of psychologism. So is the rational-choice approach, so popular in social studies.

Freud claimed that all social conflicts ultimately result from Oedipus' complex: men would revolt against authority because they identify their boss with their father. That women too can revolt against the establishment, and that one may object to a social ill without letting personal animosities occlude one's judgment, did not occur to Freud. In any event, the psychoanalytic interpretation of social facts is now out of fashion.

Searle (1995) attempted to explain social matters in terms of "intentionality," an ambiguous term as we saw in Section 6.2. But he has confined his attention to trivialities, such as the fact that the value of paper money relies on a convention, and "cases where *I* am doing something only as part of *our* doing something." A single category, such as intentionality, economic utility, or social class, cannot suggest theories explaining macrosocial facts, such as poverty, inflation, inequality, discrimination, authoritarianism, unemployment, or war, all of which are systemic.

Psychologism has been rampant in social studies since the 1870s, when neoclassical microeconomics gained the ascendancy and spilled over to all the social disciplines under the name of "rational choice theory". Indeed, this theory postulates that individual action is the root of all social events, and that all individuals, regardless of their place in society, act so as to maximize their expected utilities. Now, the concept of social utility is clearly a psychological one, since it is defined as the product of the *subjective* utility or gain by the *subjective* probability (or intensity of belief) of an action. Since neither feature is accessible to objective measurement, both are usually assumed a priori – hardly a scientific procedure (see Bunge 1996).

By contrast, experimental economics is the scientific study of the economic behavior of individuals, such as the relation between productivity and incentives of various kinds, and that of work satisfaction with the ability to influence management decisions. However, such studies – such as those of Daniel Kahneman and Ernst Fehr – concern individuals rather than economic systems, such as firms and national economies. Hence, the classing of such research as experimental *economics* is incorrect: it belongs in social psychology even though most of its practitioners are active in economics departments. So far, experimental economics proper is not even a research project, even though it should be an interesting and useful one, particularly if were to avoid psychologism.

The rejection of psychologism does not involve the denial that social actors are largely motivated by their beliefs. In 1928 the sociologist W. I. Thomas pointed out that people react not to social facts in themselves but to the way they perceive them (Merton 1968, 19–20). Hence it is true that (some) social being is partly in the eye of the beholder. It is also true that we act on our values and beliefs – but always within social contexts that precede and usually overwhelm the individual, as Marx noted. And these contexts are bound to react upon the individual actor. As a result, some of his actions will be frustrated, while others will have unintended consequences. In other words, the consequences of an action "are not restricted to the specific area in which they are intended to center and occur in interrelated fields explicitly ignored at the time of action" (Merton 1976, 154).

Naturalism predisposes the student of society to adopt an individualist approach, that is, to start from individual action. But every action happens to occur in a preexisting social context. Think of job-hunting in a period of economic recession. There is little that the isolated individual, however smart and well informed, can do to alter the social situation. Only organizations, such as states, political parties, and big corporations, can alter societal facts. And such organizations cannot be understood by individual psychology, for this discipline studies brains, not social systems. In short, naturalism cannot cope with social matters because these are artificial.

6.8 Naturalized Linguistics, Axiology, Ethics, Law, and Technology

Let us now peek at four naturalization projects, starting with *linguistic naturalism*, or biolinguistics. This is the thesis that language is natural and moreover instinctive, whence linguistics is basically a natural science (Pinker 2003). At the same time, most of the advocates of this view have adopted Cartesian mind-body dualism – which, incidentally, confirms the need to distinguish naturalism from materialism.

Part of this naturalization project is the bold assertion that we are born knowing universal grammar (UG), the intersection of the grammars of all the particular grammars. Unfortunately no one has bothered to explicitly state the rules of UG, and geneticists have not found the presumptive UG gene(s). Nor is there any reason to expect such findings, for languages are highly conventional: the word-object (or sign-signified) relation is artificial, and so is the order of grammatical categories. For instance, why should it be more natural to use *river* rather than *potamós*, or conversely, to denote a water stream in English? And why would the subject-verb-object (SVO) order, the most common in English, be more natural than the subject-object-verb (SOV) order that prevails in Japanese? There is no evidence than either is more adaptive than the other. By contrast, arguably the VSO and VOS orders are more *logical* than their alternatives because, when formalizing sentences in the predicate calculus, one starts with the more salient predicate. Thus, one formalizes "*b* said *c*" as *Sbc*. But of course rationality is anything but natural, as suggested by its recency.

Far from being a natural item like digestion and walking, speech is social. This is because, unlike animal vocalization, speech is a symbolic communication tool and, as such, subject to cultural evolution. But of course one may forget all this if one conveniently forgets the existence of historical linguistics and denies the legitimacy of sociolinguistics – the way Chomsky and his school does. But such oversights block the understanding of the language changes brought about by conquests and mass migrations. For example, the transformation of Old English (or Anglo-Saxon) into English would be incomprehensible to one who did not know of the deep social changes brought about by the Norman Conquest. Likewise, the mass arrival of European immigrants into the US helps understand some of the differences between American and British English. By contrast, we still ignore the causes of the so-called Great Vowel Shift that occurred between Chaucer (who pronounced A, E, I, O, and U the way Germans and Spaniards did and still do) and Shakespeare. If anything, this shift was maladaptive, hence unnatural.

All this is not to deny the legitimacy of the biological *aspect* of the study of speech, as distinct from language – to use de Saussure's important distinction between *parole*, a process in living humans, and *langue*, an abstract object resulting from pretending that linguistic expressions exist by themselves. For example, speech learning (its ontogeny) and speech defects, such as stammer and dyslexia, are certainly biological features, since they occur in brains. But such brains are embedded in social networks, which are built, reformed and destroyed with the help of communication tools, mainly the "natural" languages – social artifacts all of them.

Hence one should do biosocial linguistics rather than biolinguistics. However, let us move on to other facets of the naturalization program.

Axiological naturalism holds that our basic values are natural rather than conventional, as well as inter-subjective rather than subjective. This would be so because all human beings have roughly the same basic needs, which in turn is due to the commonality in our biological make-up. Moreover, this should hold for all living beings, since we are all in the same "tree of life", and since valuation is a condition of survival. So, values emerged about 3.5 billion years ago along with life. Consequently, there is nothing fallacious about value-naturalism. On the contrary, what is fallacious is G. E. Moore's famous denunciation of the "naturalistic fallacy." Cognitive neuroscientists remorselessly perpetrate this "fallacy" when, on the basis of brain imaging studies, they conclude that evaluating the goal of a human action is "correlated" with activity in the medial orbitofrontal cortex (e.g., Hare et al. 2008). Biology has trounced axiological idealism.

I find axiological naturalism compelling for biological values, particularly well-being, but not for social, moral and aesthetic values. The latter, like acquired tastes, are strongly dependent on experience, social standing, and social environment. In social, moral and aesthetics matters, one person's likes are often another's dislikes. So, there are subjective preferences in addition to objective ones. Moreover, most economic, political and cultural values are unnatural, in that they depend on work, tradition, outlook, and expectation. For example, there is nothing natural about social convention, price (contrary to value in use), or reputation.

However, the basic assumption of axiological naturalism, that values, or at least some of them, are "real" rather than subjective, stands – with the proviso that reality be understood to include not only nature but also society as well as the self, which is largely shaped by the social environment. (See Boudon 2001; Railton 2003.)

Ethical naturalism is the thesis that moral norms are either natural or reducible to natural science. The former view may be dubbed naïve, and the latter sophisticated ethical naturalism. The ancient Greek and Roman Stoics, like Zeno the Stoic, and the modern emotivist moral philosophers, from Hume to the logical positivists, were naïve ethical naturalists, since both groups recommended "following nature." When some early Darwinists suggested that evolution had a moral arrow, Thomas Henry Huxley (1983), nicknamed "Darwin's bull-terrier," reacted spiritedly, and claimed that morality is utterly artificial, since it aims at suppressing selfishness, which is natural because it follows from the self-preservation instinct. Contrary to the Stoics, Huxley famously stated that "the ethical progress of society depends, not on imitating the cosmic process, still less in running away from it, but in combating it."

Yet naïve ethical naturalism has recently been partially vindicated by experimental research in two unexpected fields: primatology and behavioral economics. In fact, observing chimpanzees and bonobos, Frans de Waal (1996) found that, far from being consistently aggressive, these close relatives of ours are on the whole good-natured – which explains why they form rather stable groups. To be sure, the apes cheat and fight for power and for access to the alpha-male's harem; but they often cooperate to achieve their goals, and they are not cruel. In short, they do not behave the fiendish way pop anthropology used to tell us. But they are not good Samaritans either.

Contrary to chimpanzees, adult humans would seem to be "strong reciprocators" imbued with a sense of fairness. In fact, working with humans, some behavioral (or experimental) economists, from Daniel Kahneman to Ernst Fehr and his colleagues at the University of Zürich, have corroborated what any parent and teacher knows but the orthodox economists and the self-styled evolutionary psychologists ignore, namely that most of us are basically fair (Fehr and Fischbacher 2003; Gintis et al. 2005). (By contrast, chimpanzees are basically selfish: Jensen et al. 2007.) Moreover, this is an aspect of human nature, for the feeling of inequity has a neural "correlate," namely the insula. Indeed, in humans the activity of this region of the prefrontal cortex is the more intense, the more unfairly a situation is perceived by the subject (Hsu et al. 2008; Purves et al. 2008, 615). In short, inequity-aversion is generalized. However, it is not inborn, as will be seen in Section 10.1.

Thus, contemporary psychology has confuted the pessimistic Judeo-Christian view of human nature, whereas it has confirmed Adam Smith's on moral emotions as the root of our attitudes to fairness and distributive justice. But, like all natural dispositions, the sense of fairness is altered by education and reflection, as shown by the individual differences in reactions to inequity. For example, practical considerations may force someone a trade-off between equity and efficiency – as in medical triage. In this case the putamen, which responds to inefficiency, competes with the insula, and a third region, the caudate/septal subgenual region, is activated as well (Hsu et al. 2008). A moral of this story is that the sense of fairness involved in distributive justice is not just an academic subject, let alone a mirage (Hayek): it is deeply embedded in any brain that has not been irreparably damaged by either standard economics or by a conservative ideology. A more general lesson is that our moral ideas have strong emotive roots that may intertwine with cognitive roots. This finding confutes utilitarianism and constitutes a partial vindication of emotivism – only partial because the latter overlooks the cognitive component.

What I have called *sophisticated* ethical naturalism is the project of reducing moral norms to the natural sciences, in particular human biology (Edel 1944). Elliott Sober and David Sloan Wilson (1998) have bravely tackled this ambitious project with the help of evolutionary biology. In particular, they have embraced William Hamilton's "rule" (hypothesis), according to which altruistic behavior occurs spontaneously in all species when a certain condition is met. This condition is that the cost c of the altruistic behavior be less than its benefit b discounted by the coefficient r of genetic relatedness between actor and beneficiary: $c < r.b$. Some interesting confirmatory cases have been cited, but they do not include humans.

In humans social relatedness takes precedence over biological relatedness. For example, we are likely to help genetically unrelated spouses, friends, and partners more than we help our genetic relatives. (In these cases $r = 0$, so that the Hamilton condition for altruistic behavior becomes "c< 0", which is meaningless.) Domestic violence is no less frequent or lethal than street violence. Ditto war: Remember that three of the main heads of state involved in the First World War, namely the German Kaiser and the British and Russian emperors, were cousins. By contrast, a mafia "family" is not based on kinship. In short, among humans neither conflict nor cooperation are fated by genetic relatedness: interest (or value), social bond, and social conflict trump genetic relatedness.

Ironically, whereas altruism is likely to puzzle biologists and orthodox economists, it does not greatly surprise psychologists or social scientists. The latter know that giving can be more pleasurable than receiving, and that sharing is a condition for keeping one's standing or reputation in any social network. But, as Robert Luis Stevenson noted long ago, normal people are half-and-half, or egotuists rather than fully egoists or fully altruists. We need to be somewhat selfish to subsist, and somewhat unselfish to coexist.

Morality cannot be successfully naturalized because moral facts are social rather than biological, as they occur only in social contexts: Robinson faced no moral dilemmas before encountering Friday. For example, volunteering and crime are moral facts because they are prosocial and antisocial respectively. Second, all morals belong to a tradition and exist in some social context. That is, moral norms are invented and applied by human actors with definite interests and in definite social circumstances. Change either actor or circumstance, and a different set of moral deliberations is likely to be involved.

This is why morals differ from one society to the next, why they have changed through history, and why there can be moral progress, as in the cases of the abolition of slavery and capital punishment (see, e.g., Westermarck 1906–1908). In conclusion, morality cannot be naturalized because it is a chapter of the social maintenance manual. But it can be "materialized" in the sense that it can and must be regarded as a feature of the social coexistence of material agents (persons) in material systems of the social kind.

What holds for morals also holds, mutatis mutandis, for ethics, or the study of morals. Ethics analyzes and evaluates rules of behavior, and as such it is a social technology, not a branch of natural history. And, unlike social crafts, social technologies are expected to make deliberate use of scientific laws. Now, a scientific law, if applicable at all, is technologically ambivalent, for it countenances two distinct and even mutually opposed technological rules or prescriptions (Bunge 1967a). Indeed, a law L involving a means M and a goal or consequence G of practical interest has the form

Law If antecedent M occurs, consequence G follows (or is likely to follow). [1]

A law of this form suggests two mutually dual technological rules:

Rule 1 To attain G, put M into effect. [2a]
Rule 2 To avoid G, prevent M from happening. [2b]

Note that, whereas the above law schema [1] is value-free, the rules [2a] and [2b] based on it (though not deduced from it) are value-laden. Moreover, the choice between seeking *G* and avoiding *G*, hence between applying [2a] and using [2b], will involve moral deliberations if the consequence *G* is likely to affect individuals other than the actor. Hence the naturalization project involves a logical circle: The attempt to derive any moral norm from science involves moral norms.

Biology alone cannot account for human nature because, as the British archaeologist Gordon Childe put it in 1936, man is the animal that makes himself. Hence only a combination of biological evolution with social evolution may explain the emergence, reform, and repeal of social, moral and legal norms. In particular, it would be foolish to explain constitutions in purely biological terms. The same holds for international codes such as the United Nations Charter, as well as for social facts in general. Which is why the expression "naturalized social science" is an oxymoron.

Legal naturalism holds that just laws are natural. But history shows that there is no such thing as natural law: all legal codes are artificial. And, like all artifacts, the so-called positive laws are bound to be imperfect, though also perfectible through research, debate, and struggle. Unfortunately, the contemporary debate over natural law has been distorted by the legal positivists: in their eagerness to defend rationality and secularism, as well as the status quo and the corresponding legal order, they hold that one must choose between the two schools, natural law and legal positivism, and that the former is just a vestige of tradition, whereas science, or at least rationality, is on the side of the "might makes right" school. Actually no such choice is compelling, because there is a *tertium quid*, namely legal realism. This school holds that the law is more than a tool of social control: it can also be a tool of social progress, in particular an adjunct of any policies aiming at attaining social justice (see Bunge 1998; Pound 1954). The disjunction of law and morality, claimed by legal positivism, is wrong because laws regulate our behavior toward others, and as such they are objective moral rules to be evaluated by their consequences (Wikström and Treiber 2009).

What holds for social technology, in particular ethics and the law, also holds for the other branches of engineering: they too are artificial activities centered on the design of artifacts. What is also true, if paradoxical, is that artificiality is of the essence of human nature: Humans, however primitive, use or even invent and fashion artifacts of many kinds, from tools and weapons to institutions and words. Only the artifacts made by nonhuman animals, such as spiders, honeybees, beavers, crows, and weaverbirds, are natural, in the sense that the corresponding animals make them instinctively, without first making any drawings and with no need for formal schooling. The human material artifacts, and their corresponding crafts and technologies, are just as unnatural as poems, songs, rules of etiquette, legal codes, religions, and scientific theories. Since naturalist anthropology cannot account for the artificial, it misses the all-round artificiality that distinguishes humans from other animals, no less. Hence it is not fit for human consumption.

Let us finally glimpse at a special naturalization program: the attempt to reduce all the sciences of man to neuroscience.

6.9 Neuro This and Neuro That

Neuroscience has recently displaced physics as the sexiest science, and it is successfully completing the naturalization of psychology that Hippocrates and Galen began in antiquity. Indeed, cognitive and affective neuroscience have replaced at the same time behaviorism, which was scientific but extremely narrow, and psychoanalysis, which was very broad but pseudoscientific. Fortunately, cognitive neuroscience has not eliminated social psychology, but has merged with it. Indeed, traditional social psychology, which was born in the 1930s, is currently being enriched with cognitive neuroscience, to produce social cognitive neuroscience (see, e.g., Cacioppo et al. 2006).

It was perhaps unavoidable that the success of the naturalization program in capturing the mental would inspire what may be dubbed *neuroimperialism* – the attempt to explain every bit of behavior in neuroscientific terms. Indeed, in recent years we have witnessed the birth of neuroeconomics, neurohistory, neurolaw, neuroethics, neuromarketing, and neuropoetics. How much of this is legitimate science or technology, and how much empty promise? Let us see.

Neuroeconomics is the study of the economic behavior of individuals in the light of cognitive and affective neuroscience (Camerer 2003). This emergent discipline has shown certain differences between planned and impulsive buying: whereas the former is guided by the prefrontal cortex, the latter is under the strong influence of subcortical systems – as was to be expected. Results of this type are interesting but limited, because they concern individuals rather than firms and markets, and because laboratory findings are a poor guide to real-life behavior.

Neurolaw is the neuroscientific study of what jurists call *mens rea*, the criminal mind. It has been known for decades that juvenile offending peaks at about the age of 17 because this is when boys form new attachments beyond the control of home and school, and when their brains become awash in testosterone, while their prefrontal cortices are still immature. But neurolaw cannot explain why the crime rate is far greater in the United States than in neighboring Canada, which in turn has a higher rate than in Western Europe, India, and Japan. Nor does neurolaw explain what drives persistent offending, that is, adult crime. These limitations of neurolaw are due to the fact that crime, unlike poetry or mathematics, is identical with antisocial behavior, and understanding such behavior calls for an exploration of the social environment (see, e.g., Wikström and Sampson 2006).

Neuroethics is the study of the way the brain internalizes moral norms, as well as the study of the pathological causes for the failure of such moral enculturation. The classical case is of course that of Phineas Gage, who lost his moral conscience as a consequence of a severe injury to his frontal lobe. But this was a very exceptional case: in the vast majority of cases, immoral behavior results from either incomplete socialization or social pathology rather than from neural deficiency.

How about the seven deadly sins: could neuroscience explain anger, greed, sloth and the rest, as suggested on a recent program of the History Channel? This is extremely unlikely, for two reasons. One of them is that every list of sins is culturally determined. For example, aggression, enslavement, theft and social insensitivity do

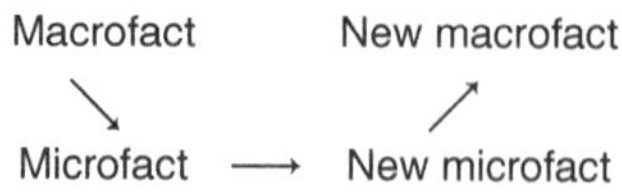

Fig. 6.1 A macrofact, such as a flood or a revolution, elicits emotions that mobilize individuals who cooperate to bring about a change in the environment. Inspired by Coleman (1990, 10ff)

not occur in the Christian list of sins. And the worst aggressions, the military ones, are not committed in anger; likewise, thieves are not motivated by anger or greed: most of them steal to feed their families in times of unemployment or social disintegration. Focusing on the brain as the source of social ills amounts to diverting attention from their structural source.

The reason neuroscience alone cannot cope with social problems is that it explores what goes on inside people's brains, not among them: Only the social and biosocial sciences are equipped to deal with social relations. Therefore the neuroscientific approach to all kinds of human behavior should be welcomed with caution, for every person is a member of several social circles or systems, which constrain behavior in some respects while stimulating it in others. Thus, while it is true that the stock market transactions are influenced by fear and greed, these emotions do not have their source in inner brains: they are generated by macrosocial facts, such as natural or political disaster, inflation, unemployment, social unrest, war, and industrial innovation. Consider the familiar scenario: disaster strikes a community, whose members get together to remedy some of the consequences of the disaster: See Fig. 6.1.

In conclusion, naturalism should be expanded to make room for the economy, the polity, and culture. This expansion is called "emergent systemic materialism", to be examined in the next chapter.

Closing Remarks

The great merit of naturalism is that it undermines magical thinking, in particular supernaturalism. But naturalism is limited, and therefore weak, for it does not account for the specificities of the social, the moral, the legal, the scientific, the technological, and the artistic. This is why it has failed to explain any large-scale social process, from the emergence of culture to the rise and fall of empires. After all, the whole point of being human is to "violate the natural order" – to enact the paradox that human nature is largely artificial.

Furthermore, according to naturalism, human nature is biological, whence social reform is ineffectual or worse, as Pinker (2003, 294), Tiger (2008), and others have claimed. (Obviously, Noam Chomsky has not noticed that his nativism and anti-evolutionism conflict with his leftist politics.) Thus, ironically, naturalism, which was scientifically and politically progressive between the Renaissance and the Enlightenment, is nowadays being invoked to support conservatism. Worse, some of the most vocal champions of secularism and science rightly fight creationism and parapsychology, but overlook equally pseudoscientific but far more harmful doctrines, such as the standard economic and political science theories. And animal-rights advocates, naturalists too, are typically unconcerned about human-rights violations.

In short, the naturalization project is sound in natural-science subjects but it is deceptive in social-science subjects: here at best it attempts to depoliticize politics, and at worse it disguises a regressive political agenda. The limitations and excesses of naturalism suggest that this worldview should be expanded to encompass the artificial and the social, which are almost wholly unnatural. (They have natural sources or roots but they are made, not found.)

In other words, before the emergence of modern man, the universe was wholly lawful. After that event, in a tiny corner of the universe, namely the inhabited regions of our planet, there have been rules (or norms) in addition to laws. For example, the mathematical concepts are subject to the rules of logic; social systems "obey" laws, both natural and social, plus social conventions and norms, many of which are local; and computers operate in accordance to both the laws of electronics and the programs that their users impose on them. Any realistic ontology must go beyond naturalism and make place for the artificial and the social.

The proposed enlargement of naturalism (or physicalism, or vulgar materialism) should not be mistaken for sociologism, or the attempt to account for everything human in purely social terms, forgetting that humans are social *animals*. In other words, the flaws of naturalism are not corrected by exaggerating the unnatural aspect of human nature, but by blending the two aspects. I submit that, when combined with the systemic approach and the scientific method, materialism has all the advantages of both naturalism and sociologism, and hopefully none of their shortcomings. However, this kind of materialism deserves a chapter of its own – the next one.

Chapter 7
Materialism

The word "materialism" is ambiguous, for it designates both a moral doctrine and a philosophy. In ordinary language, "materialism" is synonymous with "hedonism", or the pursuit of pleasure and material possessions. On the other hand, philosophical materialism is the worldview according to which everything real is material. The two doctrines are logically independent: hedonism is coherent with immaterialism, and philosophical materialism is compatible with lofty moral values. Epicurus, the most famous materialist of antiquity, was known to be extremely austere. And nowadays some of the most greedy tycoons and corrupt politicians like to denounce materialism and preach spirituality, which they conflate with religiosity.

Unsurprisingly, materialism has been reviled for more than two millennia because it undermines religion, a mainstay of conservative governments. For example, Francis Collins, the current director of the US National Institutes of Health, recently held that "the claims of atheistic materialism must be steadfastly resisted." He did not say why, perhaps because he does not know what "materialism" really means. Presumably and hopefully Dr Collins, who has a solid research record, would disapprove of any NIH investigator who were to repeat John Eccles' (1951) claim that the immaterial mind is the cause of brain activity. Indeed, scientists are expected to check their religious and idealist beliefs at the door of their labs.

Philosophical materialism has a large overlap with naturalism, since both reject supernaturalism and agree that the world, or reality, is composed exclusively of concrete objects. But they differ with regard to the characteristics of matter. Indeed, naturalism involves a very special concept of matter, namely the one investigated by physics, chemistry, and biology. Consequently, naturalists deny the existence of other kinds of matter: thinking, social, and artificial, in particular semiotic or signifying matter, such as this sheet of printed matter.

Thus, naturalism is less comprehensive than the systemic and emergentist materialism to be advocated in this chapter. That is, one can be a naturalist without being a materialist. For example, one may claim that the basic components of reality are not material things but facts (Wittgenstein), states of affair (Armstrong), or processes (Whitehead) – which are so many unwitting ways of dematerializing the world. Conversely, one can be a materialist without being a naturalist, as is the case with the materialists who, like the present writer, emphasize the specificity of the social and the artificial, both of which go beyond the natural even though their

M. Bunge, *Matter and Mind*, Boston Studies in the Philosophy of Science 287,
DOI 10.1007/978-90-481-9225-0_7,

ultimate components are natural. In conclusion, the relation between naturalism and materialism is one of partial overlap: neither includes the other.

In any event, admittedly materialism, though ancient, is still immature. This is partly because it has been banned by most schools since Plato's time, and has consequently remained mostly in the hands of amateurs, from the physician Büchner, of *Force and Matter* (1855) fame, to Lenin, the professional revolutionary who wrote *Materialism and Empiriocriticism* (1907). But not even the famous Harvard logician Willard Van Orman Quine, who wrote about ontology for most of his long academic life, and called himself a physicalist, succeed in advancing materialism. Moreover, Quine was anything but a strong materialist, for he claimed that the "existential" quantifier ∃ covers both conceptual and real existence – a case of logical imperialism, not physicalism (recall Section 6.2). He also failed to elucidate the concept of the mental, which he disliked, although he devoted many fleeting and imprecise remarks to it (see, e.g., Lycan and Pappas 1976). Much the same may be said about his erstwhile student David Lewis, the self-styled materialist who worked mainly on possible (that is, imaginary) worlds.

Since the term "materialism" is polymorphous, it is advisable to qualify it – that is, to distinguish some of the members of the family of ontologies it stands for. Let us start with the oldest form of materialism, namely physicalism.

7.1 Classical Materialism

Classical materialism equals mechanism: it is the view according to which the world is a collection of bodies. Unsurprisingly, mechanism claims that mechanics is necessary and sufficient to explain the world. A stronger version of mechanism holds that all complex bodies, such as our own, are machines. But this doctrine, advanced by La Mettrie (1748), had very few followers before the emergence of cybernetics, information theory, and Artificial Intelligence in mid-twentieth century. To believe their enthusiasts, the universe is an information system, and humans are basically self-programmed computers. (Recall Section 4.2, and see Chapter 12.)

Mechanistic materialism is the oldest secular worldview: it was invented at the same time in India and in Greece more than 2,500 years ago alongside with atomism (see, e.g., Charbonnat 2007; Lange 1905; Plekhanov 1967). Lucretius immortalized it in his beautiful philosophical poem *On the Nature of things*. The medieval nominalists, in particular William Ockham, as well as the Western followers of Averroes, were materialists of sorts, although they did not specify the precise nature of the furniture of the world. The mechanism inherent in the Scientific Revolution was definitely materialistic, and the first scientific worldview to boot. During the Enlightenment materialism was expanded to mind and society by such influential thinkers as d'Holbach, Helvétius, and La Mettrie. It was rightly regarded as subversive, and accordingly banned. It took the French Revolution of 1789 for the official censorship on materialism to be lifted. Still, how many professors dare call themselves materialists in our enlightened and liberal societies?

During the following century mechanistic materialism was nearly forgotten in France, but widely popularized in Germany and Great Britain: remember the popular books by Ludwig Büchner and John Tyndall. But materialism did not storm the academic citadel, which was occupied, even in the New World, by Kantians (subjective idealists) and Hegelians (objective idealists).

In the last century materialism became academically respectable in two unlikely places: first in the USA, with George Satayana and Roy Wood Sellars, and later in Australia, with David Armstrong, J. J. C. Smart, and U. T. Place. However, arguably, materialism has always been practiced by all scientists, even by those who held religious beliefs, since immaterial entities played no role in their theories or experiments. For example, although Newton was a devout Unitarian, God plaid no role in his equations of motion.

However, materialism is still a minority view in the philosophical community, largely because it has suffered a bad press from birth. In fact, since Plato's time, materialism has been slandered as being crass and immoral, and few if any professors have dared teach it. To this day, the ontological ideas of Hobbes, Gassendi, d'Holbach, Helvétius, Priestley, Feuerbach, Engels, and Haeckel are being systematically distorted, reviled, or just ignored in history of philosophy courses. (For instance, Randall Collins devotes to materialism a single page of his 1,000 page-long treatise on the sociology of philosophies.) And what little materialism is taught is its earliest version, namely vulgar or eliminative materialism, that is, physicalism. Even the mechanistic materialism of Descartes' two treatises is usually silenced, whereas Wittgenstein's shallow aphorisms, as well as Heidegger's nonsensical utterances, are often regarded as pearls of wisdom.

There are several reasons, some good and others bad, for rejecting or ignoring materialism. Six sound reasons used to be the following. Materialism is a collection of disjoint and sketchy theses rather than a well-constructed and consistent theory; it has remained analytically crude and, in particular, alien to modern logic; it has underrated the power of ideas; no materialist has accounted for mathematics; few if any materialists have gone beyond utilitarianism in matters of value theory and ethics; and most materialists have been amateurs rather than professional scholars. In any case, none of the materialist contemporaries of Plato, Leibniz, Berkeley, Kant, or Bolzano attained the sophistication and eminence of these idealist philosophers.

But of course the bad reasons invoked to ignore or reject materialism have been far more effective, though hardly philosophical. One was and still is that materialism undermines the fear industries, in particular fear of God, death, and scientific progress. Another illegitimate reason for rejecting or ignoring materialism has been the intimate relation of dialectical materialism with the Marxist-Leninist version of communism. However, the first reason is unjustified because idealism, in particular subjectivism, can be just as godless as materialism: recall that in his first *Critique* Kant stated that "God is a mere idea."

The second reason for rejecting materialism is unjustified too, because there have been plenty of conservative materialists, from Thomas Hobbes to Nietzsche to the social Darwinists to the contemporary evolutionary psychologists, according to whom "biology is destiny". Something similar holds for idealism. For example, there have been both liberal and fascist neo-Hegelians, such as Benedetto Croce and Giovanni Gentile respectively. In short, there is no *logical* relation between materialism and sociopolitical ideology: neither implies the other. (By contrast, realism, an epistemological school, is strongly correlated with political progressivism, because social reform assumes that society exists outside the subject's mind, and that political action consists in attempting to alter social reality.) Materialism should be judged by its theoretical merits, not by its accidental political associates, who are distributed over the entire political spectrum, from Marx to Nietzsche.

However, let us examine the honorable reasons for ignoring or rejecting materialism, because this examination may suggest an entire program for updating materialism. To begin with, let us go back to materialism's next of kin, namely naturalism. Materialism shares many crucial theses with naturalism. So much so that Dewey, Hook and Nagel (1945) – the most influential American philosophers of their day – were at pains to distinguish the two philosophies. The only difference they detected is that they regarded the psychoneural identity theory as speculation in need of scientific test – a fair assessment at the time, since cognitive neuroscience was still embryonic.

The matter of mind has always been the main bone of contention between naturalism and materialism. Whereas the materialist holds that the mental functions to be brain functions, a naturalist may believe in the immaterial mind. Consequently, whereas the materialist is irreligious, the naturalist may allow a place for a religion – one with a laid-back God and without afterlife, though. This difference has a political counterpart: Whereas there has never been an Epicurean emperor, there was one Stoic emperor, Marcus Aurelius; and one of Nero's ministers was the Stoic philosopher Seneca. Shorter: Naturalism is dissident but politically harmless, whereas materialism is bound to weaken any faith-based regime – which is why it has had such a bad press since Plato's days.

In sum, classical materialism, like naturalism, is physicalist, or physicochemicalist, and thus open to the accusation of crassness, that is, of being unable to appreciate the higher values. But it had the virtue of clarity and, in particular, of avoiding the pseudo-subtleties of Hegelian dialectics, to which we turn next.

7.2 Dialectical Materialism

The next version of materialism to gain some popularity was dialectical materialism, sketched by Frederick Engels in 1877. This ontology was intended as the materialist counterpart of Hegel's "logic". Like Hegel's, it was centered in the concept of conflict ("contradiction"), and became part of the Communist ideology in the twentieth century. Consequentgly it experienced the ambivalent reception of that ideology: adopted as official dogma in the Soviet sphere, and reviled or ignored in the same dogmatic fashion elsewhere. Materialism, both Marxist and non-Marxist, was practiced by prestigious British and French historians in mid-twentieth century, and later on by some important British and North American anthropologists and archaeologists, starting with Gordon Childe and Marvin Harris, both of whom reached wide readerships.

Dialectical materialism is often summarily accused of being an ideology, and accordingly dismissed out of hand. Actually it is a serious if flawed ontology. Karl Marx and Friedrich Engels cobbled together dialectical materialism out of the materialism of the eighteenth century and Feuerbach's, on the one hand, and Hegelian dialectics on the other (see, e.g., Cornforth 1954; Engels 1940, 1954; Shirokov 1937; Wetter 1958), At first sight this synthesis has the virtues of classical materialism and of Hegel's dynamicism (or process metaphysics). On closer inspection, dialectical materialism turns out to be confused as well as neither fully materialist nor consistent with modern science. Let me substantiate these charges.

The following are only some of the most outstanding confusions of dialectical materialism (Bunge 1981). One: the lumping of logical contradiction with ontic opposition and strife, and the ensuing talk of "dialectical logic", which would encompass ordinary logic as a sort of slow-motion approximation. Both the conflation of contradiction with ontic opposition and the (interrupted) project of building a dialectical logic are understandable in the context of Hegel's system, where all existents are ideal, whence "everything real is rational, and everything rational is real". But these confusions are inexcusable in the light of mathematical logic, whose laws apply only to predicates and propositions, and of factual science, whose laws apply, if only approximately, exclusively to material things.

A second confusion of materialist dialectics is the so-called third law of dialectics, namely the "transformation of quantity into quality and conversely". Such transformation is just impossible. What is meant by that clumsy phrase is that certain quantitative changes give rise to new qualities (properties), and that in these are followed by new quantitative patterns. For example, when a village grows to become a city, the kind of political governance changes, as a consequence of which population growth may be controlled. The preceding is just an example of emergence, the occurrence of radical novelty as a feature of the combination of two or more things. (The so-called complexity theory deals with emergence in the case of systems composed of a large number of items.)

The other two "laws" of dialectics state that every existent is the "unity of opposites," and that all change comes from the "contradiction" or "struggle" of such opposites. Both hypotheses are false. The first is false, as shown by the existence of elementary (indivisible) quanta, such as electrons and photons. Besides, the statement in question leads to an infinite regress: indeed, every one of the two "opposites" in question should in turn consist of two opposites, and so on without end. The second hypothesis, about the source of all change, is falsified by all the cases of cooperation both in nature and in society. For example, atoms combine to form molecules, and these to form liquids or solids; and persons cooperate to form families, business firms, gangs, and other social systems. In fact it may be surmised that every existent is either a system or a component of such. This is of course the central hypothesis of the systemic ontology (Bunge 1979a). However, let us return to the confusions of dialectical materialism.

A third important confusion is that between "changing" and "relative". This confusion occurs in the recurrent statements by Engels and Lenin, that "all knowledge is relative" – meaning temporary rather than eternal. This confusion is harmless when incurred by Planck, when he wrote about "the progress from relative to absolute", meaning from covariant to invariant properties and laws – that is, from items that hold relative to some reference frame to those that – like the basic laws and the speed of light in a vacuum – hold for all reference frames. But, when bound with Marx's thesis that knowledge is created by society as a whole, rather than by individuals or teams, the said confusion leads easily to anthropological and epistemological relativism – one of the postmodernist scourges.

Finally, all dialectical materialists, especially Lenin, have consistently confused materialism, which is basically an ontological thesis, with realism, which is a family of epistemologies. Yet one may be at the same time realist and immaterialist, or nonrealist and materialist. In fact, there are four possible combinations of these schools:

MR (e.g., Democritus)	$M\overline{R}$ (e.g., Nietzsche)
$\overline{M}R$ (e.g., Aquinas)	$\overline{MR}$ (e.g., Kant)

Let us now examine my accusation of inconsistency. Far from being monistic, dialectical materialism is dualist in both its philosophy of mind and its social ontology. In fact, it regards the mind as immaterial, and society as composed of a "spiritual superstructure" resting on a "material infrastructure". The former thesis is clear in Lenin's denunciation of Joseph Dietzgen's assertion that thoughts are material: "if such inclusion is made, the epistemological contrast between mind and matter, idealism and materialism, a contrast upon which Dietzgen himself insists, loses all meaning" (Lenin 1947, 251).

Incidentally, Lenin's political heir criticized such contrast, and unwittingly reinvented neutral monism, or the double-aspect view, that had been advanced by Spencer as the alternative to both materialism and idealism. According to Stalin, "the mental and the material are two different forms of the same phenomenon" – as Rosental and Yudin (1945) inform us in the entry "dualism" of their authoritative dictionary of Soviet Marxist philosophy. But two decades later Stalin's name was erased from the same work and, with it, the confusion of materialism with neutral monism (Rozental and Yudin 1967). However, let us go back to Stalin's teacher.

When commenting on Hegel's *Logic*, Lenin (1981, 182) took a further step in the same immaterialist direction: he moved from dualism to trialism. Indeed, he stated the very same thesis that Popper (1967) was to popularize half a century later under the name "world 3": that "there are *actually*, objectively, *three* members: (1) nature; (2) human cognition = the human *brain* (as the highest product of this same nature), and (3) the form of reflection of nature in human cognition, and this form consists precisely of concepts, laws, categories, etc. [original emphases]." Adding The One to this triad would result in Plotinus' spiritualist cosmology. Nor was that the only time that Lenin anticipated Popper: he also pointed out, in 1908, what Popper announced in 1952: that Berkeley was a precursor of Mach. No doubt, these are sheer coincidences. But they should force Popper's admirers to be more critical.

Unsurprisingly, psychoneural dualism was the official Soviet philosophy of mind, and the psychoneural identity (or materialist) thesis was rejected as a piece of "vulgar materialism," as Jarochewski (1975, 168) informs us. The present writer's defenses of psychoneural identity in the Russian journal *Filosofskie naukie* (1979) and its Hungarian counterpart (Bunge 1982b) were severely criticized for being critical of dualism. Interestingly, whereas my Soviet critic was presumably a Marxism expert, my critic in the Hungarian journal of Marxist philosophy was an eminent neuroscientist, Janos Szentagothai, who also happened to be a devout Christian.

Still, some dialectical materialists have rightly criticized many spiritualist and irrealist philosophies. For example, Engels criticized a number of confusions and errors of Eugen Dühring's. But this was an obscure independent scholar and an easy target. It would have been far more useful if Engels had taken on the most influential philosophers of his time, such as Schopenhauer, Comte, and Mill, as well as the far more popular paraphilosophers, such as Nietzsche and Spencer. And in his *Materialism and empiriocriticism* Lenin (1947) criticized the idealist misconceptions of a number of eminent physicists of his time. While some of his criticisms were on target, Lenin's procedure was dogmatic rather than scientific: it amounted to making vehement statements of the form "That is wrong because it contradicts Engels," and accusing idealists of being "lackeys of the bourgeoisie." Regrettably most Soviet philosophy professors followed this example, first set by religionists two millennia earlier.

In sum, dialectical materialism is fuzzy rather than exact, and the bulk of its intelligible theses is inconsistent with modern science. Worse, although they profess to love science, dialectical materialists typically proceed in a dogmatic fashion. Even great scientists, such the physicist John D. Bernal and the biologist J. B. S. Haldane, fell under the spell of the charlatan Trofim Lysenko just because he was Stalin's protégé. Which shows, incidentally, that scientific experience is insufficient immunization against pseudoscience. Only scientific knowledge together with rigorous philosophical (in particular methodological) reflection can vaccinate us against that plague.

7.3 Historical and Australian Materialisms

Let us now survey two materialists schools with a more restricted purview than the ones discussed earlier in this chapter: historical and Australian. Historical materialism, originally proposed by Marx and Engels without any dialectical nonsense, is the social ontology and philosophy of history according to which the prime movers of society are biological need and economic interest rather than ideas.

There are two main versions of historical materialism: apolitical and political. Any students of society who start by asking how people earn their living, rather than what their beliefs, ceremonies, and political order are, qualify as historical materialists. Thus Ibn Khaldûn and the members of the French *Annales* school, such as Fernand Braudel, Marc Bloch, and Lucien Febvre, as well as the American

Immanuel Wallerstein, may be said to have practiced historical materialism, just as much as the British Marxists Eric Hobsbawm, Edward P. Thompson, and Eric Wolf (see, e.g., Bloch 1949; Braudel 1969; Hobsbawm 1997).

A great merit of the *Annales* historical materialists was that, far from being economicists, they practiced what they called *histoire totale*. It may also be called *systemic historiography*, for it goes from geographic environment, demography, and foreign trade, to matters of daily life such as sex and taste in food, whether in the small Pyrennean village of Montaillou or in the vast Mediterranean basin: theirs, like Ibn Khaldûn's, was systemic materialism. Incidentally, Joseph Stalin, a self-styled historical materialist, was actually a historical idealist, for he held that people are moved primarily by political, moral, and spiritual considerations (Wetter 1958, 219). The same holds for the Marxist Antonio Gramsci, who was far more interested in ideas and in politics than in the so-called material basis of existence.

Historical materialism had two great merits. First, it was first advanced at a time when the dominant secular philosophies were neo-Hegelianism, a branch of objective idealism, and neo-Kantianism, a version of subjective idealism. Furthermore, Marx and Engels suggested looking at the economy first, and regarding everything social in a historical perspective, and therefore as transient rather than eternal. This proposal was both original and enormously fertile. In the opinion of Geoffrey Barraclough (1979, 64), a non-Marxist scholar, that is "the only coherent theory of evolution of man in society, and in that sense the only philosophy of history, which exercises a demonstrable influence over the minds of historians today." The Marxist influence on archaeology, particularly evident in the work of Gordon Childe and Soviet archaeology, was equally strong and positive (Trigger 2006) – in stark contrast with the Marxist censorship of advances in physics, chemistry, biology, psychology, and sociology.

Max Weber is widely believed to have been the anti-Marx because he claimed to follow the lead of the idealist philosopher Wilhelm Dilthey, the champion of the *Verstehen* approach. But Weber's substantive work was largely inconsistent with his declared philosophy (Bunge 2007a); and his importance has been greatly exaggerated, for he never personally conducted any empirical research, and he missed all the great social movements of his own time: nationalism, militarism, democratization, trade-unionism, socialism, feminism, secularization, the transformation of industry and everyday life through science-based technology, and above all the emergence of the German Empire, as well as globalization through imperialism. In my view, Weber's current popularity owes more to his enormous erudition and to his conservative opinions than to his original contributions to social science. And, because of the myth that he was the anti-Marx, Weber was a beneficiary of the Cold War, the same political process that fractured and eventually destroyed the *Annales* school.

However, Marxist historical materialism was marred by three serious defects. One was that, particularly in the case of Engels, it was not based on original research. For example Engels' brilliant *Origins of the family* (1884) relied exclusively on the fieldwork of Lewis Henry Morgan (1877), a pioneer of evolutionary anthropology. A second shortcoming of Marxist historical materialism was that it indulged in prophecy: it claimed that socialism was inevitable, and thus it unwittingly invited political passivity. We have learnt at great cost that nothing in evolution, whether social or biological, is inevitable: there are accidents of all kinds as well as unpredictable innovations, such as nuclear weapons, informatics, and the shift to the right of the working class.

A third flaw of Marx and Engels' historical materialism is that it was not consistently materialist, since it assumed that society is split into two tiers: the material (or economic) infrastructure, and the spiritual (or politico-cultural) superstructure (Engels 1954). I submit that a consistent materialist will regard the whole society as a material system, though one consisting of equally material subsystems, among them its culture (Bunge 1981). In traditional societies, kinship relationships and ethnic "identities" are decisive, and consequently people associate into tribes, clans, or ethnic groups. Modern societies, by contrast, are constituted essentially by subsystems of three kinds: the economy, the polity, and the culture. A consistent materialist conception of society regards all three subsystems as concrete things. In particular, cultures (in the sociological sense) are material systems because they are composed by concrete persons that produce or exchange cultural goods, from poems and theorems to blueprints and recipes, through concrete communication channels (Bunge 1981).

In its Marxist version, the infrastructure-superstructure split comes with the additional postulate that the former moves the latter and, moreover, acts as an unmoved mover. But in a scientific ontology there are no separable parts, even though, of course, they are distinguishable. A scientific social ontology will treat the three above-mentioned subsystems as interacting. In particular, the economy will not only act on the polity and the culture, but will also be subject to political and cultural inputs. For example, in contemporary society every political movement will either favor or harm business, and industry will be fed not only by labor and capital, but also by technology, which in turn depends on science. Just think of the impact of electronics on all the branches of the economy. See Fig. 7.1.

Finally, a word on the materialist philosophy of mind proposed by the Australian philosophers Ullian T. Place (1956), Jack Smart (1963b), and David Armstrong (1968). All three argued for what they called "central-state materialism". This is the old "identity theory", according to which all mental processes are brain processes (see O'Connor 1969). In particular, consciousness is "a process in which one part of the brain scans another part of the brain" (Armstrong 1968, 94).

This view came unexpectedly as a breath of fresh air, at a time when most philosophers were still diffident of metaphysics, suspicious of materialism, and

Fig. 7.1 (**a**) The Marxist view of society. (**b**) The systemic view of society

sympathetic to behaviorism and its philosophy, namely logical positivism. The latter regarded the mind-body problem as a pseudoproblem, since it equated materiality as "possibility of sensation." Herbert Feigl (1967) was the only survivor of the Vienna Circle to take the problem seriously and to call himself a monist. But he was no materialist: Feigl adopted the neutral monism (or double-aspect theory) that he had learned from Bertrand Russell. He took this position because he approached the mind-body question as an epistemological problem, not an ontological one. And from his viewpoint there were two equally legitimate "perspectives" or modes of description of mental experiences: the egocentric and the intersubjective ones. But neutral monism does not adopt a definite position, for it does not tell us what the mind is. Therefore it does not help the researcher who wants to explain vision, illusion, or pain, for to explain is to unveiling mechanism. And a mechanism is the process that makes a concrete system what it is (Bunge 2006a). However, let us go back to the Australians.

The limitations of Australian materialism can easily be seen with hindsight. Firstly, it was confined to the philosophy of mind: until recently it had nothing to say about space or time, causation or chance; neither about life, society, or artifact. It did not even propose a precise and correct definition of "matter", beyond telling us that it is what physics investigates. Smart (1963a, 651) even conjectured that space-time points might be the ultimate entities of physics – even though they have no physical properties.

Secondly, and as a consequence, Australian materialism did not place the mind-body problem in its wider metaphysical context. Therefore it used only ordinary-knowledge notions of thing, state, process, or causation. For example, Place held that mental events are "composed" of physical events, but he did not offer a clear concept of process composition.

Thirdly, the Australian materialists equated "material" with "physical", and consequently they regarded brains and computers as physical systems, whereas in fact the former are biological and the latter artificial. Hence they were not eager to learn from neuroscience, evolutionary biology, developmental psychology, or social psychology. (Smart (1963b) even adopted La Mettrie's thesis that "man is a machine.") Accordingly, they rejected emergence and levels organization, and retained the mechanistic faith that biology would eventually be reduced to chemistry, and in turn chemistry to physics: their philosophy of science was radically reductionistic and therefore more a matter of faith than scientific fact. But all that is past: Australian philosophy is nowadays in the grips of possible-worlds metaphysics. Even Armstrong (1997) has given up on matter: he holds that the basic constituents of world are states of affairs – as if these could exist apart from material things. In short, Australian materialism has dematerialized for lack of scientific nourishment.

Still, the Australian materialists must be credited with having revived interest in materialism, and with having discredited psychoneural dualism. In short, they had all the virtues and shortcomings of mechanistic (or vulgar) materialism. We shall argue that scientific materialism overcomes the latter's limitations in scope, depth, and precision.

7.4 Scientific Materialism: Emergent, Systemic, and Science-Based

I take scientific materialism to be the fusion of materialism with scientism, or the thesis that whatever can be studied is best investigated using the scientific method (Bunge 1977a, 1979a, 1981; Bunge and Mahner 2004). Thus scientific materialism is a case of philosophical syncretism: see Fig. 7.2.

Scientific materialism emerged in name the 1850s. It was the work of Ludwig Büchner, Heinrich Czolbe, Jakob Moleschott, and Karl Vogt (see, e.g., Engels 1940, 1954; Gregory 1977). All four admired natural science, but only Moleschott trained as a scientist, and none of them went beyond popularizing mechanistic materialism and empiricism. Their writings, particularly Büchner's *Kraft und Stoff* – dubbed the Bible of materialism – were extremely popular. They effectively discredited religion, spiritualism, and the Naturphilosophie of Goethe, Hegel, and Schelling: they advanced naturalism and the standing of science in popular culture. But they were not original; they were mechanistic at a time when mechanism started to decline even in physics; and they did not propose any research projects. For materialists to have an impact on science they must exhibit scientific credentials and must seek to solve some of the philosophical conundrums that slow down the progress of science in their time.

Contemporary scientific materialism, as I see it, has five distinctive traits. Evidently, the first of them is *materialism*: it holds that all existents are material; it is *dynamicist* though not dialectical: every thing is changeable, but nothing is the union of opposites, and cooperation is just as important as strife; it is *systemic* though not holistic; consequently, it is also *emergentist*, since a peculiarity of systems is that they possess properties that their components lack; and it is *scientistic* and therefore critical and work in progress rather than finished product.

Many naturalists share three or four of these principles. But the second thesis, that there is nothing outside nature *or society*, is distinctive of scientific realism. This tacit extension of reality, to include the made or artificial along with the found or

Aristotelianism	Thomism	Kantianism
∧	∧	∧
Naturalism Rationalism	Aristotelianism Christianity	Empiricism Idealism
Materialism	Dialectical materialism	Positivism
∧	∧	∧
Naturalism Sociologism	Materialism Dialectics	Empiricism Scientism
Logical positivism	Phenomenology	Scientific materialism
∧	∧	∧
Positivism Logicism	Idealism Irrationalism	Materialism Scientism

Fig. 7.2 Examples of philosophical syncretism. Warning: These are just cartoons. Truth is more complex. For example, Aristotle was not fully naturalistic

natural, emphasizes emergence and suggests a multi-layered universe and the concomitant limitation of the reductionist strategy (see, e.g., Bunge 2003a). The earliest systemic materialist was the influential Baron Thiry d'Holbach (1770, 1773); and the first systemic materialist to emphasize emergence along with scientism was Roy Wood Sellars (1970).

Since scientific materialism is an ontology, it is relevant to all the research fields with the sole exception of the formal sciences. Indeed, because neither logic nor mathematics deals with the real world, they are all free from ontology. This assertion looks obvious, yet it has occasionally been denied, so that we should examine the ontological presuppositions of the factual sciences, also called empirical sciences and *Realwissenschaften* – natural, biosocial, and social. Let us start by tackling three hopeless monsters: materialist logic, mathematics, and semantics.

Materialist logic and mathematics: Logic and mathematics would be the most general sciences of being. The first intimation of this doctrine is found in Aristotle. Two millennia later, at the time when logical positivists were going through their physicalist phase, the unjustly forgotten Swiss mathematician Ferdinand Gonseth (1937) stated that logic is "the physics of the arbitrary object". This opinion has a grain of truth: All of the many logical theories are, indeed, the sciences of the arbitrary object, which is why they are topic neutral. But Gonseth's formula overlooks the facts that (a) there are as many logical theories as logicians care to invent, whereas in principle there is only one truest physical theory for each category of physical objects; (b) unlike physics, which involves empirical tests, logic is impregnable to them; and (c) energy, the universal physical property, is not a property of abstract objects, just as consistency, the universal metalogical desideratum, cannot be a feature of any physical objects.

The eminent mathematician Saunders McLane (1998), the cofounder of category theory, proposed a more subtle thesis: that mathematics is the science of patterns. This is true but, as with existence, one should distinguish formal from material patterns (or laws). For one thing, most mathematical patterns – for example, those of category theory – have no known counterparts in the real world. For instance, the statement that a given diagram is commutative has no physical meaning: it only says that two maps can be composed to form a third one. Secondly, all the real patterns involve properties, such as momentum, valence, and heredity, which formal objects cannot possibly possess. For example, it makes no sense to ask what the spin, the energy, the evolutionary ancestors, or the price of a number or of an operator are. Yet McLane's thesis does hold a grain of truth, namely this: The formalization of real patterns (natural or social laws) involves formal patterns. In other words, the scientific application of mathematical concepts requires the addition of semantic hypotheses, that is, formulas of the form "Construct C represents factual item F". However, we have just run into our next subject.

Semantic materialism is the nominalist thesis that there are no constructs: that there are signs (such as words and numerals), not concepts; sentences, not propositions. An indicator of nominalism is calling *sentential* calculus what the rest of us call *propositional* calculus. A second is the construal of predicates as collections of individuals, i.e., classes – that is, the confusion of predicates with their extensions, hence the identification of properties that, though different, are possessed by the same objects (or by the same n-tuples of individuals), as is the case of "price" and "quantity" in the universe of goods of a given type. A third example of nominalism is Quine's (1953) stubborn refusal to countenance meanings. A fourth is the thesis (Hilbert and Bernays 1968, I:21) that number theory deals with numerals (*Ziffern*). But of course symbols, such as numerals (the names of whole numbers), are not mathematical objects: as Frege reminded us, unlike concepts, signs have physical and chemical properties, not conceptual ones. All mathematicians admit tacitly the distinction between concepts and their symbols when they write down designation rules, such as "Let $\mathbb{R}$ designate the real line". Incidentally, nearly all members of this set are nameless, for $\mathbb{R}$ is non-denumerable, whereas the set of names is countable. Hence a strict nominalist should expel the real line from his universe of discourse.

Since semantics is the science of meaning and truth, both of which are rather abstract concepts, it is not obvious how to go about reducing them to psychology. (However, see Chapter 15 for the naturalization of factual truth.) By contrast, it is clear that the applications of semantics are not independent of ontology. For example, the application of any theory of reference calls for a decision concerning the nature of the referents in question. For example, a materialist will assume that the mass concept refers to material bodies, whereas an idealist might claim that it refers to our idea of a body, which in turn is an idea in itself, not an autonomous physical entity.

Consider, for example, any two quantum-mechanical "observables" (dynamical variables) represented by the non-commuting operators A and B (such as position and momentum, or the components of the angular momentum) such that $AB - BA = iC$, where C is a third operator. (This is the mathematical root of Heisenberg's "uncertainty principle".) The Copenhagen interpretation of this formula is that A and B cannot be *observed* simultaneously, or that their exact values cannot be *measured* at the same time. But this interpretation (or semantic hypothesis) is false, since no mention of instruments of observation or measurement is made in either postulating or deducing the value of the difference between AB and BA. The realist interpretation of the formula in question is quite different: the properties represented by A and B are objectively spread out rather than sharp: such fuzziness inheres in nature, as argued earlier (Section 3.4). No wonder that we cannot know the precise values A of and B at the same time: they just do not exist at the same time. What a measurement device can do is to increase the sharpness (or "definition" or determinateness) of A at the price of that of B, or conversely.

The second pivotal concept of semantics, that of truth, is actually a whole family of concepts: formal (or mathematical) truth, factual (or empirical truth), moral truth, and artistic truth. These distinctions are both semantic and epistemological. But they also pose a problem to the materialist. Indeed, the consistent vulgar materialist, such as the nominalist, will tend to reject not only the distinctions in question but also the very concept of truth, because he claims to use symbols rather than concepts, and symbols can be handy or cumbersome, but not true or false. The scientific materialist, by contrast, has no problem with constructs: he holds that they are human creations, not self-existing ideal entities, even though there is nothing to prevent us from feigning that they enjoy independent existence. But the materialist should not object to thinking of propositions as if they existed independently of the corresponding mental processes. Such distinction, as well as the relation between truths of both kinds, are indeed worth making, and we made them in Chapter 6. Our immediate task is to tackle the next-door neighbor of semantics: the theory of knowledge.

A *materialist epistemology* is an extension of its naturalist counterpart (Section 7.2 above). It postulates not only that cognition is a brain process, but also that it cannot be fully accounted for in neuroscientific terms because human brains do not exist in a social vacuum: we all learn from others, and we all fish in the common pool called "human knowledge". The social embedment of cognition is such that the isolated brain does not function normally: it is likely to hallucinate. Suffice it to recall the sensational experiments on sensory deprivation performed in 1951 by Donald Hebb (1980) – and recently used by the CIA on prisoners at Guantánamo. In short, cognition is a brain process strongly influenced by the environment. Consequently epistemology ought be socialized as well as naturalized. Materialism encourages this development, whereas idealism and naturalism discourage it. More on the social below.

7.5 Materialist This and That

Materialist methodology prescribes, similarly to its naturalist counterpart, that nothing immaterial should be invoked in the design, construction, and operation of test devices, such as measuring instruments. For, if it were, then the corresponding empirical operations would be unreliable. For instance, unexpected pointer readings would be accounted for in terms of entities, such as malignant spirits, other than those being observed. In science, empirical anomalies are explained in terms of natural factors, such as drafts, vacuum leaks, chemical or bacteriological contamination, faulty experimental design, or witness bias. The resort to immaterial factors is regarded as unscientific. Hence materialism is irrefutable by empirical means. This does not show that it is dogmatic: it only shows that scientific research presupposes materialism.

And yet, once in a while it is claimed that science reveals the direct action of mind on matter. The best known of such claims is of course parapsychology, which purports to investigate telepathy, telekinesis, and similar "psi phenomena". But of course no solid evidence for such claims has ever been produced (see Kurtz 1985). Moreover, no such evidence should be expected if cognitive neuroscience is taken seriously, for this discipline assumes and confirms the psychoneural identity thesis, according to which mental processes are brain processes, hence just as undetachable from the body as metabolism. Therefore, the tolerance of parapsychological research, recommended by empiricists, can only lead to endowing them with a respectability that they have not earned.

A second apparent violation of methodological materialism is the Copenhagen version of the quantum theory of measurement. Indeed, according to it, a measurement is not completed until the corresponding observation of the measuring instrument has been carried out; and since such observation (instrument reading) is a conscious act, one must conclude that the consciousness of the observer plays a critical role in the objective behavior of the mensurandum (London and Bauer 1939, 41).

The best-known example of such alleged active role of the observer's consciousness in the physical world is Schrödinger's cat. It is claimed that, as long as the lid of the box is not opened, the cat is in a superposition of the live and dead states: the observation act would project this superposition onto either of the said states. But a camera placed inside the box would record the evolution of the cat's state of health without the observer's intervention. Furthermore, quantum mechanics does not contain any variables representing observers, let alone their mental states. This can be ascertained by axiomatizing quantum mechanics. The result of this operation is an objectivist (realist) and materialist, hence ghost-free, interpretation of the theory (Bunge 1967b; Pérez-Bergliaffa et al. 1993).

But of course such arguments won't deter anyone bent on subordinating science to an unscientific philosophy: He could argue that the results of classical measurements too depend on the observer's consciousness, since they involve observations. He could go even further, and claim that every time a measurement is performed, and one of the theoretically possible values turns up, the unrealized possibilities are consigned to parallel universes, in each of which the counterpart of the terrestrial observer performs a similar measurement (Everett 1957). But this is a piece of science fiction, because it violates all of the conservation laws, in particular that of energy; and also because the parallel universes are in principle inaccessible from each other, in particular from ours, and thus their existence is an act of faith rather than a testable scientific hypothesis (Bunge 2006a, 97–98). Dealing with minds leads naturally to psychology.

Materialist psychology denies the existence of freestanding minds, souls, or spirits, whether human or divine. Eliminative materialism, as we saw in Section 7.1 above, goes much further: it also denies the very existence of mental processes in humans and other animals. (Thus the irrepressible John Herman Randall Jr (1958) wished to "kill 'mind' just as James had destroyed 'consciousness'.") Radical behaviorism, or stimulus-response psychology, held this view, which ran its course in the 1960s, when cognitive psychology and cognitive neuroscience were resurrected.

Standard cognitive psychology is dualist, not materialist, for it overlooks the brain. By contrast, cognitive (and affective) neuroscience adopts the psychoneural identity hypothesis, that all mental processes are brain processes. (More precisely, the collection of mental processes is included in that of brain processes. In obvious symbols, $M \subset B$, rather than $M = B$.) Regrettably, many upholders of this materialist hypothesis express it clumsily, in ways that suggest adhesion to mind-body dualism, such as "the brain causes the mind" (John Searle); and, more commonly in the scientific literature, "This brain system subserves (or instantiates) that mental function". Such clumsiness stems from insufficient analysis and lack of a comprehensive ontological framework.

Social (or better sociological) materialism is the view that every society is a material system. This is the view of the historical materialists, both Marxist and non-Marxist, as well as that of the cultural materialists like Harris (1979) and Trigger (2003a). Note the difference between sociological materialists and physicalists: the former hold, if usually in a tacit manner, that society, though material, is a predominantly *extraphysical* system, since its components – persons and artifacts – are not physical. (The neologism "extraphysical" denotes things that are beyond the ken of physics. Organisms and social systems are extraphysical but not supraphysical, because they are not free from physical constraints, such as energy conservation.) The social relations too are extraphysical, even though some of their carriers, such as communication channels and processes, are physical. Caution: whereas the Marxist wing of historical materialism exaggerates the importance of the economy, Harris' cultural materialism exaggerates the biological.

Linguistic materialism retains de Saussure's classical distinction between *parole* or speech, and *langue* or language: whereas the former is a social fact, the latter is a system of signs or symbols. Hence, while pure linguists study languages as abstract systems, field linguists, neurolinguists, psycholinguists and sociolinguists study real speakers and linguistic communities. Although idealists like Chomsky and his followers ignore empirical linguistics, actually the two branches of this science are mutually complementary.

Legal (or juridical) materialism is another special case of social (or sociological) materialism. Like its epistemological partner, legal realism, legal materialism holds that the legal codes are operation manuals for social coexistence, which requires the efficient functioning and coordination of the various components of society. Legal materialism opposes legal naturalism (or natural law), in regarding legal norms as artificial or man-made. But it also opposes legal positivism (the might-makes-right school) in recalling that there is nothing absolute about the *Grundnorm* (basic norm) of a legal order, since whatever is made can also be unmade. Legal idealism (e.g., Dworkin 1986) claims that the law trumps all – an assertion that overlooks the pressures from above (the power elites) and from below (the labor unions and liberal and socialist parties). By contrast, legal materialism holds that the law is observed only when it does not overtly inconvenience the economic, political and cultural powers that be, and when it does not make the daily life of ordinary folks too hard – otherwise people will attempt to break the law. As with legal norms, so with moral imperatives, to which we now turn.

Moral materialism argues that morality is the core of our social-survival kit: that moral norms should facilitate existence and coexistence. It is consequently secular rather than religious: moral materialism enjoins us, the way Spinoza did, to regard fellow humans, rather than immaterial extraterrestrials, as non-negotiable: to make bearable and useful sacrifices, such as volunteering for good causes, instead of wasting resources on rituals and ceremonies designed to placate or bribe imaginary powers.

Moral materialism also enjoins us to set up the material conditions for the rule of moral norms, for it is hard to behave altruistically on an empty stomach, to disobey immoral orders emanating from tyrants, or to resist the temptation to exploit or oppress the weak from the heights of political power. The noble slogan of the French Revolution, *Liberté, egalité, fraternité*, will not be implemented unless the power elites are defanged – which is to say, unless power is distributed equitably throughout society. In sum, let us keep spreading lofty moral seeds while preparing a ground propitious to their germination.

Ethical materialism is the view that moral problems are social problems that arise when scarce resources are handled by persons with unequal power (Section 1.7). Examples: the allocation of benefits and burdens among members of a social system, such as a family, a business firm, or a political organization. Since moral problems are social problems, they can only be solved by social action; and, according to scientism, they are best solved in the light of the best social knowledge rather than by moral intuition or political ukase (Bunge 1989; Railton 2003).

Metaethical materialism holds that moral norms neither drop from Heaven nor are etched in stone. Contrary to the Moses legend, metaethical materialism holds that moral norms are both man-made and subject to change along with other features of social life. These changes are prompted by an examination of the consequences deriving from earlier moral norms – which is a reason for advocating consequentialist and experimental rather than deontological and dogmatic ethics. For example, the *Déclaration des droits de l'homme et du citoyen* (1789) was a product of the French Revolution, which replaced the feudal order with the bourgeois one. In turn, that ethico-political manifesto exerted a strong influence on political life around the world, leading eventually to universal suffrage, which had not been contained in that charter.

But eventually political rights were found insufficient, and socioeconomic rights, such as unemployment and health insurance, were incorporated. Thus ethics, like engineering, has evolved in the light of reflection and experience. But, unlike engineering, ethics has also been an object of political contention. For example, in the most advanced societies sick people are treated as patients rather than customers; and the death penalty has been abolished not only on moral grounds but also because statistics has shown that it is not a crime deterrent.

Furthermore, metaethical materialism, just like metaethical realism, does not object to the so-called naturalistic fallacy. Indeed, although it does not perpetrate the fallacy of trying to deduce *ought* from *is*, metaethical materialism asserts that (a) we jump over the is-ought chasm every time we perform an action guided by a norm or rule; (b) it is desirable to both justify and criticize moral norms in the light of social science; and (c) all imperatives, in particular the moral ones, are translatable into declaratives, as when converting "Do X!" into "You have the duty of doing X."

Finally, *political materialism* argues that politics, far from being pure, is the arm of economic and cultural (in particular religious) interests. Consequently it also holds that, to be affective, political action must be perceived as advancing the interests of some section of society – which is why there are or have been aristocratic, bourgeois, middle-class, working-class, agrarian, Catholic, Evangelic, Islamic, and other political parties. It follows that any political movements that were to pursue lofty but impractical goals, such as universal love or enlightenment, are bound to fail: to stand a chance of victory, a political movement must be seen as advancing some strong material interests. In other words, political realism depends on political materialism. However, this does not entail that politicians must jettison generous ideals: it only entails that that such ideals must have practical roots and must be pursued by realistic means. For example, universal literacy benefits not only the literate but also the businesses that need workers capable of reading instructions and commercial ads – or at least the prices of the items advertised in illustrated commercial catalogues. (More in Bunge 2009.)

7.6 Hylorealism

Realism is the view that the external world exists by itself (ontological component), and that it can be known, if only partially (epistemological component). But realism does not commit itself as to the nature of the world. This is why it must be distinguished from materialism, although both are often confused. This confusion is rather natural because both doctrines share the ontological principle that reality exists independently of the inquirer – for example, that, as Balzac put it in one of his novels, flowers came before botany.

Like materialism, realism comes in several flavors. The main varieties of realism are naïve and scientific. Naïve realism holds that things are really the way we see them: that our brain (or our mind) just "reflects" the things out there. Scientific realism, by contrast, says that cognition is constructive rather than passive "reflection": that it involves the creation of constructs, such as those of zero, electron, nation, and universe, that go beyond appearances and beyond intuition; and that the visible world is best explained in terms of invisible entities, such as atoms, photons, genes, neurons, and governments.

The constructive nature of scientific research is shown not only by philosophical analysis: it is also a result of cognitive neuroscience. In fact, the popular view that the human brain is passive, that it only responds to the momentary demands of the environment, is false: the brain is in constant spontaneous activity, even during sleep. This is why, although its weight is only about 2% of the total body weight, the human brain consumes about 20% of the total body energy budget. Even perception, the most basic mental process, is anything but a fully stimulus-bound process. As William James (1890, II:103) put it in his classic *Principles*, "*whilst part of what we perceive comes through our senses from the object before us, another part* (and it may be the larger part) *always comes* [. . .] *out of our own head*" (original italics). We shall return to this topic in Section 9.3.

In sum, although realism and materialism are logically independent from one another, each is incomplete and shallow without the other. This is due to the nature of both our sensory equipment and the external world: because our senses are limited, they penetrate only skin-deep into reality. This forces us to think beyond percepts: to hypothesize imperceptible entities, as well as to design indicators and tools incorporating such indicators, so that we may check which of the entities we have imagined actually exist in the real world. This why scientists practice hylorealism even while preaching positivism.

7.7 Spirituality in a Material World

In ordinary language, "spiritual" is taken to be the opposite of "material". In biological psychology and the corresponding philosophy of mind there are no such things as immaterial spirits or souls. In these disciplines there are only human brains and their functions. In this perspective, everything spiritual is material, in being a brain process. Even the higher cognitive activities, such as abstract thinking, are not only cerebral: they are also heavily interlaced with sensory and motor processes (Mahon and Caramazza 2009). In sum, pure thought is embodied, not disembodied. Moreover, everything spiritual is rather expensive, because mental activity is a metabolically high-maintenance activity, and one that requires long years of apprenticeship – and a modicum of economic and political freedom.

The cases of religious worship and theological reflection and disputation are more complicated than that of academic pursuits, because they may involve intense emotions and costly commitments and because, far from being disinterested pursuits, people engage in them expecting rewards both terrestrial and other-worldly. Just think of the atonement and the social approval sought by the ordinary religious believer: the stakes, whether celestial or mundane, are, or used to be, far higher than those involved in disinterested activities. Only the Tibetan Buddhists have succeeded in maximizing the expected utility of prayer, by cranking their prayer machines without thinking or feeling anything other some physical weariness.

Closing Remarks

In conclusion, materialism is plural: it is a whole family of doctrines, from the blandest – naturalism – to the hardest – physicalism, or eliminative materialism. Like atheism, naturalism is a negative outlook: it says that there is nothing outside nature, but it does not say what is inside it. By contrast, materialism holds not only that the supernatural is mythical, but it also commits itself as to the nature of the universe. For example, physicalists hold that all things are physical, whence all factual sciences are physical or reducible to physics. Physicalism, in particular mechanism, was plausible between the Scientific Revolution and the Enlightenment, although biology had already started to weaken it. But from ca. 1850 on physicalism became increasingly untenable in the face of the maturation of biology and social science. And yet most contemporary philosophers hold on to the obsolete identities "Material = Physical = Corpuscular," and "Materialism = Physicalism." For example,

Kim (2006) writes of the mental as "supervening" on the physical – not the biological. (Recall Chapter 5.)

Nowadays only a handful of philosophers are physicalists – in particular those who believe that the brain is a physical system rather than a biological one. They do not look outdated just because of an equivocation, namely the popular belief that all materialism is physicalist, just as some of the most innovative biologists in the past, such as Claude Bernard and Jacques Loeb, called themselves "mechanists" when criticizing vitalism, although actually they were physico-chemicalists. Likewise only sociobiologists believe nowadays that sociology, economics, and history are natural sciences. Most social scientists know that the social is irreducible to biology, if only because there are social values, such as merit, in addition to natural values, such as health. However, the students of society often admit unnatural items for a wrong reason: for believing either that the source of social action is the immaterial mind, or that "society" is the plural of "individual", and that social life is driven solely by individual interests.

In the meantime, the nonphysical sciences have been advancing at a fast pace despite the idealist (in particular neo-Kantian) strictures and the naturalist delusion. The ideas of emergence and level of organization have been gaining converts in practice even if often tacitly, that is, without philosophical sophistication. For example, any contemporary biologist knows that the whole multicellular organism has properties that none of its cells possesses, and that cells do things that molecular biology does not even attempt to explain. Scientific materialism accommodates all of these novelties, whereas naturalism and physicalism do not, even though they are far closer to the truth than the radical fringe of idealism, namely hermeneutics, which shares the cabalistic myth that words create things.

Active natural scientists, with the sole exception of the dwindling band of dualist psychologists, are materialists even if they do not call themselves such. Indeed, no living physicist, chemist, or biologist dares claim that the things they study interact with immaterial agencies, such as disembodied souls, ghosts, or deities. Shorter: scientific materialism has the support of natural science. Moreover, it may be argued that it is a necessary condition for the progress of all the sciences.

But materialism is far from being sufficient. For example, until mid-nineteenth century psychology did not advance beyond Galen, because it was purely speculative; Soviet biology went backward from the moment that Lysenko's fantasies became official creed; and the rational-choice speculations in social studies have not solved any problems in this field despite focusing on material interests. For materialism to help science, it must combine with the scientific method. Next we must tackle the hardest and most prickly problem faced by all materialists: the nature of mind.

Part II
Mind

Chapter 8
The Mind-Body Problem

The most popular view about the nature of the mind is that it is immaterial, hence separable from the body. Moreover, it is still widely believed that we are alive ("animated") as long as we have souls (*animae*), and that we die when these leave us. Saul Kripke (1971) revived this prehistoric myth by claiming that mental processes cannot be neural because the brain-mind association is contingent rather than logically necessary – hence people in alternative worlds might not need brains to think. This is all that modal (or possible worlds) metaphysics manages to tell us about mind. Psychologists tell us much more, of course, but many of them are still not interested in the brain, which cognitive neuroscientists regard as the organ of mind: whether or not they are dualists, those psychologists behave as if they were. Indeed, they act as dualists whenever they write about the neural "substrate" or "correlate" of this or that mental function, which is like saying that the legs constitute the substrate or correlate of walking.

However, mind-body dualism has been challenged numerous times since antiquity. For example, in the late Renaissance, when a professor gave his inaugural lecture at the restive University of Padua, his students asked him what he thought about the nature of the soul, its relation to the body, and its prospect of immortality: They wanted to know whether the professor was a spiritualist or a materialist. Cesare Cremonini, the most famous and best-paid philosophy professor of his time, taught Alexander of Aphrodisias's naturalism, in particular his view of the mortality of the soul – at the same time that, as late as the 1620s, he kept teaching Aristotle's astronomy and refused to look through his colleague Galileo's telescope (Renan 1949).

At about the same time, Descartes wrote two basically naturalistic works, his *Traité du monde* and *Traité de l'homme*, which he did not dare to publish for fear of the Inquisition. And a century later, La Mettrie, Helvétius, Holbach and other *philosophes* reasserted the conviction of Hippocrates and Galen, that the brain is the mental organ. Their works were widely read despite being banned and, in the case of Helvétius' *De l'esprit*, burned by the public executioner. David Hartley's materialist view of the mind was tolerated, but Joseph Priestley had to emigrate to the young U.S. for holding that the soul perishes with the brain.

In the nineteenth century, Broca and Wernicke created neuropsychology, and talk of physiological psychology became popular in medical schools. But the

M. Bunge, *Matter and Mind*, Boston Studies in the Philosophy of Science 287,
DOI 10.1007/978-90-481-9225-0_8,

philosophy professors still refused to listen. Likewise the behaviorists, who used to rule the American psychology departments roughly between 1920 and 1960, decreed the irrelevance of the brain to behavior. And the psychoanalysts, who reigned in American psychiatry until recently, fantasized about the soul and its mastery of the body.

The cheapest solution of the mind-body dilemma is to deny the existence of beliefs, desires, fears, hopes, intentions, and the like. This is the so-called eliminativist stand: to claim that all such mentalistic categories belong in folk (or prescientific) psychology (see, e.g., Stich 1999). A number of influential American philosophers, such as Willard Van Orman Quine, Richard Rorty, and Paul Churchland, have taken this stand, which is nothing but a rephrasing of behaviorism, and may be regarded as a primitive reaction to psychoneural dualism.

By sticking to overt behavior, the eliminativists or behaviorists defined their ensign, behavior, as "an animal's reaction to a stimulus," or even as "what an animal does." Thus they put breathing, sweating and other automatic actions in the same category as goal-directed actions such as maze-solving and nest-building. And they overlooked everything that the brain does when we do not behave, in particular ideating and emoting. No wonder that the students of animal behavior still do not agree on what constitutes behavior (Levitis et al. 2009). Fortunately, psychologists kept cultivating psychology as the scientific study of the mind conceived of as a (sub)set of brain processes (Hebb 1980, 2). Still, this redefinition of the task of psychology raises two questions: Which brains, and which processes? However, we must postpone these questions to Section 9.3.

The result is that this, the hardest of all the Big Questions, is being actively investigated in the lab. This, in spite of the philosophers who, like McGinn (2004), declare the mind to lie beyond the reach of science. Why bother studying the quickly growing technical literature on cognitive neuroscience, the scientific approach to the mind-body problem? True, some philosophers of mind, in increasing numbers, claim to adopt the neuroscientific approach to the problem. But most of them use a single example: "Pain = Firing of *C*-fibers." This is wrong because these nerves only conduct stimuli from receptors of a certain type: feeling pain is a conscious process occurring in the deep recesses of the brain. And we have become so used to equating academic freedom with the license to repeat obsolete or even nonsensical stuff, that, while we still heap scorn on those colleagues of Galileo's who refused to look through his telescope, we take seriously the philosophers of mind who refuse to look at the contemporary science of mind, namely cognitive neuroscience.

Other philosophers of mind are not so much narrow-minded as confused for lack of a broad and clear ontology. Thus John Searle (2007, 40 ff.), who has published extensively on this subject, tells us that he opposes both materialism and psychoneural dualism. Yet he also claims that mental states are *caused* by brain processes at the neuron level. States of one kind caused by processes of another? This talk of upward causation sounds dualistic to me. Moreover, it is reminiscent of the nineteenth-century vulgar materialist Karl Vogt, who famously claimed that "the brain secretes thought just as the liver secretes bile." There is an elementary ontological confusion here: By definition, processes are sequences of states, and only events are supposed to cause events (more in Chapter 14). For instance, not LSD by itself, but taking LSD, causes hallucinations.

A psychoneural monist of the materialist kind would state that all mental states are brain states, and that the same holds for mental changes of state, that is, events and processes. For instance, she would say that feeling disgust is the same as a certain process in a brain network that includes the insula. She would also state that all brain processes are caused by other processes, either in the brain or acting on it. And she might also know that this has been a central thesis of materialism, Indian as well as Western, for at least 2,500 years, and that only vulgar or eliminative materialists (that is, behaviorists) reject it. All of which suggests that the mind-body problem can be treated profitably only within a broad ontological framework, and with some information about its history. However, in the following we will be able to give only an inkling of all this. (More in Bunge 1980a; Bunge and Ardila 1987; Hebb 1980; Purves et al. 2008.)

8.1 Introductory Dialogue

PHILOSOPHER: You just blushed!

SCIENTIST: I suppose I did.

PHILOSOPHER: Why did you blush?

SCIENTIST: Because of your comment on the philosophical naiveté of scientists.

PHILOSOPHER: Sorry. Anyway, you just showed unwittingly a clear example of the power of mind over matter. Indeed, your embarrassment, a mental process, caused your blushing, a physiological process.

SCIENTIST: Not so fast. What makes you deny that embarrassment, or any other mental process for that matter, is a physiological process?

PHILOSOPHER: Because we describe it in psychological terms.

SCIENTIST: Ah, but this is only because of tradition and for the sake of brevity. I could explain.

PHILOSOPHER: Go ahead, we have time.

SCIENTIST: Well, to begin with, the embarrassment in question happened in my brain, not elsewhere in my body, let alone out of it.

PHILOSOPHER: How do you know? After all, it is mere conjecture that mental events are brain events, right?

SCIENTIST: Conjecture, yes. Mere conjecture, no, because it has been well confirmed by electrophysiological and lesion studies, that the emotions are processes in brain networks that include both cortical and subcortical regions.

PHILOSOPHER: Suppose I grant you this. What do we gain by conceiving of the mental as neural?

SCIENTIST: We gain understanding. You understand a process when you expose its mechanism. And all mechanisms happen to be processes in material things, such as brains.

PHILOSOPHER: And what would be the mechanism of the process we are discussing?

SCIENTIST: Schematically, it would be the following causal chain: Hearing your remark (auditory area) – Understanding it (Wernicke area) – Feeling anger or shame (a circuit containing the hypothalamus, orbitofrontal cortex, or other brain regions) – Activation of the brain's motor circuitry – Dilatation of face capillaries. I grant, though, that the details have yet to be worked out. But the big picture is there, and that picture is a whole research project, not just a story.

PHILOSOPHER: Yet you just described a case of downward causation, from a high level to a lower level one.

SCIENTIST: If you wish. But I prefer to call it a top-down process occurring in the head. And the point is that this process occurs in the brain, and is describable in purely neuroscientific terms.

PHILOSOPHER: I dispute this: the very word "embarrassment" is a psychological term that does not occur in the neurophysiological vocabulary. And surely the same applies to the rest of the psychological vocabulary.

SCIENTIST: True, but the point is that the mental processes can be explained, at least in principle, in neurophysiological terms. However, the neurophysiological processes called "mental" are qualitatively different from all the others: after all, they are recent arrivals in evolution. Moreover, they do not occur in a social vacuum. Indeed, embarrassment, like shame and compassion, is a social emotion.

PHILOSOPHER: So, you eliminate neither the mental nor psychology. You retain a distinction between levels, and allow for psychology, and perhaps even for sociology.

SCIENTIST: Absolutely. We identify mental with neural of a very particular kind, and we fuse psychology with neuroscience – and sociology as well when it comes to social learning and action. In short, we perform ontological reduction at the same time that we effect epistemic coalescence.

PHILOSOPHER: I'll chew on that.

SCIENTIST: You do that. But remember that the mind-body problem is scientific as well as philosophical and theological. Recall also that no idea can be discussed fruitfully in a vacuum. In particular, a materialist idea of the mind should be discussed in the context of a comprehensive materialist ontology. Proceeding otherwise would be philosophically naive.

PHILOSOPHER: Touché.

8.2 Science, Philosophy, and Religion Intersect

Some of the most interesting conceptual Big Questions lie at the intersection of science, philosophy, and religion. The problems of the nature and origin of the universe, life, mind, society, science, and religion are in this category; so are the problems of what is good and what right, and what must be done to identify and solve the moral and social problems that raise human action and inaction. Any of these problems is a counterexample to the positivist thesis that those three domains – science, philosophy, and religion – are mutually disjoint.

But although the said trichotomy fails for problems, it does hold for the approach to those problems. Indeed, whereas theologians resort to either revelation or the exegesis of so-called revealed scripture, scientists investigate problems both empirically and theoretically: They seek new knowledge to solve problems, whether old or new. Hence the soothing thesis, that science and religion can coexist peacefully because they would be concerned with "non-overlapping magisteria," as the great Steven Jay Gould (1997b) claimed, is false.

Religion and science are ontologically at odds, because the former asserts, whereas the latter denies, the existence of supernatural entities, as well as of disembodied souls. They are also methodologically incompatible with one another because, while scientists search for truths, religionists claim to have already found them. Which is why there is no free search for new truths under a theocracy – nor under a totalitarian regime.

Something similar holds for the relation between science and philosophy. Indeed, a number of problems are at once scientific and philosophical. Examples: What are matter, causation, chance, life, and mind? Although at first blush physics and chemistry have solved the first problem, actually it still poses a number of philosophical questions, such as those discussed by Michael Frayn in his famous play *Copenhagen*. For instance, do microphysical events happen only when some experimenter elicits them? Likewise, biologists have yet to answer the question "What is life?," and psychologists the question "What is mind?" to everyone's satisfaction.

How should one go about tackling such hybrid problems? On these questions, the philosophers split into two camps: the pro-scientific and unscientific. Whereas the former seek guidance from science, the unscientific ones engage in unbridled speculation. Unsurprisingly, the unscientific philosophers constitute the vast majority. After all, ignoring science is much easier than learning some of it.

I am not counting just the well-known enemies of science, such as Berkeley, Vico, Rousseau, Fichte, Schelling, Hegel, Nietzsche, Bergson, Croce, Gentile, Husserl, and Heidegger. Nor am I counting the philosophers indifferent to science, such as Moore, Wittgenstein, Strawson, Kripke, David Lewis, and Habermas. I am also including among the unscientific philosophers some great and influential thinkers who are usually regarded as proscientific, such as Locke, Hume, and Kant. The reason is that all three believed that all we can know are phenomena or appearances, characterized by secondary properties or qualia, such as color and taste.

Thus, Locke (1690 Bk IV, sec. iii, 28), warned that "we can have no distinct knowledge" of the motions of bodies beyond our experience, because we do not understand how they cause sensations in us, other than by the intercession of "an infinitely wise Agent." So, "we are utterly incapable of universal and certain knowledge" of the bodies around us. Little did Locke suspect that his skepticism concerning the power of science was behind his own times, since the Scientific

Revolution was already well advanced. In particular, he was unaware that Newton's magnum opus (1687) – which contained precisely some of the laws of motion that Locke had decreed unknowable – appeared the same year that he finished his *Essay*. Luckily, not even Locke's immense intellectual authority could prevent the triumphal march of Newtonianism. However, his skepticism regarding the power of science did eclipse the important work of his near-contemporary Thomas Willis, the early modern neuroanatomist who regarded the brain as the organ of emotion, perception, and memory (Zimmer 2004). Which is one more example of the harm that superficial philosophies, such as phenomenalism, can cause. For example, in astronomy the phenomenalists defended the geocentric view of the planetary system; in physics they attacked the atomic theory; and in psychology they favored behaviorism.

Worse, even most of the philosophers who profess love of science do not use it to tackle any of the big problems in question. For instance, some positivists have claimed that the mind-body problem is a pseudo-problem, and others have repeated the double-aspect doctrine. Lenin and his followers have held that ideas are the opposite of matter. Ludwig Wittgenstein claimed that only a human being (rather than her brain) can be said to perceive or be conscious – because that is the way psychological predicates are used in ordinary language. Hilary Putnam proposed that the mind is just a set of computer programs – a view adopted by Daniel Dennett and other philosophers of mind. Karl Popper defended psychoneural dualism and revived Plato's analogy: mind is to brain what pilot is to boat. Still others, such as John Searle, maintain that the brain causes the mind – which is like stating that legs cause walking, rather than walking being the specific function of legs.

The variety of bizarre philosophical opinions about the nature of mind stems from the tacit, alas mistaken, conventions that the philosophical imagination should not be constrained by any scientific findings, and that philosophical problems can be tackled one by one rather than in bundles. I happen to hold the views that philosophy should be intimately bound with science, and that none of the Big Questions involving facts can be successfully handled except in the light of precise empirically testable theories about the nature of reality and the knowledge of it (Bunge 2006a).

In sum, the problem of the nature of mind has always been of great interest to scientists, physicians, philosophers, theologians, shamans, and dabblers in the esoteric. As with other important problems, whereas a few philosophers have made useful suggestions, most have hindered the serious investigation of it – for instance, by claiming that it is insoluble, or that it must be approached via language, or via computer technology. Let us glance at three influential views of the mind, every one of which has been enthusiastically endorsed by some philosophical school.

8.3 Classical Psychoneural Dualism

There are currently three main conceptions of the mind: psychoneural dualism, computerism, and the psychoneural identity thesis. Let us peek at them. (More on dualism in Armstrong 1968; Bunge 1980a; Bunge and Ardila 1987; Ingenieros

1946; Kim 2006; Lovejoy 1955.) Psychoneural dualism is of course the ancient opinion that matter and mind are distinct entities or substances; that the one can exist without the other; and that they may interact, but that neither can help explain the other. Dualism has been defended by famous philosophers, such as Plato, Descartes, and Popper, as well by a few eminent neuroscientists, among them Jackson, Sherrington, Penfield, Sperry, and Eccles; and it is a component of all religions and primitive cosmologies, as well as of psychoanalysis and New Age. Its great advantages are that it looks obvious; it seems to explain effortlessly every bit of human behavior; and it inheres in the dogma of the survival of the soul after death.

Moreover, at first glance psychoneural dualism is supported by logic. Indeed, consider the following well-known argument, in the light of Leibniz's law, that two objects are identical if, and only if, they have the same properties.

1. I have direct knowledge of my mental states.
2. I do not have direct knowledge of my brain states.

Therefore, by Leibniz's law, my mental states are not identical with my brain states.

This argument is fallacious, because having or lacking direct knowledge is not a property of the items in question, namely mental states and brain states. Indeed, "the 'property' ascribed in premise (1), and withheld in premise (2), consists only in the subject item's being recognized, perceived, or known as something-or-other. But such apprehension is not a property of the item itself, fit for divining identities, since one and the same subject may be successfully recognized under one name or description, and yet fail to be recognized under another (accurate, coreferential) description. Bluntly, Leibniz' law is not valid for these bogus 'properties' " (Churchland 1984, 42).

Furthermore, psychoneural dualism is open to a number of fatal objections. Let us recall a few of them.

1. *Dualism is conceptually fuzzy*. Indeed, the very expression "mental state" is at best shorthand, because every state is a state of some concrete (material) thing at a given time. (For instance, the state of a hospital patient at a given time is roughly indicated by the values of its vital signs at that instant.) And the expression "mind-body interaction" is an oxymoron because, by hypothesis, the immaterial mind is impregnable to physical stimuli, just as matter cannot be directly affected by thoughts or emotions. The very concept of an action is well defined only with reference to material things. Think, e.g., of the action of the terrestrial gravitational field upon an aircraft, of the action of nitric acid upon a copper penny, or of that of coffee, wine, anesthetics, or cocaine on the brain.
2. *Dualism is experimentally irrefutable* since one cannot manipulate a nonmaterial thing, as the soul or mind is assumed to be, with material implements, such as lancets and pills. In other words, only material objects are changeable and can act upon material tools such as measuring instruments. The brain is such an object but the nonmaterial mind is not, as shown by the utter failure of psychics, mediums, and parapsychologists.

3. *Dualism considers only the adult mind.* Hence it is inconsistent with developmental psychology, which shows how cognitive, emotional and social abilities develop (grow and decay) along with the brain and the individual's social context.
4. *Dualism is inconsistent with cognitive ethology*, in particular primatology, which shows that we share some mental abilities with our evolutionary next of kin. It is also inconsistent with comparative psychology and cognitive archaeology, for these sciences suggest that our mental abilities have evolved as a result of biological and social changes. True, some of this evidence derives from sheer anthropomorphism. But there is also solid evidence from anatomical and physiological studies. After all, there would be no human physiology, from Galen's dissections of monkeys and Claude Bernard's classical animal experiments to our days, without studying animals, even phylogenetically as distant from us as fruit flies and worms.
5. *Dualism violates physics*, in particular the law of conservation of energy. For instance, energy would be created if a decision to take a walk were an event in the nonmaterial soul. Moreover, dualism is inconsistent with the naturalistic ontology that underpins all of the factual sciences. This makes brainless psychology an anomalous solitary discipline. It also deprives the science of mind of the panoply of surgical and pharmaceutical tools that allows it to treat successfully the mental disorders that do not respond to psychotherapy.
6. *Dualism confuses* even investigators who are contributing to its demise. In fact, in the cognitive, affective and social neuroscience literature one often reads sentences of the forms "*N* is the neural *substratum* (or *correlate*) of mental function *M*," and "Organ *O subserves* (or *mediates,* or *instantiates*) mental function *M*" – as if functions were accidentally attached to organs, or were even prior to them, and organs were means in the service of functions. Why not say simply "Organ *O* performs function [or does] *F*"? After all, one does not say that legs *subserve* walking, guts *mediate* digestion, and so on. Nor does one ask what came first, the nose or smelling.

   Why not say simply that the brain feels, emotes, cognizes, intends, plans, wills, and so on? Talk of substratum, correlate, subservience and mediation is just a relic of dualism, and it fosters the idea (functionalism) that what matters is function, which can be studied independently of stuff. But there is neither walking without legs nor breathing without lungs. In general, there is neither function without organ nor organ without functions. The smile of the vanishing Cheshire cat belongs in fiction, not in science. True, most cognitive neuroscientists use the expressions "neural correlate" and "neural substrate" without assuming that there is something "over and above" brain function. But this is no excuse: they should learn to say what they mean.
7. *Dualism isolates psychology* from most other disciplines, insofar as none of them admits the stuff/function dichotomy. Imagine generalizing the correlate talk: "the planetary correlates of Kepler's orbits," "the light correlate of refraction," "the molecular correlates of chemical reactions," "the fluid correlates of streams and eddies," "the organismal correlates of evolution," and so on. Much the same

applies to "substrate" talk, as in the common phrase "the neural substrate of decision-making." Something similar holds for "subservience" talk: it is a relic of both dualism and finalism and, more particularly, of Plato's idealist principle (*Laws* X: 896) that "the soul is the first origin and moving power of all that is, or has become, or will be."

8. *Dualism is barren at best, and counterproductive at worst.* In fact, it has spawn superstitions and pseudosciences galore, from beliefs in the supernatural and the afterlife to parapsychology, psychoanalysis, and memetics. And dualism has slowed down the progress of all the disciplines dealing with the mind, in particular biological psychology, neurology, psychiatry, psycho-neuro-pharmacology, and neuroengineering. But it has not deterred some of these experts from studying or even recommending the use of psychotropic agents for military and riot-control purposes.

In short, psychoneural dualism is scientifically and philosophically untenable. Worse, it continues to be a major obstacle to the scientific investigation of the mind, as well as to the medical treatment of mental disorders. Yet, it is still very popular, particularly in its computer version, which involves hardware/software dualism. However, we shall postpone a discussion of this fashionable form of dualism to Chapter 12.

8.4 Mind Over Matter?

The cognitive control of behavior, as in shoveling, driving, drawing, writing, and self-restraint, has traditionally been regarded as proof of the power of mind over matter, or "downward causation" (Campbell 1974b). The psychosomatic and placebo effects are often regarded in the same way. In particular, the health benefits of optimism and the sickening effects of low self-esteem and social stress have been said to exemplify the power of mind over matter.

But of course in a scientific perspective there can be no Mind → Matter causation, if only because it would violate momentum and energy conservation. There are only (a) same-level causal relations between neural events, such as the joy or sadness caused by learning something; and (b) bottom-up and top-down causal relations among neural and non-neural bodily events, such as the dullness caused by a blow in the head, and the quickening of the heartbeat caused by a sight of the beloved.

In particular, the placebo effects involved in abating pain and favoring recovery are nowadays explained in purely scientific terms, albeit only sketchily, partly thanks to the discovery of nerves connecting the cerebral cortex to the immune system (see, e.g., Cacioppo et al. 2006, and the journal *Brain, Behavior, and Immunity*). For example, the healing or distressing power of a symbol (Symbol → Meaning → Behavior) can sometimes be explained as a case of classical conditioning (Benedetti 2009). On the other hand, placebo analgesia may involve endogenous opioids generated by the brain. The placebo effects involving reward expectation, as in the pain relief elicited by the mere sight of a stethoscope on a white coat, is a process in a complex brain circuit. Finally the social pains and pleasures, such as envy and *Schadenfreude*, have been shown to involve the same neural circuit as those caused by painful physical stimuli (Takahashi et al. 2009). In sum, placebo effects are real: they are links in physiological causal chains. But their benefits are limited. For example, placebo treatments can reduce the discomforts and pains caused by cancer, but they cannot stop tumor growth.

In short, what used to be regarded as cases of Mind → Body action turned out to be Brain → Brain, or Brain → Rest of the Body processes. This shift, from Freudian psychosomatic fantasy to psycho-neuro-endocrino-immunology, was a late result of the pioneering experiments of Walter Cannon (of homeostasis fame) and Hans Selye (the father of stress research). They and their followers showed conclusively that the nervous, endocrine, and immune systems constitute a supersystem – one more triumph of systemic materialism. The importance of this synthesis for pharmacology and medicine, in particular psychiatry, should be obvious.

Ironically, psycho-neuro-endocrino-immunology explains why shamans and logotherapists are occasionally effective, as well as why social exclusion can be literally sickening. Thus, the morbidity and mortality rates of the poor are higher than those of the rest of us, not only because they cannot meet all their needs and wants, but also because they experience negative social evaluation, which lowers immunity (e.g., Kemeny 2009). This result confirms the earlier finding, of social psychologists, that relative inequality can hurt more than absolute poverty. However, let us go back to the mind-body problem.

In recent years, neuroengineers invented devices which, when implanted in the brain of a paraplegic, or in that of an immobilized monkey, allow the animal to control a computer cursor or a robotic arm. Dualists interpret this technological feat as evidence for the power of mind over matter. But of course to the bioengineer what happens is that one part of the body, namely a region in the prefrontal cortex, activates the neural prosthesis and makes it execute the desired movements. Likewise the analgesic effect of placebos and the (unreliable) healing power of prayer can be understood as effects of processes in the brain upon other parts of the body. It is matter over matter all the way down. It could not be otherwise since causal chains are sequences of events, and every event is a change in a material thing. See Fig. 8.1.

The explanation of the resolve and fortitude of the Vietnamese Buddhist monks who in the 1970s set themselves on fire to protest against the war is similar. But of course this case is far more complex than the control required by daily life activities, for it involves free will and a degree of abnegation and learned emotional regulation that surpasses anything that ordinary people can achieve. Incidentally,

Prefrontal cortex ⟶ Motor cortex* ⟶ Finger**

Fig. 8.1 The causal chain between the decision to move a finger and the actual finger motion passes through the activation of the motor cortex. The *star* symbolizes an event, and the *arrow* a nerve pulse. In the case of patients with neural prostheses, the last link is a computer cursor or a mechanical hand

emotion regulation, whether deliberate or automatic, is a hot topic in psychology (see Gross 2007).

The materialist monist does not regard these as cases of the power of mind over matter, but of the control that parts of the brain, in particular the prefrontal cortex, can exert over the rest of the body. Hence, the cases in question are examples of psychoneural identity, not counterexamples to it. So much so, that the cognitive control of behavior has become a standard chapter of cognitive neuroscience, as well as the scientific basis of neuroengineering (see DiLorenzo et al. 2008). Just try to base neuroengineering on an alternative philosophy of mind, such as Descartes', Hume's, Kant's, Hegel's, Husserl's, Wittgenstein's, or Popper's versions of mind-body dualism.

Consider the neuroengineer who designs or implants a motor neural prosthesis that may enable a person paralyzed from the head down to move a cursor on a screen, or a mechanical arm, just by imagining this action. That technologist or technician pushes an electrode into the motor cortex of a brain, not into an immaterial mind. This reorientation has been accompanied by a slight change of terminology: Neuroscientists and bioengineers speak of *top-down*, rather than *downward*, causation. The causal chain in question is not interrupted by an immaterial chasm: it is fully material. Materialist monism underlies and motivates neuroengineering and the neural prosthesis industry, whereas psychoneural dualism discourages them. It is nice to know that the right philosophy can bring health and money in addition to promoting or blocking social progress. See Fig. 8.2.

Finally, let us remember the converse of top-down causation, namely bottom-up, or molecule-to-brain causation. This happens every time we drink a cup of coffee or a glass of wine. Indeed, caffeine and alcohol are consumed precisely because of their effects on cognition and mood respectively. Although the functions of these substances are notorious, their mode of action is still under investigation (see, e.g., Iversen et al. 2008). Which shows that scientists, unlike most philosophers of mind, are not functionalists: they are not satisfied until they discover the mechanism underneath the function of interest.

In short, the mind-mind and mind-body causal arrows are viewed as follows by dualists and materialist monists:

Event type	*Dualism*	*Materialist monism*
Same-level	$\text{mental}_1 \rightarrow \text{mental}_2$	$\text{neural}_1 \rightarrow \text{neural}_2$
Downward	mental $\rightarrow$ bodily	neural $\rightarrow$ bodily
Upward	bodily $\rightarrow$ mental	bodily $\rightarrow$ neural;

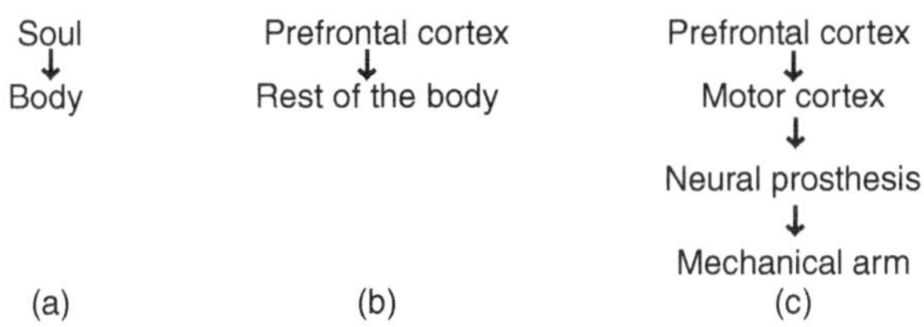

Fig. 8.2 Top-down causation. (**a**) Dualist dogma. (**b**) Materialist hypothesis. (**c**) Neuroprosthesis

8.5 Dualism is Hazardous

The consistent dualist will reject any treatment of mental disorders involving such material means as neurosurgery and psychotropic (or neuroactive) pills. He will also reject the neural prostheses that neural engineers are designing to help people who have suffered severe brain injuries. But, because of the efficacy of such means, the dualist neurologist, clinical psychologist or psychiatrist may be willing to sacrifice his philosophical integrity, and prescribe them to tamper with the allegedly immaterial minds. After all, he has to support his family.

Regrettably, the diagnosis of mental disorders is still at a rather primitive stage. This is so to the point that every new edition of the standard work in the field, the *Diagnostic and Statistical Manual of Mental Disorders*, offers new criteria. For example, it seems hard to distinguish between manic-depression and dementia praecox, as well as to identify schizophrenia. The reason is sometimes said to be that some of the key symptoms of different disorders are the same. But this situation is common in medicine: the symptom-cause relation is many-to-one rather than one-to one. Such ambiguity is characteristic of all inverse (or ill-posed) problems, whether in medicine, engineering, or physics (see Bunge 2006a).

How might this situation be improved? I submit that the remedy is to replace the usual *symptomatic* (or *clinical*) *typologies* with *somatic* (or *neuroscientific*) *classifications*. (The classes in a classification proper are pairwise disjoint, which is not the case of types.) To take a familiar example: a vulgar digestive discomfort may be caused by a malfunction of the stomach, the intestine, the liver, or the gall bladder. Only a thorough examination of these organs might spot the cause and thus suggest the proper therapy.

I submit that the case of mental disorders is parallel: their diagnosis will remain uncertain as long as psycho-neural dualism prevails, for this philosophy of mind counsels focusing on mental symptoms or phenomena, rather than starting with them but tracing them back to objective brain facts. For example, kleptomania and drug abuse are very different behaviors, yet both disorders are successfully treated with naltrexone, a drug that blocks the same opioid receptors, which suggests that they involve the same brain circuits. Symptoms do not solve the diagnostic problem: they only pose it.

Accordingly, I propose supplementing the extant psychological typologies with neuroscientific classifications, where the diagnostic statements are not like "panicky," "bulimic," "obsessive-compulsive," "pathologically indecisive," or "excessively uninhibited" but "possible amygdala trouble," "possible hypothalamus problem," "possible disturbance of the caudate nucleus," "possible prefrontal cortex damage," and "possible orbitofrontal damage" respectively. Note that in all these cases it is assumed that mental disorders are organ disorders rather than either whole-brain or molecular abnormalities.

In any event, mental disorders are not whole-brain diseases. For example, Parkinson's disease is a disorder of the substantia nigra; depression involves serotonin imbalance; Alzheimer's disease starts as β-amyloid plaques in neurons; multiple sclerosis is due to severe myelin loss in peripheral nerves – and so on.

In short, all psychiatric disorders are organic: none are purely mental or functional. Hence they should be classed in neuroscientific terms. This implies that genetic psychiatry, in particular the search for single genes "responsible" for mental disorders, does not look promising, at least for now. It takes more than a crooked or missing gene to bend the mind, or any other function of a highly complex cell system. The searchlight should be trained on the middle of the levels scale. By the way, the functionalist view of the mind, favored by most contemporary philosophers of mind, should be dropped as being both scientifically shallow and medically hazardous.

This does not imply that we should adopt the so-called Swiss-Army penknife model, favored by phrenologists and evolutionary psychologists. This is the opinion that the brain is made up of mutually independent modules, each of which carries out a single mental function. Neuroscientists know that this is not so: that every brain component interacts with several other subsystems in the body. They also know that in some cases two very different mental disorders are due to the disturbance of a single brain organ. For example, both the Huntington disease and the obsessive-compulsive disorder are due to disturbances of the caudate nucleus, one of the neural assemblies comprising the basic ganglia. Still, it is true that in every neural circuit there is a dominant component. For example, panic may be induced by either a visual, auditory, or haptic stimulus, but the amygdala will be a part of all varieties of panic. Hence the rule: To treat recurrent panic episodes, check the amygdala. Likewise, to treat recurrent stomach trouble of any kind, check the gut even if the ultimate culprit proves to be the liver, the gall, or even the cerebral cortex. (See more on localization-cum-coordination in Chapter 9.)

The car mechanic proceeds in like manner: He classes a car's troubles into problems with ignition, cylinders, transmission, electric circuitry, and so on, and tries to guess parts damage from the driver's complaints. This is why car mechanics are, on the whole, more efficient than clinical psychologists and psychotherapists: they do not detach symptom from organ. They move from symptom to dysfunction to part trouble. This is why they prefer deep essentialist classifications to superficial symptomatic typologies.

(Essentialism is the ontological thesis that things have properties of two kinds: essential and accidental. The methodological counterpart of essentialism is this: Only the essential properties should be used to define classes. For example, the periodic table of the elements rests on the essential properties of atoms, which are the numbers of its protons and neutrons. An atom does not change species if it only loses or gains one or two electrons. Regrettably, the eminent evolutionary biologist Ernst Mayr wrote some influential pages against essentialism; and, believing biology to be an autonomous science, he did not take the trouble of checking how species are defined in other sciences, such as chemistry.)

To conclude, philosophies need not be socially useful, but they must not be socially harmful. I submit that the dualist philosophies of mind are hazardous to mental health because they divert the researcher's and the therapist's attention from the brain to an immaterial and therefore inaccessible item. This is not to deny the beneficial effects of some talking therapies: After all, words are physical stimuli, and some of them may enhance self-healing brain processes, while others may induce the patient to correct bad habits. In other words, some words are as good as deeds because they affect the brain, not the mythical soul.

8.6 Explaining Subjectivity Objectively

Subjectivist philosophers renounce the real world and attempt to describe it in terms of subjective experience. Thus, the founder of phenomenology discovered the delights the *Lebenswelt* or life-world (Husserl 1970) only after spending his whole academic life immersed in himself ("egology"), and claiming that the essence of things can be grasped only by "bracketing out" the external world (i.e., pretending that it does not exist). Thereupon he stopped his search for transcendence, and proposed that everything, even mathematics, should be traced back to every-day life terms – a sort of cheap sociologism.

Of course. neither Husserl nor his increasingly numerous followers have ever carried out their epistemically reactionary program. For instance, they did not even attempt to explain in plain language how rain forms, fire starts, metabolism proceeds, or economic crises originate. In any case, the later Husserl did not invent anything: He unwittingly rewrote the phenomenalist project that Hume had sketched two centuries earlier, and Kant, as well as the positivists, adopted. This project consists in *explaining objectivity in subjective (or first-person) terms*. For instance, Ernst Mach attempted to account for the physical world in terms of sensations.

This is exactly the converse of what scientific psychology, in particular cognitive, affective and social neuroscience, has been doing: *Explaining subjectivity in objective (or third-person) terms*. Take, for instance, the phantom limb experience undergone by uncounted war veterans. No one denies that this experience is subjective, since only their sufferers can have it. But it is real: the amputee treats the phantom part as real, and often really suffers pain, which she locates in a place that she no longer owns. Being real processes, these must be explained in objective (or nonexperiential) terms.

In fact, phantom limb experiences were first explained in terms of the "body schema" etched on the somatosensory cortex, an engram that persists after the amputation. Ronald Melzack (1989) found this theory too vague, as well as incapable of accounting for all the data – such as that persons born without a limb may feel pain, whereas some people with whole limbs are unable to feel pain in them under harmful stimuli. Subsequently Melzack proposed his own theory: He conjectured that we are born with a "neuromatrix" of the body self, which is subsequently modified by experience; and that this particular neural network, far from accepting all inputs from the body, filters them and gives us the feeling of having a whole body of our own.

This hypothesis accounts for the extreme cases of congenital insensitivity to pain, and the spontaneous occurrence of pain (neuropathy). Further research may modify this conjecture, but any alternative is likely to keep its core idea, that our brains construct our body images and, in general, all our experiences and, above all, the unity of experience.

Obviously, the project of explaining subjectivity objectively is in tune with both the Scientific Revolution and the Enlightenment, for it is materialist and realist rather than idealist and subjectivist. The execution of this project requires shifting attention from brainless minds to "minding" brains, because to explain is to unveil mechanisms, and these are the processes that make material systems tick (Bunge 2006a). Indeed, as a recent textbook puts it, "The overarching goal of the field of cognitive neuroscience is to explain mental processes and behavior in terms of the structure and function of relevant regions of the brain and the rest of the nervous system" (Purves et al. 2008, 57).

Attaining this goal has been a dream of the materialist philosophers since antiquity. Drop this philosophy, and that great project cannot even be formulated; remove the brain, and no mind remains; ignore the neural "basis," "substrate," or "correlate" of mental processes, and no hope of understanding them can be reasonably maintained. Hence any brainless philosophy of mind must erase two and a half millennia of mind science.

Closing Remarks

Of all the big and venerable scientific questions, the mind-body problem is the hardest and the one surrounded by the thickest theological barrier and philosophical fog. No wonder, because the soul has been the traditional property of priests, shamans, charlatans, and philosophers who refuse to learn modern psychology, and yet proclaim confidently that there is no such problem, either because Plato solved it, or because there is no mind – or because the mind is mysterious.

For example, Colin McGinn (1993, 36) shares the opinion of John Tyndall, a nineteenth century physicist quoted by William James (1890, I:147) that "[t]he passage of the physics of the brain to the corresponding facts of consciousness is unthinkable." But "thinkable" (or "conceivable") is a psychological concept, not an epistemological one like that of the plausibility (or verisimilitude) of a hypothesis in the light of a given body of knowledge.

Besides, philosophers are not expected to resort to arguments from authority, the way McGinn does, particularly when the alleged authority earned his reputation in a far removed field. In the case of the mind-body problem, philosophers are expected to build on cognitive neuroscience. This science is the only (provisional) authority that a philosopher of mind should admit. But McGinn's four-pages long bibliography includes only two scientific references, and none at all to the proponents of psychoneural identity, which happens to be the philosophical propeller of cognitive neuroscience.

Let us find out what contemporary mind scientists think about the matter of mind.

Chapter 9
Minding Matter: The Plastic Brain

The philosophy of mind is the chapter of ontology that deals with the most basic and universal traits of the human mind, It can be either traditional (prescientific) or contemporary (science-oriented). The traditional philosophy of mind is, in a nutshell, the hypothesis that everything mental happens in the immaterial mind. This is the view of shamans and priests, as well as that of psychoanalysts, psychics, and New Age cultists. Idealist philosophers, linguists indifferent to real speakers, as well as many psychologists, concur. The famous philosopher Hilary Putnam (1975, 291) put it memorably, "We could be made of Swiss cheese and it wouldn't matter." And the founder of possible-worlds metaphysics denied the identity of mental states and brain states because one might *imagine* a brain state that is not a mental state, as well as a mental state existing without the corresponding brain state (Kripke 1971, 162–163). Thus, the fact that zombies are conceivable is used to prop up a philosophy of mind indifferent to the science of mind.

By contrast, medical researchers and materialist philosophers since antiquity, as well as cognitive neuroscientists since Broca and Wernicke, have postulated that all mental facts happen in brains. Simpler: We think and feel with our heads – which is why beheading has always been regarded as the best defense against dangerous thinking. The methodological partner of this materialist philosophy of mind is of course the strategy that may be condensed into the rule: *To understand the mind, study the living brain*. This is exactly what cognitive neuroscience has been doing ever since Donald Hebb (1949) wrote its research project (see, e.g., Purves et al. 2008). For example, memory traces consist of groups of sparsely distributed neurons. This is why fear memories, which are engraved in the amygdala, can be literally killed by ablating portions of that brain subsystem.

Admittedly, brainless psychology can correctly describe some mental phenomena. But it can explain none, because genuine explanation involves revealing mechanisms (Bunge 1967a; Machamer et al. 2000); furthermore, all mechanisms are processes in material systems (Bunge 2006a). The medical implication is that serious mental disorders are brain diseases that call for therapies disrupting neural processes. However, since living brains are immersed in a social environment, cognitive and affective neuroscience must be fused with social psychology. The result is social cognitive neuroscience (see, e.g., Cacioppo 2006).

M. Bunge, *Matter and Mind*, Boston Studies in the Philosophy of Science 287,
DOI 10.1007/978-90-481-9225-0_9,

The thesis that the brain is the mental organ can be upheld only if the brain is assumed to be relentlessly active, even in sleep, in the absence of external stimulation, and while performing "mindless" (routine) tasks: if it is seen as capable of stimulating and reorganizing itself – that is, endowed with spontaneity, self-motivation, self-winding, and self-programming, as well as with the ability of repairing itself.

Admittedly, assuming that only an understanding of the brain can help understand the mind involves far greater investments in ingenuity and experimental work than adopting any of the simple rival views, from the old soul psychology to fashionable computationalism, because the human brain happens to be the most complex system in the world. As Ken Hill, of artificial intelligence fame, said, "If the brain were simple enough for us to understand, we'd be so simple we couldn't."

9.1 Pychoneural Identity

Let us now take a quick look at the most important alternatives to classical dualism, namely the pychoneural identity theory, or rather hypothesis. (In the exact sciences, a theory proper is not a stray conjecture but a hypothetic-deductive system.) This is the reductive conjecture that all mental processes are neural processes. More precisely, *For every mental process M, there is a process N in a brain system, such that* $M = N$. For instance, seeing is the specific function of the visual system; feeling fear, a specific function of the system centered in the amygdala; deliberating and making decisions are specific functions of the prefrontal cortex, and so on. (See, e.g., Bindra 1980; Bunge 1980a; Changeux 2004; Hebb 1980; LeDoux 2002; Mountcastle 1998; Zeki 1993.)

In the present context, a *function* is understood as a process in a concrete thing, such as the circulation of blood in the cardiovascular system, and the formation of a decision in the prefrontal cortex. And a *specific function* of a system *S* is one that only *S* can perform. For instance, the brain performs very many functions, but only the brain can think. Functionalists delete the *S*'s in the previous account, or they conceive of organs as mere means for the performance of functions. An example of this teleological mode of thinking is the title of the book by Popper and Eccles: *The Mind and its Brain*. Why not *Walking and its Legs*, *Digestion and its Gut*, and the like?

The functionalists incur the same mistake Aristotle accused Plato of making: that of speaking of motion in itself rather than of moving bodies. By contrast, the cognitive and affective neuroscientists attempt to locate the mental processes in the brain, and to disclose their mechanisms: they work on the physiology of the mind, just as other scientists work on the physiology of digestion. See Table 9.1.

Regrettably, the psychoneural identity hypothesis is often ill formulated, even by people who claim to hold it. In one popular version, "thought and consciousness are *products* [*Erzeugnisse*] of the brain" (Engels 1954, 55). In another, "the brain *causes* the mind" (Searle 1997) or, in general, "neural anatomy causes mental function." Both are unwitting formulations of epiphenomenalism, the dualist

Table 9.1 Physiology of mind in a nutshell

Organ	Specific function(s)
Amygdala	Feeling fear
Auditory system	Hearing
Brain stem	Wakefulness
Entorhinal cortex	Map of the spatial environment
Hippocampus	Short-term memory, spatial orientation
Insula	Disgust
Nucleus accumbens	Pleasure
Olfactory bulb	Olfaction
Prefrontal cortex	Cognitive control, executive functions
Visual system	Vision
Valuation & decision	Medial prefrontal cortex
Whole brain	Consciousness

doctrine that mental events are the passive byproduct of brain activity, much as noise is a byproduct of engines.

The production version of epiphenomenalism is mistaken, because a product is generally thought of as stuff separable from its source and incapable of reacting upon it – for instance, the bile is a product of the liver. The causal version of epiphenomenalism is wrong too because the causal relation holds between events, not between things and events occurring in them. For example, rotation is not an effect of the wheel, but just what the wheel does – its specific function in addition to skidding, getting overheated by friction, making noise, etc. Neither is the wheel the "physical correlate" or "substrate" of rotation. It is equally absurd to say that the wheel "subserves" or "instantiates" rotation, or that it is the "material substratum" of rotation. One says instead that wheels turn. Besides, and contrary to epiphenomenalism, if the brain were to cause the mind, then the latter should in turn react upon the former, since every event initiates a new causal chain (Lachs 1963).

The mental functions are neither products nor effects of the brain: they are what the brain can do. Likewise, muscle contraction is neither product nor effect of the muscle: it is just the specific muscle activity. (See Bunge 1980a for a formalization of the concept of a specific function.) The monistic psychoneural hypothesis is simply that mental processes are processes in brain systems, such as the language "areas" in the cerebral cortex.

This hypothesis has been central to medical psychology since Hippocrates and Galen, as well as a target of theologians and idealist philosophers. And of course it is the central assumption of cognitive neuroscience, both of which attempt to "map the mind onto the brain." However, let me haste to warn that this map has turned out to be very different from the one-to-one chart of a territory: see Section 9.4 below.

To emphasize that materialist monism asserts a certain identity, and not a mere correlation or parallelism, let us reformulate it in a more explicit if pedantic fashion: *For every mental process M, there is a process N in a brain system, such that $M = N$.* Equivalently: *For every mental function F there is a brain system B that performs F.* Medical corollary: *If B is injured or absent, F is disturbed or fails to occur.*

The psychoneural identity hypothesis in this formulation is empirically testable. Indeed, it is possible to alter neural systems through pharmacological or surgical means, or else TMS (transcranial magnetic stimulation), and measure the resulting changes in behavior. Unlike MRI (magnetic resonance imaging), which reveals certain mental functions, TMS alters some of them on top of measuring some of their traits.

9.2 Supervenience and Emergence

Some philosophers of mind, such as Hilary Putnam and Saul Kripke, discuss the mind-body problem in the light of sheer fantasies. Some of the most popular are those of the brain in a vat, Dry Earth, zombies, and the "transporter" in the movie *Star Trek*. Let us glimpse at the latter, used by Jaegwon Kim (2006, 8–9) to illustrate what he calls "supervenience of the mental on the physical", as well as to buttress the physicalist or reductionist version of materialism.

The said "transporter" is an imaginary device that instantly dismantles a person, collects all the information about its molecular components and their relative positions, and transmits this information to another location, where another device uses this information to synthesize an identical copy of the original person. I submit that this fantasy is just that, for it rests on a wrong concept of living matter, one according to which all that matters about a system is composition and structure, regardless of mechanism and interaction with the environment.

The composition-structure model of an organism, or indeed of any material system, is deficient because it is static: it overlooks the fact that, to be alive, a thing must undergo myriads of chemical reactions – some simultaneous and others sequential – that are regulated and timed with exquisite accuracy. Some reactions cannot start until others have been completed, simply because they "use" products of the latter. If only for this reason, the very idea of instant life is naïve, to put it mildly. This holds even for the simple chemical reactors used in the chemical industry.

Consequently, putting together in the right positions all the chemicals that constitute a living thing won't "animate" the system, not even subjecting it to electric discharges, the way Mary Shelley fantasized when imagining Frankenstein. Producing synthetic life takes far more than that. Fortunately, the psychoneural identity hypothesis does not stand or fall with science-fiction fantasies: it rests on, and in turn motivates, solid neurocognitive research – something that philosophers of mind rarely consult. (For example, the 10-pages long bibliography of Kim's 2006 book does not list a single paper in cognitive neuroscience.)

I have yet to see a scientist earnestly discussing impossible science-fiction devices. For that matter, I have yet to see the word *supervenience* in a scientific publication. Scientists prefer the word *emergence* when dealing with qualitative novelty. As we saw in Section 5.7 , "emergence" has two mutually complementary senses: (a) a peculiarity of a system that none of its constituents possess, such as the ability to make decisions, that only certain multi-million systems of neurons in the

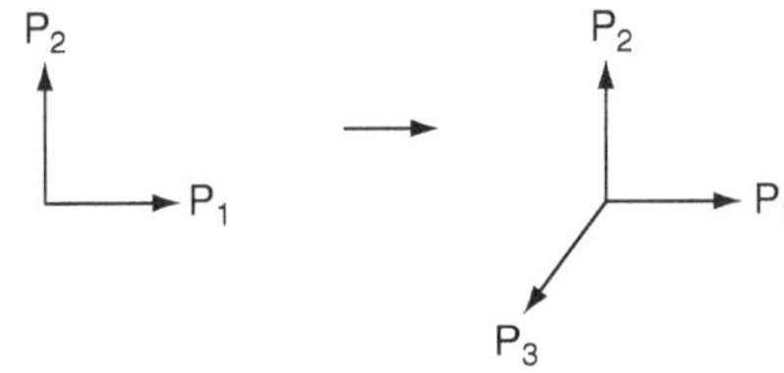

Fig. 9.1 A new property P_3 emerges in a thing with properties P_1 and P_2

prefrontal cortex has; and (b) a radically new property that arises in the course of a process, such the ability to "read" other individuals' minds, a novelty arisen in both the developmental and the evolutionary process. Both concepts, the static and the dynamic, can be represented by the sprouting of a new axis (or coordinate) in the state space for things of the given kind: See Bunge 1977b and Fig. 9.1.

Saying that a new property P_3 arises, or that it emerges from previously absent properties of things of some kind *K*, is short for "In the course of their history, things of kind *K* acquire property P_3." The reason is of course that, as Aristotle argued against Plato, there are no properties without bearers: Every property is the property of some thing or other. (More in Chapter 14.) In the case of living things, the history in question may be either the life history (or ontogenesis) or the species' history (phylogenesis). Thus, one may investigate at what age and under which circumstances do infants normally learn to think of other people's mental processes, or which of our remote ancestors acquired this ability.

In any event, whereas the concepts of emergence are clear and in frequent use in the sciences since George Lewes introduced it in 1874, and particularly since Lloyd Morgan (1923) popularized it in biology and psychology, the notion of supervenience is neither well defined nor common outside contemporary philosophy of mind.

9.3 The Plastic Brain

The best known hypotheses about the nature of the brain are that it is basically fixed from birth, and that experience can only fine-tune or modulate it (nativism); that it is in a blank slate at birth and thereafter totally at the mercy of its environment (empiricism, behaviorism, and computationalism); or else, that it is plastic: that it changes as it learns, forgets, invents, plans, and decides, whether spontaneously or under external stimulation.

The first hypothesis goes well with Charles Sherrington's view, that the main function of the nervous system is to coordinate the different parts of the organism. Because he focused on the homeostatic function of the brain, Sherrington (1964) overlooked its mental functions, and leaned toward psychoneural dualism. The second hypothesis is consistent with Ivan Pavlov's postulate, that the brain is "the organ of the most complicated relations of the animal to the external world." Unsurprisingly, he upheld the psychoneural identity hypothesis (Pavlov 1955).

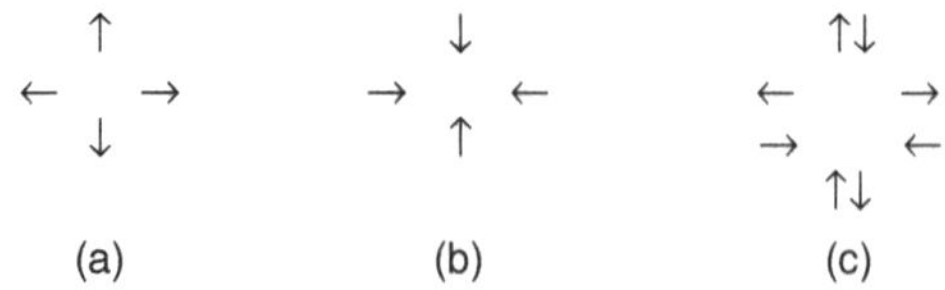

Fig. 9.2 Three views of the brain: (**a**) rigid and self-sufficient (nativism); (**b**) plastic and passive (empiricism); (**c**) plastic and interactive (cognitive neuroscience)

The third hypothesis is that of cognitive neuroscience since Hebb (1949). Like Pavlov, Hebb (1980) defended psychoneural monism; but, unlike Pavlov, he tried to educate his fellow psychologists instead of attacking them. See Fig. 9.2.

The third hypothesis incorporates the partial truths of both genetic and environmental determinisms, in particular the hypothesis that the brain regulates both the internal milieu and the animal's interactions with its environment. However, this third view goes way beyond both Sherrington's and Pavlov's, for it emphasizes the unrelenting self-generated or spontaneous neuronal activity, even during sleep.

A decisive evidence for this view is the constructive nature of episodic memory (Bartlett 1932; Tulving 2002). That is, when recalling episodes, we often combine different memories rather than faithfully reproducing what really happened. To put it in neural terms, the recall of an event often consists in the aggregation of different memory traces. This is why the testimonials of eyewitnesses are unreliable. This is also a reason that the computer model of the mind is wide of the mark.

Another compelling evidence for the hypothesis of spontaneous brain activity is that 60–80% of the energy budget of the brain is spent on inter-neuronal communication, whereas responding to the momentary demands of the environment may take as little as 1% of the total energy budget (Raichle 2006).

Of course, the rate of energy expenditure depends on the nature of the task, and it is likely to be greatest during the early learning stages, while the animal "gets the hang of it." For example, the neurobiological study of the hippocampus of rats while learning a task, and then recalling it, led György Buzsaki and his coworkers to this conclusion: "During learning, the temporal order of external events is instrumental in specifying and securing the appropriate neuronal representations, whereas during recall, imagination, or action planning, the sequence activity [of the neurons concerned] is determined by the intrinsic dynamics of the network" (Pastalkova et al. 2008).

"A fundamental property of the brain is plasticity, the ability to change in response to experience and use" (Feldman 2009, 34). Such changes are functional (physiological) or structural (anatomical), and while some of them are caused by external stimuli, others occur spontaneously. Learning, forgetting, imagining, inventing, decision making, and the alterations our memories suffer over time, constitute perhaps the most compelling evidence for the hypothesis that parts of our brains are *plastic* rather than either elastic (leaving no traces) or rigid – incapable of either learning or forgetting anything.

Ramón y Cajal intuited, and half a century later Donald Hebb (1949) theorized, that the basic mechanism of learning and forgetting is synaptic plasticity – changes in the strength of inter-neuronal connections. In 1948 Jerzy Konorski confirmed the occurrence of plasticity, and coined this term; and in 1966 Terje Lømo confirmed experimentally the Cajal-Hebb hypothesis. Seven years later he and Timothy Bliss strengthened those connections through irradiation with high-frequency electromagnetic waves. (The technical name for this synaptic facilitation is LTP, short for long-term potentiation.) Lately Attila Losonczy added an anatomical change, namely the alteration of dendritic spines, as a plasticity mechanism. (See Craver 2009 for an analysis of this large multilevel and multidisciplinary research project.) At about the same time, traveling waves ($v \approx 10$ cm.s^{-1}), a long range and faster inter-neuronal communication mechanism, were discovered in several regions of the brain.

Neural plasticity research, which has spanned an entire century, is a clear illustration of both the long-term, interdisciplinary, and international nature of basic research on the so-called Big Questions, as well as of the intimate intertwining of hypothesis and experiment. Furthermore, the same research has shown that learning and forgetting involve processes on several levels, from the release and reception of molecules of various kinds – light like calcium and heavy like glutamate – to the synthesis and degradation of proteins, morphological changes in dendritic spines, and the assembling and dissembling of cell assemblies – as Hebb (1949) called the systems composed by several neurons.

Synaptic plasticity has refuted the nativist, in particular genetic, conception of the brain as strictly determined by inheritance, and replaced it with the *epigenetic* conception. The latter "postulates that the connections between neurons are established in stages, with a considerable margin of variability, and are subject to a process of selection that proceeds by means of trial and error" (Changeux 2004, 185). In view of the large and increasing body of evidence for this dynamic view of the mental organ, the persistent popularity of its opposite in its various versions – Dawkins' genetic determinism, Chomsky's innate ideas, and computationalism – is puzzling.

The findings on plasticity suggest the following view of the brain of the higher vertebrate (mammal or bird). Such brain contains subsystems of two kinds: with constant and with variable connectivity – or committed and uncommitted respectively. The former are in charge of routine functions, whereas the others are capable of discharging new functions, that is, of learning. Every mental function is the specific function of a neural system that is plastic or was plastic before the animal learned the task in question to the point that it became a routine. The smallest neuron assembly capable of performing a mental function may be called a *psychon*. In short, every mental process is a process in a psychon or in a system of psychons. (For details see Bunge 1980a; Bunge and Ardila 1987.)

The preceding answers the question of the aim and scope of psychology, which we asked at the start of the previous chapter. Psychology is the scientific study of the mind regarded as the collection of processes, other than the "household" ones, such as protein synthesis an blood circulation, that happen in the plastic subsystems of the brains of the higher vertebrates (mammals and birds), as well as of the behavior controlled by such subsystems.

9.4 Localization-Cum-Coordination

A recurrent controversy in neuroscience is that between localizationists and holists. Localizationists, like Galen and Gall, have tended to be materialist monists, whereas holists, like Jackson and Freud, have tended to be dualists. The reason for this strong correlation should be obvious: Facts are somewhere in space, whereas immaterial objects, such as souls and numbers, are either nowhere or everywhere. This is why, in his *Méditations* (1641), Descartes held that the *res cogitans* was out of space – an assertion that provoked Leibniz's derision. But in the end, being more of a scientist than a theologian, in *The Passions of the Soul* (1649), Descartes tacitly recanted substance dualism and conjectured that the pineal gland was "the seat of the soul."

Franz Joseph Gall was a neuroanatomist who, in the late eighteenth century, claimed to be able to read people's mental faculties by locating bumps on their scalps: he believed that each bump indicated the presence of a highly developed "mental organ." This doctrine, phrenology, combined Hippocrates' psychoneural monism with Galen's hypothesis that every region of the brain performs a particular mental function. Gall's main ideas, that mental processes are brain processes, and that the brain is a system of specialized subsystems, were basically sound; regrettably his methodology was wrong, for he did not bother to check his bold conjectures.

Two centuries later, the philosopher Jerry Fodor (1983) revived phrenology. Indeed, he postulated that the mind – not the brain, which hardly interested him – is a collection of mutually independent modules, every one of which is "domain specific", i.e., carries out a specific task by itself. This hypothesis, dubbed the *Swiss-army penknife*, was to become a mainstay of nativism à la Chomsky and Pinker, as well as of pop Evolutionary Psychology.

Fodor's imaginary modules did not develop, but were born fully functional: this one for vision, that for grammar, and so on. In short, Fodor's theory, like Chomsky's psycholinguistics, is at odds with the most important finding of developmental psychology: that *all knowledge is learned.* It also contradicts the plausible conjectures that modularization (specialization) is a gradual constructive process, and that every module is functionally connected to at least one other module. In other words, specialization is not inborn, but emerges in the course of individual development (Elman et al. 1998). For example, we are normally born with a visual system, but we must learn to look if we wish to see microscopic details or overall patterns. It took the genius of Cajal to notice for the first time neural networks in what others had seen a homogeneous blob.

In any event, all the known cases of cognitive "domain-specific" modules, and the corresponding deficits, have been found in adult brains. The neonate brain is ignorant; and its deficits, if any, are anatomical or architectural, not physiological or functional (Karmiloff-Smith 2006.) In conclusion, Fodor's inborn, rigid, and mutually independent modules did not stem from cognitive neuroscience, but was in the tradition of the aprioristic *Naturphilosophie* of Hegel and Schelling.

Much the same holds for its polar opposite, the holistic extravagance of Bennett and Hacker (2003), that the whole person, rather than the brain, is the bearer of mental predicates. (If this were true, quadriplegics would be mindless, beheading would not be more of an impediment to thinking that finger amputation, and neurosurgeons should start cutting at the toes.)

The "whole person" approach to disease, characteristic of "holistic medicine," is a stumbling block to neuroscientific research and neurological practice, as Norman Geschwind (1974) emphasized in his classic 1965 paper on disconnection syndromes. This seminal paper reported on his own study, together with Edith Kaplan, of a patient who had undergone the section of his corpus callosum, the bridge between the brain hemispheres. The patient in question "appears to behave as if there were two nearly isolated half-brains, functioning almost independently" (p. 23). For example, he was unable to name the object, known to him non-verbally, that he was holding in his right hand. (Other neurological patients exhibit the dual behavior: they cannot handle objects that they can name.) At first, his behavior baffled the scientists who had failed "to regard the patient a made of connected parts rather than as an indissoluble whole" (p. 224). The parts in question were the visual system and the language area. Thus Geschwind (p. 225) rejected both holism and atomism, and considered the animal "as a union of loosely joined wholes." That is, he tacitly adopted the systemic approach. See Fig. 9.3.

If the extension of the mind to the hole person were deemed to be insufficient, consider Andy Clark's (2008) "extended mind," which includes the thinker's paper, pencil, computer, and library. (Why not generalize, and regard the kitchen as belonging to the "extended gut", the gym as part of the "extended musculo-skeletal system, and so on?) This won't do, as brains cannot be replaced, repaired, or set aside like tools. One more step in the operation of mind extension might land us in panpsychism. Which illustrates the cynical principle that, given an arbitrary extravagance, there is at least one philosopher capable of inventing an even more outrageous one.

Let us now go back in time by one century. From its start in mid-nineteenth century, neurolinguistics seemed to confirm strict materialist localizationism, that is, the one brain area-one mental function hypothesis. At the beginning of the eighteenth century it was discovered that damage to one side of the brain impaired the control of movement of the opposite side of the body. In the next century Pierre-Paul Broca discovered that lesions in the left hemisphere of right-handed people impair speech production; and one generation later Carl Wernicke found that damage to a certain region in the same hemisphere impairs speech comprehension. Much

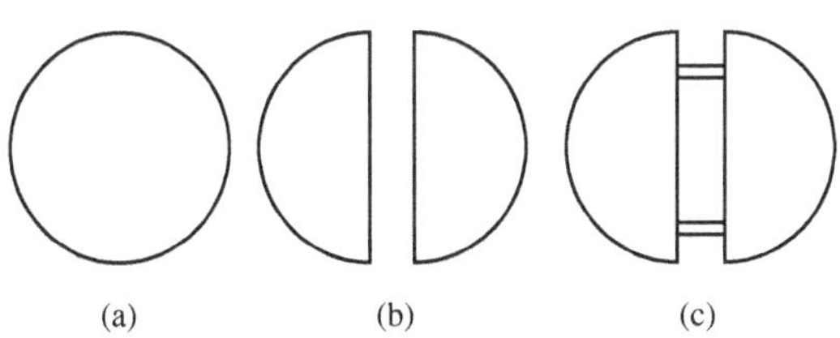

Fig. 9.3 (**a**) Holism: every thing is a seamless whole. (**b**) Atomism: All complex things are collections of mutually independent individuals. (**c**) Systemism: All complex things are either systems or interdependent system constituents

later it was found that certain lesions erase prepositions, and others articles; that, whereas some aphasias are syntactic, others are semantic, depending on the site of the accident or lesion; that, in bilinguals, brain injuries in certain sites diminish performance in one of the languages but not in the other; that certain lesions cause the loss of words naming inanimate objects but not organisms, or vice versa; that some psychotic disorders occur only in one of the languages of bilinguals, and so on (Paradis 2004). And it has recently been found that distinct regions of the visual system react to living and nonliving stimuli, even in adults blind since birth (Mahon et al. 2009). But, since such reactions are automatic, they do not constitute innate knowledge.

The study of perception in neurological patients revealed that vision too is mosaic: Although we see things as unitary entities with a number of phenomenal (secondary) properties, every one of these seems to be apprehended by a different brain system. This is why some patients cannot see shapes, while others cannot perceive color, texture, or motion, and still others recognize what they perceive but cannot say where it is, or conversely. Thus, contrary to what the Gestalt school held, the brain is firstly an analyzer, and a synthesizer only in a second phase. But we still do not know exactly how the synthesis of the various sensations is achieved: this is the famous binding problem.

At the beginning of the twentieth century, Korbinian Brodman, working with a microscope on dead brains, found in the human cerebral cortex 52 distinct areas composed of cells of different types. Half a century later Wilder Penfield and his collaborators found that patients whose cerebral cortices were subjected to weak electrical stimulation suddenly remembered long-forgotten episodes, felt scents, hummed a song, or hallucinated (Penfield and Rasmussen 1968). One of his associates, Brenda Milner, discovered that memories are localized: Visual memories in the visual region, verbal memories in the left temporal lobe, motor memories in the motor region, and so on. Thus materialist localizationism seemed confirmed both anatomically and functionally.

Numerous subsequent lesion studies led to the generalization that learning something may be a function of a number of brain system, but in most cases there is one essential (necessary and sufficient) neural circuit (Thompson 2005). So, knowledge is fairly well localized, which is why it is erased when its seat is excised. Yet none of the many localization findings weakened Penfield's faith in psychoneural dualism – as if science and religion were located in different brain regions.

More recently, work with neuroimaging techniques, in particular fMRI (functional magnetic resonance imaging), gave birth to cognitive neuroscience, and seemed to confirm materialist localizationism. For example, fear is a function of a circuit including the amygdala, and social exclusion activates the insula together with a region of the prefrontal cortex and another of the cingulate cortex. And shining light on special neurons in the fruit-fly larvae can activate hole neural circuits controlling innate behavior, such as the escape response.

This kind of work has elicited malicious comments on the part of some veterans of the much earlier, and sensationally fruitful techniques, namely lesion studies and single-unit electrophysiology: they accused the neuroimaging crowd of perpetrating phrenology. This impression was soon corrected by a plethora of discoveries made with fMRI, a method that allows one to image the whole brain at once. This work revealed that we see mainly with the primary visual area, hear with the hearing area, smell with the olfactive bulb, and evaluate and make decisions with the frontal lobes. Even a cursory perusal of the *Journal of Neural Engineering* will show that the design of neural prostheses presupposes the localization of mental functions. If holism were true, such prostheses would not work or they could be implanted anywhere in the brain rather than in rather well localized places.

Work with the same technique confirmed the difference between fluid intelligence (or problem-solving ability) and crystallized intelligence (skills and knowledge), originally pointed out by Cattell (1987). It turned out that the former is located in the prefrontal cortex, whereas the latter is a function of posterior and parietal areas (Ferrer et al. 2009).

It has always been known that our reasoning abilities improve with age (or rather with learning). But now we also know that such improvement results from the maturation of the prefrontal cortex, in particular its rostrolateral region (Wright et al. 2008). As our brain develops (or decay), so does our intelligence. Further, individual differences in fluid intelligence correspond to differences in brain structure (Gray and Thompson 2004).

However, none of this confirms modularity *à la* Gall or *à la* Fodor, for research with the same tool has also revealed that one and the same brain area may have different functions, and may participate in several distinct neural circuits. For instance, we see with the visual system and, in particular, with the primary visual cortex. But, when looking intently at something, the frontal and parietal areas too become involved (Bressler et al. 2008). Likewise, sniffing involves the temporal lobe in addition to the olfactory bulb (Sobel et al. 1998). This explains the difference between seeing and looking – and similarly between smelling and sniffing, as well as between hearing and listening. Likewise, when making a decision or controlling a behavior, affective areas combine with cognitive ones. And, although the specific function of the hippocampus is to recall places and guide our navigation between them, it also participates in imagining the future; that is, the hippocampus is shared by the system (or circuit) that imagines future events.

In general, as Dehaene and Naccache (2001, 13) write, "besides specialized processors, the architecture of the human brain also comprises a distributed neural system or 'workspace' with long-distance connectivity that can potentially interconnect multiple specialized brain areas in a coordinated though variable manner." Such long-distance connections among remote brain subsystems become manifest every time transcranial magnetic stimulation (TMS) at any one site spreads to distant areas (Bestmann et al. 2004).

This is to be expected if one regards the brain as a system of more or less strongly bound subsystems, as a consequence of which hitting any one target affects also distant regions. But it sounds paradoxical (counter-intuitive) if one presupposes that the brain has been intelligently designed, each component of which performs a single function. But it has not: the brain has been clobbered together by an opportunistic craftsman, namely evolution. This is one of the many lessons of evolutionary biology: that intelligence is a product of an utterly unintelligent process that produces imperfect organs and leaves debris behind.

The coordination of the various subsystems of the brain is a necessary condition for the occurrence of conscious states (see, e.g., Singer 2009). It also underlies the cognition-emotion connection. This explains, for instance, that the response of patients with frontal brain damage to emotional stimuli is weaker than that of normal subjects (Damasio 1994). In lay terms, cognition and emotion, though separate, are connected and modulate one another (Phelps 2006).

The emerging consensus is that the integration of cognition and emotion occurs mainly in the lateral prefrontal cortex, and that "the neural basis of emotion and cognition should be viewed as strongly non-modular" (Pessoa 2008, 148). An obvious methodological consequence is that the names "cognitive psychology" and "affective neuroscience" were obsolete at birth: "cognitive neuroscience" should have used from the start. Even better, the original name, "physiological psychology", coined around 1880, should have been kept. Its synonym, *psychobiology*, is even better because shorter and yet more comprehensive.

In short, there is functional localization together with interaction, and often coordination as well. This hypothesis may be called *moderate localizationism*, and it is an instance of the systemic ontology, according to which every thing is a system or part of one (recall Section 1.3). The methodological counterpart of this ontology is the research strategy summarized in the slogan *Distinguish but don't detach*. For example, look for brain nuclei or modules, but do not isolate them from the rest, for in fact they are likely to be anatomically linked to other modules. Thus, the amygdala, famous for being the anxiety center, is connected to most cortical areas. And the hippocampus, once believed to be the primary organ for memory encoding and retrieval, is now known to share this task with the prefrontal cortex.

In general, in the brain there is functional integration or synthesis together with anatomical segregation or specialization. For this reason, large-scale neuroimaging (using mainly fMRI), which alone can identify the entire neural circuit performing a cognitive process, should be combined with single-cell electrical recording – plus the classical psychological study of whole animals in their social context for good measure (see Bunge and Kahn 2009; Logothetis 2008).

Moderate localizationism may be summarized as follows. Every brain subsystem carries out at least one specific function – that is, a function that it alone can perform. Thus, only the visual cortex can see, only the hypothalamus can regulate appetite, only the insula can feel disgust, only the amygdala can feel fear, only the hippocampus can "store" places and routes, and so on. But every such specialized organ needs the support of other parts of the body. Likewise we can only walk with our legs, but only as long as the heart, the lungs, the brain and many other organs cooperate. In the brain, like in any other system with heterogeneous components, division of labor requires coordination along with specialization. See Fig. 9.4.

To find out whether "area" A of a brain discharges task or function F, one inactivates A either temporarily (e.g., by cooling it down below 20°C) or for good (by surgical ablation). If A ceases to perform F, one concludes that A is *necessary* for F. But one may obtain the same result by inactivating an upstream part. For example, by just temporarily sewing the left eyelids of a kitten, Hubel and Wiesel famously

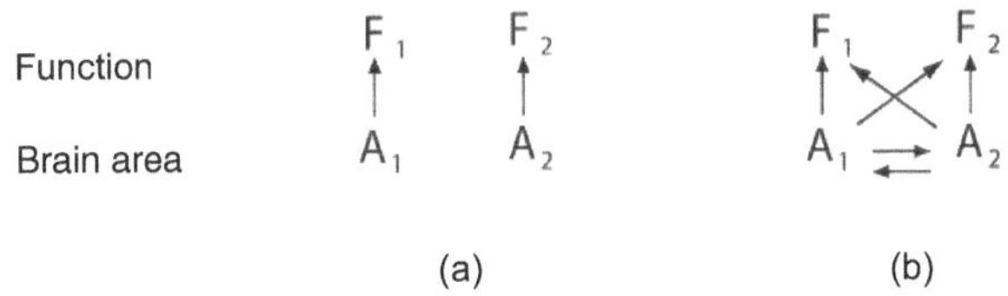

Fig. 9.4 Relations between brain areas and their functions. (**a**) Radical localizationism: one-to-one map. (**b**) Moderate localizationism: interactions between areas plus multiple functions

inactivated the right half of its visual cortex, and showed that the normal development of this part of its brain had been arrested in an irreversible manner: Even after the stitches on the eyelids were removed, the animal did not recover binocular vision, because half of its visual area had degraded due to lack of sensory stimulation.

At about the same time, Donald Hebb and his students showed that sensory deprivation decreases the ability to solve problems and causes hallucinations. Severe restriction of movement, too, increases illusory pattern perception and delusory imagination, to the point that the subject may become hypersensitive to painful medical procedures, may imagine conspiracies, and develop superstitions (Whitson and Galinsky 2008). The lesson for political manipulation should be obvious.

In short, the human brain is the organ of mind and it is always extremely active, but it does not function normally in isolation. It is also divided into regions with different specific functions; but these regions, far from being self-contained and mutually independent modules (or "Darwinian algorithms"), can combine to the point of constituting systems capable of "generalized intelligence." We shall return to this subject in Section 10.5.

9.5 Advantages of Psychoneural Monism

Let us now list some of the virtues of the materialist conception of mind.

1. It is no less than the hypothesis that drives cognitive, affective, behavioral and social neuroscience, which is at the cutting edge of contemporary psychology and psychiatry.
2. It can explain, at least in principle, all the mental phenomena known to classical psychology, and then some. For example, we have "mirror neurons" in the premotor cortex that are stimulated by the perception of certain actions of other people. Allegedly, they allow us – as well as monkeys – to imitate effortlessly some skilled movements of others (e.g., Rizzolatti and Craighero 2004). Besides, these neurons would constitute the "neural basis" (mechanism) of learning by imitation (Prather et al. 2008). It has also been speculated that mirror neurons are involved in forming the "theories of mind" (guesses about others' mental processes) that we invent to explain the behavior of others. However, this motor theory of action understanding – only the latest of this kind in the space of two centuries – has been subjected to severe criticism (e.g., Hickok 2009). So, the

least that can be said for it is that it is empirically testable, whereas dualism is not.

3. It has garnered a number of astonishing findings, such as that mood can be medically controlled by regulating the dopamine level; that the brain has one system for seeing the environment and another for the visual control of movement; that a furious charging bull can be suddenly stopped in its track by a radio pulse acting on an electrode implanted into its brain; that compulsive behavior can be elicited by the same pills that control Parkinson's tremors; and that the feeling of trust, so basic to all human bonds, can be enhanced with nasal spray delivering oxytocin, a "bonding" hormone involved in sex, labor, caregiving, and lactation (Kosfeld et al. 2005).
4. It can tackle problems that could not even be posed in brainless psychology, such as those of locating the engrams for words of certain categories, and finding out the action of chemicals on mood, cognition, and social behavior. For instance, neurolinguists have found that lesions in specific regions of the cerebral cortex cause the loss of certain speech functions (see, e.g., Paradis 2004). And psychologists have found that administering certain hormones or neurotransmitters alters such basic behaviors as child care.
5. It is breaking down artificial barriers between traditional psychological disciplines, such as cognition/emotion and individual/social.
6. By conceiving of mental diseases as brain disorders, materialist monism has helped replace ineffective shamanic psychiatry, in particular psychoanalysis, with increasingly effective, though still rather primitive, biological psychiatry (see, e.g., Shorter 1997)
7. The psychoneural identity hypothesis fits into the materialist (or naturalistic) ontology inherent in modern science, which does not include disembodied spirits or organless functions, yet does admit, if tacitly, the huge qualitative diversity of the world, and even the need to distinguish several levels of organization. In particular, psychoneural monism undermines the idealist fantasy that the world is mental, for, if it were, every human brain would include the universe. (Incidentally, the fantasy that everything is in the mind was held not only by Berkeley but also by Kant, albeit far less clearly.)

In sum, psychoneural monism does not have the flaws of its competitors; it is also consistent with the ontology that underlies all of the natural sciences; and, most importantly, it is the hypothesis that guides cognitive neuroscience. Still, some philosophers still claim that it cannot account for qualia, intentionaliy, and consciousness. Let us see.

9.6 The Qualia Objection to Psychoneural Identity

Qualia, or raw feels, are the sensations of color, taste, smell, touch, and the like. We know that they "reside exclusively in the sentient body" (Galileo 1953, 312). The corresponding predicates, such as "is red", are phenomenal, not physical. Hence physicalism cannot account for them. (However, recall from Chapter 7 that physicalism is only the most primitive version of materialism.) Does this disqualify the attempt to analyze qualia in non-phenomenal terms, such as wavelength and neuronal firing rate? Let us see.

We have first-person knowledge of qualia. Bertrand Russell would say that we know them by acquaintance; and his opponent, Henri Bergson, counted qualia among "the immediate data of consciousness". The set of qualia at a given moment is often called "phenomenal consciousness". But this is obviously a misnomer, because even protozoa detect stimuli of certain kinds, yet no scientist would argue that they are conscious. For example, the unicellular organism *Euglena Viridis* detects light, but does not know *what* it "sees." And some bacteria "feel" the terrestrial magnetic field thanks to tiny bits of magnetite in their body – which is more than even the most exalted philosophers of mind can do.

One of the most common objections to the hypothesis that mental processes are brain processes is that, whereas physical things may detect physical stimuli and react to them, they could not possibly experience qualia or raw feels, such as seeing red or an after-image, feeling pleasure or pain. For instance, if this book were set on fire it would not feel pain. Nor can it share in the pleasure it may give some of its readers.

Admittedly, qualia are very special reactions to physical stimuli, quite different from the reactions of physical things. For example, one does not remember a blow the same way a horseshoe retains the shape the ironmonger gave it: there are as many kinds of memory as species of things with memory – as physicists, ethologists, and other scientists discovered long ago. Cognitive neuroscientists too know this, which is why they study animal memory, not ironware memory. Most philosophers of mind concede that science explains memory, or may eventually explain it. Thus David Chalmers (1996) calls this an *easy* problem. (For dissenting opinions, consult any memory expert, from Bartlett (1932) to Schacter (2001) and Tulving (2002).)

By contrast, the same philosopher and his many followers believe that explaining pain is a *hard* problem, perhaps an intractable one. Why? Kim (2006, 221) explains: "what makes pain pain is the fact that it is experienced as painful – that is, it hurts." In case the reader smells a typo or two, I invite him to look up page 15 of the same book: What is distinctive of pains is that that "they hurt." But if pain hurts, then presumably sight sees, hearing hears, memory remembers, thought thinks, speech talks, sadness saddens, death kills, the world worlds, nothingness nothings, and so on. Existentialists and other nonsense mongers clap. The rest of us moan.

Is the "hard" problem soluble? Of course not if posed in cryptic terms. Kim (op. cit.: 223) and other philosophers of mind do not take this view. They claim that one thing is certain: Brain science cannot solve that problem. This would be because the concept of pain "does not even occur" in brain science. But it does: it occurs in cognitive neuroscience, neurology, anesthesiology, psycho-neuro-immunology, and psycho-neuro-pharmacology. To check this statement, just peruse any of the journals devoted to these sciences.

But Kim does not even mention any studies on the neurophysiology of pain, in particular migraine, chronic pain, phantom pain, the inability to feel pain, or placebo analgesia. Nor does he mention that, half a century ago, some chronic pain patients underwent cingulotomies (lesion of their cingulate cortices), as a consequence of which their distress abated although the pain intensity did not, because it is a process in a different component of the so-called pain matrix.

Another absent from Kim's account of pain is Ronald Melzack's classical finding in 1957, that puppies raised in isolation are "unmindful" of pain: although they feel noxious stimuli, such as flames, they are not afraid of them and consequently do not learn to avoid or expect them. Nor does the psychobiology investigation of social pains, such as humiliation and envy, occur in Kim's account, although it has been a hot topic for some years (e.g., Lieberman and Eisenberger 2006). Why bother with brain science, if one knows a priori that it cannot possibly know "what it is like to experience pain." Don't ask what this locution means, for you will betray ignorance of the black art of metamorphosing confusion into mystery, and nonsense into theory.

And yet in 2006, when Kim's book appeared, the Decade of Pain Control and Research, declared by the U. S. Congress, was in its sixth year – though with decreasing Federal funding. At the same time, hundreds of pain research centers and clinics were in operation around the world, and a number of specialized journals were being published – among them *Anesthesiology*, *Cephalalgia*, *European Journal of Pain*, *Headache*, *Journal of Pain Research*, *Pain*, and *Pain Research and Management*.

This is not to suggest that a full scientific understanding of pain has been achieved. But it is known that we have a whole "pain matrix" (system), and that one of its components, the anterior insula, is more active, the longer a person has suffered pain. It is also well known to neuroscientists, psychologists and physicians that pain is a topical research and medical problem. This is too serious a problem to be left in the hands of philosophers who oscillate between two flagrant errors: that pain is in the immaterial mind, and that it is identical to the firing of the proverbial *C*-fibers.

Qualia are a nuisance for physicalists and computationalists (or functionalists), unless they deny their existence, as Dennett (1991) did. By contrast, qualia should not bother emergentist materialists, who know that living matter has peculiar (emergent) properties (see, e.g., Bunge 1980a, 2006a). For example, ciliates swim away from acids, whereas copper coins won't; and amoebas will engulf bits of food whereas ferns won't. Likewise, nervous tissue tends to circumscribe stimuli (through lateral inhibition), whereas elastic media propagate them. An electric circuit transforms an electromagnetic stimulus into a response of the same kind, whereas a neural circuit responds in a qualitatively different fashion. But physicalists (vulgar materialisrts) do not care for the specific functions of nervous systems, and computationalists care more for symbols than for natural things. So neither of them is interested in the biological study of qualia.

This is related to the discovery of the ancient atomists, and emphasized by Galileo, Descartes, and Locke, that physical objects do not possess secondary (or phenomenal) properties, or qualia, such as color, smell, loudness, and taste. They have only primary properties, such as composition and energy. By contrast, qualia, or raw feelings, are only in the mind: they emerge in the brain when it perceives external objects, and occasionally also in the absence of external stimuli.

According to both materialist monism and cognitive neuroscience, qualia are brain processes. However, this does not entail that qualia are physical. As the founders of psychophysics discovered in mid-nineteenth century, there is a radical difference between a color and the corresponding electromagnetic radiation, as well as between a thermal sensation and heat, or between loudness and the amplitude of the corresponding sound wave. Now we know that the difference is that in the first case there is a brain, and perhaps even a first-person report, both of which are absent in the second case.

The distinction between primary (objective) and secondary (subjective) properties poses serious problems to phenomenalists and physicalists (vulgar materialists) alike. Indeed, the former cannot accept primary properties, and the latter cannot account for secondary properties. Consequently phenomenalists are bound to either ignore or distort all the sciences but classical (brainless) psychology, while physicalists are bound to admit only physics and chemistry, and thus to deny the very existence of qualia. (More in Bunge 2006a.)

Physicalists (or naturalists) deny qualia even though they experience them every time they perceive something. Non-physicalist materialists are not afraid of qualia: they assume that these occur only in brains, so that they must be tackled by cognitive neuroscience. In fact, any recent textbook on neuroscience contains chapters devoted to vision, hearing, olfaction, taste, etc. In other words, qualia are being accounted for precisely in the materialist fashion that dogmatic and antiscientific philosophers, such as Husserl (1970, 134ff.), had condemned a priori. In other words, subjectivity is being explained in objective terms (more in Bunge 2006a). In particular, the psychology of perception is explaining phenomena (appearances), a task that no philosophical phenomenologist has even attempted.

The materialist who includes qualia in his inventory of the world is likely to be an emergentist materialist. That is, he will emphasize that systems, such as the brain, possess properties that their constituents (e.g., neurons and neuron assemblies, such as cortical columns) lack. Such global properties are said to be *emergent*, because they arise (or else disappear) in the course of processes, such as those of self-organization, morphogenesis, and dissolution, that are conspicuous in development and evolution – as we saw in Chapter 5.

A related objection to psychoneural identity is that in daily life we describe mental processes in non-neuroscientific terms. For example, one says that so-and-so fell in love, rather than describing in detail the very complex brain process he underwent – a process we know only in broad outline anyway. But this is common to occurrences of all kinds rather than a peculiarity of the mental. Thus, a driver may describe her car's breakdown in a simplistic way, in a more complicated one by her mechanic, and in a still more sophisticated manner by a car engineer. Likewise, the explanation of a cardiovascular accident, such as a stroke, involves items, such as fat deposits in the arteries, and beta-blockers, which do not occur in its clinical description. Philosophers of mind rightly call this the contrast between folk psychology and scientific psychology.

A seemingly more technical fashion of formulating the objection in question is this: For an identity between mental phenomena and brain processes to exist, the former must share all the properties of the latter – but they don't. Indeed, whereas mental phenomena have secondary properties, brain processes have only primary properties: the former are subject-dependent (subjective), whereas the latter are objective. But this objection is invalid, for it could be raised against all scientific descriptions. For instance, a weak electrical discharge causes a shock that the patient may described as a painful tickle, while a biophysicist would explain it in terms of the action of an electric current on living tissue.

To take a simpler example, and one that has bedeviled semanticists since Frege for over a century: What is the difference between the Morning Star and the Evening Star? They *are* the same planet, namely Venus, but they *look* different, because both the atmosphere and the subject have changed. In fact, seeing Venus in the morning is not the same experience as seeing it in the evening, although the physical object looked at is one and the same. Most philosophers are baffled by this example because they believe that they are dealing with a single referent, namely Venus, while actually there are three referents: Venus, subject, and atmosphere. Venus is what I call the *central referent* of the propositions "I saw the Morning Star" and "I saw the Evening Star." These propositions summarize "I saw Venus in the morning [or through a cool and clean atmosphere]" and "I saw Venus in the evening [or through a somewhat warmer and polluted atmosphere]" respectively (Bunge 1974a).

This is just an example of the contrast between the scientific description of a subject-independent occurrence, and the intuitive or ordinary-knowledge description of a human experience. In other words, we have to do with *two descriptions of two different facts*, one that involves a knowing subject, and another that does not. If the object in question is a mental process, we have these two different facts: a brain process seen by a scientist from the outside, and the same process as experienced and described by the owner of the brain in question. If the brain happens to be that of a cognitive neuroscientist who observes her own mind, she too is likely to give two different descriptions: the objective one in terms of primary properties, such as the rate of consumption of either glucose or oxygen, and the egocentric one in terms of feelings, images, and the like.

In summary, there are qualia and they are exclusive to sentient beings; but they can be explained in objective (subject-independent) terms, namely as features of brain processes. In general science is expected to explain subjectivity (or experience) in objective (nonexperiental) terms – and this is what psychology is all about. But this argument won't persuade the ordinary-language philosopher who, like Stoljar (2006), believes that science, in particular physics, ignores "the physical", whence it would be Quixotic to attempt to reduce the experiential to the nonexperiential. It would be impolite to disturb their slumber.

9.7 Reduction and Merger

The factual identity postulated by the psychoneural identity hypothesis is of the same kind as the identities "Heat = Random atomic or molecular motion", "Light = Electromagnetic radiation whose wavelength is comprised between 390 nm and 740 nm", and "Rusting = Combination of a metal with oxygen." Fiction writers and speculative philosophers are of course free to invent worlds where such factual identities do not hold, where everyone gets anything for free, and where society rewards nonsense. But responsible people do not mistake conceptual possibility, or conceivability, for factual possibility, or lawfulness; and they do not regard the ability to invent fantasy worlds as evidence for their real existence.

There is a huge philosophical literature on reduction, but the subject is far from having been exhausted. For one thing, it is seldom made clear what is the object of reduction: thing, property, or construct? For instance, it is often stated that Water = H_2O. But this equation is false. What is true is that a water body, such as a drop of dew or a lake, is *composed* of H_2O molecules; and composition is only one of the features of a system. A water body has also a structure – essentially the hydrogen bonds shown in Fig. 5.4; in addition, a water body has typical mechanisms, in particular random molecular motion on one level and flow on the next.

Second example: it is usually said that "Temperature = Kinetic energy". But this equation is falsified by any macrophysical body, such as a spoon. It would be truer to say that "Temperature = Average kinetic energy of a system of atoms or molecules in random motion." It will be noted that the first example concerns a thing, and the second a construct (a function) representing a physical property. In my terminology, the first is a pseudo-example of ontological reduction, whereas the second is wrongly alleged to be an instance of epistemological reduction. Let us elucidate these two technical terms. (See details in Bunge 1973, 1977c, 2003a; Bunge and Ardila 1987.)

I suggest that reduction can be either *ontological*, as in "$M = N$", or *epistemological*, as in "M-logy is deducible from N-logy", where M and N denote properties, such as "mental" and "neural" respectively. To state that the mental is the same as neural is to perform an ontological reduction. And to claim that psychology has or will become a chapter of neuroscience is to express hope for an epistemological reduction. Incidentally, the reduction of the mental to the neural does not amount to eliminating the former, but to explaining it. For example, cognitive neuroscientists seek to explain such subjective phenomena as feeling angry or elated in the objective terms of neuroscience. Such reduction does not eliminate the phenomenal concepts of anger and elation: psychoneural monism is not the same as eliminative materialism. Likewise, the explanation of a small-scale social event, such as explaining the formation or dissolution of a partnership in terms of the convergence or divergence of individual interests does not eliminate the concepts of social aggregation and disintegration.

Neither of the two reductions, the ontological or the epistemological, implies the other. The ontological thesis of the identity of the mental and the neural only suggests the research project of either reducing psychology to neuroscience or merging the two disciplines. In the former case one aims for psychobiology, whereas in the latter one aims for joining them. But, since the mental life of gregarious animals is strongly influenced by their social life, a good pinch of sociology must be added to cognitive neuroscience: naturalism does not work for the mind of highly artificial gregarious animals such as us. (Recall Chapter 6.)

When dealing with mental phenomena, the physicalist adopts an extreme form of reductionism: he skips chemistry and biology, equates "mental" with "physical", and hopes that one day psychology will become a branch of physics. Thus the physicalist flattens the world to the ground floor: he ignores the existence of the supraphysical levels or organization, such as the chemical, biotic, and social ones. The physicailist oversimplifies and impoverishes both reality and our knowledge of it. Moreover, he confirms the psychoneural dualist's intuition that it is impossible for a *physical* thing to think, whereas the correct question is whether anything *other* than the brain, which is a *biological* thing, can think.

By contrast, emergentist materialism overcomes the ontological and epistemological limitations of physicalism: recall Chapter 5. In particular, it denies that atoms are just aggregates of elementary particles, and cells just clumps of molecules; that "we are all embryos" (as certain American senators have held), and that "you are your neurons", as it has also been claimed. Of course, physics is the basic science, but it is not omniscient, because there are supraphysical (yet not aphysical) levels.

This suggests that there are limits to reduction: that in most cases reduction is moderate rather than radical. For example, it is not true that heat is the same as molecular motion, as can be seen from the fact that it is possible to prepare a high-speed molecular beam at near absolute zero temperature. What is true is that heat is the same as *random* molecular motion. This is a case of straight ontological reduction together with partial epistemological reduction, because it involves a probabilistic hypothesis in addition to mechanics.

The logical difference between radical and moderate reduction is this. An idea (concept, hypothesis, or theory) B may be said to have been radically (or strongly) reduced to an idea A, if B has been proved to be deducible from A without further ado. Examples: statics and kinematics have been fully reduced to dynamics, and optics to electromagnetism. By contrast, an idea B can be said to have been moderately (or weakly) reduced to an idea A if there is a third idea C such that, conjoined with A, entails B. For example, statistical mechanics is deducible from mechanics jointly with Boltzmann's subsidiary hypothesis that the initial positions and velocities of the atoms or molecules in question be distributed at random. And chemistry is reducible from quantum mechanics only when conjoined with classical chemical kinetics plus certain hypotheses about chemical bonds, such as that the covalent (or non-ionic) ones consist in the sharing of electrons. In short,

Radical or strong reduction of B to A: $A \vdash B$

Moderate or weak reduction of B to A: $A \,\&\, C \vdash B$

Mental phenomena are of course far more complicated than physical and chemical processes, but from a purely logical point of view not more so than the case of chemistry. Indeed, to explain the mental we need not only neuroscience but also some of sociological concepts, such as "aggression" and "cooperation." For example, depression is explained not only in terms of genetic predisposition and serotonin imbalance, but also with the help of data concerning problems in the patient's social life in family, workplace, and society at large.

Notice that, contrary to popular opinion, not all reduction is to lower levels. The reason is that every thing, except for the universe as a whole, is embedded in some higher-level system. For example, a child's scholastic performance must be accounted for not only in neuroscientific terms but also with reference to the child's place in his school, family, and neighborhood. Likewise, microeconomic activities cannot be explained in abstraction from macroeconomic, demographic, and political circumstances. Hence neuroeconomics, the attempt to explain economic activity exclusively in neuroscientific terms, is wrong-headed.

In sum, reduction can be of either of three types: micro, macro, and mixed. See Fig. 9.5.

In short, the program of cognitive neuroscience is this: Mental phenomena can be explained, at least in principle, through ontological reduction (mental = neural) combined with the epistemological merger or convergence of the various branches of neuroscience, and of the latter with psychology and sociology (Bunge 2003a; Bunge and Ardila 1987). Epistemological (or inter-theoretical) reduction, by contrast, is rare not only in neuroscience but also in physics (Bunge 1973a). Craver (2009) espouses this thesis, and calls *mosaic* the unity and advancement of neuroscience brought about by the merger of its various chapters. (The same author also shares my thesis that to explain something is to describe the underlying mechanism.) However, this mosaic is in the service of a deep ontological reduction: that of the mental to the neural.

In general, the study of all things requires both analysis and synthesis. This is why the advancement of modern science has always consisted of two parallel movements: the branching or splitting into specialties, and the merger or convergence of initially separate disciplines: see Fig. 9.6.

Macrolevel	Macrolevel	Macrolevel
↓	↑	↓↑
Microlevel	Microlevel	Microlevel
(a)	(b)	(c)

Fig. 9.5 Three types of reduction: (**a**) microreduction (e.g., Visual sensation = Visual center activity); (**b**) macroreduction (e.g., the dogma of transcultural psychiatry according to which society, not the brain, is to be blamed for mental diseases); (**c**) mixed (e.g., Feeling inadequate = Being both poorly skilled and socially misplaced)

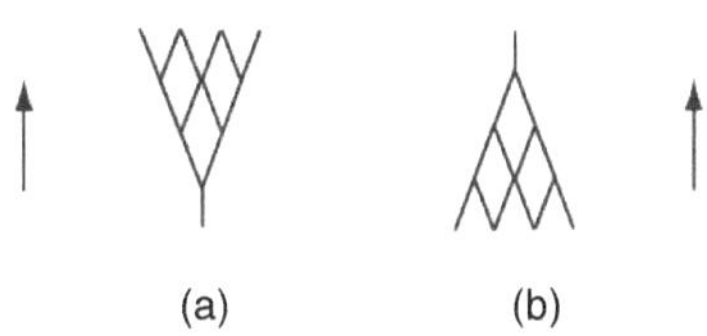

Fig. 9.6 (**a**) Analysis, divergence, or specialization. (**b**) Synthesis, convergence, or merger (From Bunge 2003a)

Concluding Remarks

The pronouncements of most philosophers about the nature of mind have been often dogmatic or confused. For example, Plato thought that the soul is immaterial and guides the body. Husserl believed in the immaterial mind; that the body is nothing but a tool of the mind; and that introspection, together with pretending that the external world does not exist (phenomenological reduction), is the only way to study the mind and even the world. Wittgenstein (1967, 105) wrote that "[o]ne of the most dangerous of ideas for a philosopher is, oddly enough, that we think with our heads or in our heads." Both the positivist Ayer and the rationalist Popper adopted psychoneural dualism as a matter of course, just because it is part of ordinary knowledge. The linguistic philosophers believe that the clue to the mind is the philosophy of language – which of course presupposes that nonhuman animals are utterly mindless. And the computer worshippers, notably Putnam, Fodor, and Dennett, have assured us that the mind is a set of computer programs that can be "realized" or "embodied" in alternative ways.

Few philosophers of mind have bothered to learn what cognitive neuroscience has to say about mental processes. Most of them have not even learnt that the brain is a biological system, not only a physical one, and consequently brain science is not a branch of physics. This is why they keep denying that a *physical* thing could have experiences, feelings, and thoughts. Although most of them regard themselves as critical thinkers, and some of them materialists of sorts, in fact they proceed in the aprioristic, hence dogmatic, fashion characteristic of philosophical idealists. Consequently, far from advancing the science of mind, they would have hindered its advancement if scientists had read them.

It is humbling to learn from Herodotus (Book Two, 2) that around 650 B.C.E. the pharaoh Psammetichus had a mute goatherd raise two newly born infants in isolation, to find out which language they spoke spontaneously. Never mind the result: What matters is that, more than two and a half millennia ago, someone knew what many a modern philosopher of mind does not: that empirical questions call for empirical research.

Chapter 10
Mind and Society

Immanuel Kant and the Baron Thiry d'Holbach were born in Germany just one year apart at the start of the Enlightenment. If Kant had lived in sparkling Paris rather than in Königsberg, and d'Holbach had stayed in dark Edsheim, his native town, they might have exchanged philosophies: Kant might have become the great materialist and realist philosopher of the century, and d'Holbach his idealist counterpart. Of course, the previous sentence is a counterfactual, and as such untestable, and therefore neither true nor false. But it is not a ludicrous fantasy, because we know that nurture and opportunity are just as important as nature.

Whereas in remote and dark Königsberg Kant had no one to learn from or contend with, d'Holbach's salon was frequented by some of the most interesting, advanced, and important thinkers of the Enlightenment: Beccaria, Condillac, Condorcet, d'Alembert, Diderot, Franklin, Helvétius, Hume, Rousseau, and Turgot among others. So, whereas Kant's mental life was basically a soliloquy, d'Holbach's was a constant stimulating and witty dialogue with his brightest and most daring contemporaries. While in dismal Königsberg it was possible, and to some extent desirable, to ignore the drab and dangerous external world, a similar flight from reality was impossible in the bustling and loud construction site of the *Lumières*.

Society reciprocated: While d'Holbach's incendiary books were popular despite having been banned, Kant's rather arcane works circulated only among a few learned men. The Baron d'Holbach, like Voltaire and Hume, but unlike Kant, was what we now call a public intellectual, one whose views are noticed because they matter. Moreover, whereas d'Holbach was scientifically and politically progressive, Kant was just as conservative as Hume. And yet, judging from the complexity of the problems that the two philosophers tackled, Kant's native intelligence and learning were superior to d'Holbach's.

The preceding, if true, reinforces the thesis that nature is of little avail without nurture, just as a severely deficient brain cannot learn much. In other words, everything mental is at once neural and social. For example, although all humans, in all cultures, are born with similar brains, one and the same stimulus may cause pleasure in one culture and disgust in a different one. Experience etches local traits on universal brains.

M. Bunge, *Matter and Mind*, Boston Studies in the Philosophy of Science 287,

DOI 10.1007/978-90-481-9225-0_10, © Springer Science+Business Media B.V. 2010

The methodological lesson is that psychology cannot perform its main job, which is to describe and explain the mental, without the assistance of the social sciences. Notice that this thesis is the dual of the rational-choice approach, according to which only individual behavior can explain the social. It is also at variance with pop Evolutionary Psychology, according to which biology explains social science.

Let us glimpse at three areas of active contemporary research: developmental cognitive neuroscience, social cognitive neuroscience, and evolution.

10.1 Development

According to genetic determinism, individual development is the automatic unfolding of the "program" inscribed in the genome. In other words, genome would be destiny. A corollary of this thesis is that identical twins and clones have identical nervous systems and therefore – via psychoneural identity – the same behavior and the same inner life, if any. But this corollary was refuted in the 1970s by studies of the water flea *Daphnia magna*, the parthenogenetic fish *Poecilia formosa*, as well as of humans: in all three cases the nervous systems of clones exhibited significant anatomical differences (Changeux 2004, 188). For example, human "identical" twins have different fingerprints and can have different handedness. As Peter Medawar (1957, 154–155) put it, the inborn differences between individuals are combinatorial: "one individual differs from all others not because he has unique endowments but because he has a unique *combination* of endowments."

Clearly, such differences emerge in the course of development, some because of different external stimuli, whereas others may be consequences of the vicissitudes of the migration of neurons and the growth of axons and dendrites. Gould (2002) rightly emphasized the importance of "contingency" (accident) in both development and evolution. Its mere occurrence disproves the extravagant opinion that there are development algorithms. In short, genome is opportunity, not destiny. To change the metaphor: DNA proposes, and the internal milieu, jointly with the environment, disposes.

Besides being somewhat erratic, animal development is notoriously mosaic – as is evolution. That is, the various parts of the organism do not mature at the same rate, because their development is controlled by different genes. For example, the reproductive system matures much faster than the nervous system. Correspondingly, the reward and reinforcement system matures before the control system, located elsewhere in the brain. This mismatch generates notorious social problems, such as that of the impulsiveness, selfishness, and irresponsibility that characterize adolescent behavior by contrast to adult behavior. Indeed, the teenager's prefrontal cortex, a newcomer in evolution and the center of cognitive control, is not well prepared to control the powerful new emotions that emerge as the brain is suddenly flooded by sex hormones. Hence teen-pregnancy, which in turn is handled differently in different societies: permissively in some, and punitively in others.

Young motherhood was the rule when people aged quickly and on average enjoyed only half of today's life expectancy. In those times risky behavior in search of instant gratification often favored learning and consequently survival as well. By contrast, in our tightly structured modern society, crafted and controlled by more mature, risk-avoiding, and therefore conservative people, innovation is carefully controlled in schools and workplaces, and discouraged when suspect of being socially disruptive.

Developmental psychologists are bound to face, again and again, the age-old nature/nurture debate: is that ability or behavior innate or learned, intuitive or rational, instinctive and universal, or social and particular? In particular, Chomsky and his followers have claimed that language is "the mirror of the mind" rather than our main communication tool, that it is instinctive, and even that we are born knowing universal grammar and more. Developmental psychologists and social scientists have always known otherwise: their empirical studies have found that speech and gesturing are learned, and that their primary function is communication (e.g., Dunbar 2003; Tomasello 2008).

Language is a tool of social intercourse to such an extent that it is newly reinvented every time it is absent. For example, creoles have been observed to develop from pidgins in the course of a single generation. And the deaf schoolchildren in Nicaragua created their own sign language, without any coaching, while playing in the schoolyard. By contrast, children who were locked up from birth for several years never progressed beyond a protolanguage devoid of syntax. We are certainly born with the ability to learn languages, as well as mathematics and theology, but this potentiality is actualized only in suitable social environments.

In short, while no one denies that all normal humans are born with the *ability to learn* almost anything, from manual skills and languages to science and philosophy, there is no evidence whatsoever that anything learnable is encoded in the genome. The genotype-phenotype connection is very indirect, whence any talk of cognitive genetics is purely promissory. Moreover, this project is seriously flawed because it skips crucial levels: cell, neuron assembly, and surroundings. Any serious studies of the emergence and improvement of cognitive abilities are undertaken by developmental psychologists and anthropologists. For example, it has been found only recently that both the ability to recollect details and to suppress (not "repress") unwanted memories emerge in late childhood.

Far from jumping from genes to behavior, these scientists study the gradual development (ontogeny) of children in various social environments (see, e.g., Karmiloff-Smith 2006). What they have found again and again is that, whereas some environments can actualize certain predispositions (by activating the corresponding set of genes), others can frustrate them. For example, Wright et al. (2008) have found the following causal chain: Childhood lead exposure → Abnormal brain development → Behavioral deficits → Greater likelihood of involvement in crime. Of course, lead exposure is not the only possible trigger of antisocial behavior: many different brain-society pairs may produce the same output. For example, growing up in a poor and violent environment may predispose to crime even more strongly than lead exposure, if only because such environment is not conducive to regular school attendance.

Another example is this. Dehaene and coworkers (2008) asked how young children and Amazonian tribesmen map numbers onto space. They instructed their subjects to place the numbers from 0 to 100 on an unmarked ruler from left to right. They found that naïve individuals used up roughly half the left half of the ruler to place the smaller integers, and crowded all the remaining numbers in the right half: they adopted the logarithmic scale – as anyone acquainted with the Fechner-Weber psychophysical law might have expected. Only the educated subjects distributed evenly, that is, they assigned the same numerical value to all the segments of equal length. The authors concluded that, whereas the logarithmic scale is innate and therefore universal, the linear scale is learned and therefore culture-bound. And they added that this solution of the problem should reconcile the nativist and the environmentalist stands. Perhaps they are unduly optimistic, because the nature/nurture debate has a large ideological-political component. (Contrary to received opinion, though, this component is ambiguous. Indeed, a nativist may claim either that all humans are born equal, or that social status is inborn.)

Another clear case that demands joint attention to nature and nurture is moral development. In particular, preference for fairness or equity, which is generalized among normal adults, is not inborn. ("Normal" = neither psychopathic nor market fundamentalist.) Presumably, such preference develops along with socialization (or enculturation), since it emerges only around the age of 7 or 8; and even then it is parochial, that is, limited to the members' social group (Fehr et al. 2008).

Moreover, our reactions to unfairness depend critically on serotonin level, which can be altered experimentally: intolerance to unfairness increases with serotonin depletion (Crockett et al. 2008). Caution: this finding does not prove that our sense of fairness, or any other social emotion, is purely a matter of chemistry. It only proves that morality is inherent in minding matter. Presumably, the sense of fairness does not fully develop in a caste society, where everyone "knows his place" from birth.

As with fairness and altruism, so with violence. It is well known that physical aggression changes during the life course, and in North American males it peaks at about 17 years, when the brain, still immature, is awash with hormones, and the adolescent enjoys new freedoms and new opportunities, and makes new friends. But we do not know yet the relative contributions of genes, brain maturation, and social environment (see, e.g., Loeber and Pardini 2008). Only one point seems clear: that antisocial behavior is exceptional, even among people born in violent societies. The case of Iceland is particularly clear, since Icelanders have a rather uniform and constant genome due to the lack of immigration in modern times. Yet, according to their sagas, Icelanders were remarkably murderous and treacherous 1,000 years ago, whereas they have been particularly peaceful and gentle during the last few centuries. Perhaps this is due to the fact that severe climate deterioration and deforestation transformed prairies and forests into glaciers, leaving no land to fight over, and forcing people to concentrate in villages and cooperate in the face of an increasingly harsh environment.

Note, incidentally that there are at least two different concepts of an innate trait, those of occurring from conception or from birth. And in either case it is preferable to ask whether a given property-cluster, rather than a single trait, is innate, because properties happen to come in clusters, as will be discussed in Chapter 14. (See Mameli 2008 for both problems.)

Lastly, developmental psychology is moving increasingly closer to comparative and evolutionary psychology. That is, students of child development are asking which other animals possess the same traits, and at what stage in evolution may those traits have emerged. One result of this convergence of disciplines, which would have delighted Ernst Haeckel, is the principle that the cognitive traits that we share with other animals tend to emerge early in human development (see Platt and Spelke 2009).

10.2 I and Us

It is hard to dispute that the brain not only controls the rest of the body but also models its surroundings and helps us adapt to it, as well as adapt it to us. One of the simplest and most persuasive proofs that the brain models its environment is this: When a rat walks around a circuit, the place neurons in its hippocampus fire at the same time. As for the role of the nervous system in adaptation and in niche construction – that is, the alteration of the environment to the animal's benefit – suffice it to recall that even the humble soil worm passes tons of soil through its guts in the course of its existence, thus unwittingly aerating the soil and enhancing its fertility.

There is more: The behavior of a gregarious higher vertebrate is unintelligible if the animal is extricated from its social environment, because much of that behavior consists in interacting with conspecifics. If in doubt, remember how severely impaired the mental life of autistic children is. Yet most of psychology, even nowadays, is asocial: it ignores the individual's social context, which is like studying lungs in a vacuum. Social psychology corrected the social blindness of classical psychology. In particular, Vygotsky (1978), Luria (1976), Humphrey (1983), and Cole (1996) investigated the social root and social function of the higher mental functions, and showed some of the cultural differences in thinking. In recent years, social neuroscience has started to reveal some of the neural mechanisms of the social aspect of human life (e.g., Cacioppo et al. 2006).

Take, for instance, self-consciousness, which we use not only to control our own behavior, but also to understand that of others. In fact, my knowledge of others, and my ability to get along with them, or to attempt to modify their behavior, derive to a large extent from an analogy with myself: I model others after myself, and so I can empathize with them and foresee some of their actions and inactions. When this ability is severely impaired, as in the case of Asperger patients, the results are maladjustment and unhappiness.

In turn, self-consciousness is to some extent a product of social intercourse: I am the more self-conscious, the more intensely I hope or fear that my behavior will be watched and judged by my fellow humans. In pathological cases, such as that of Jean-Jacques Rousseau, self-consciousness degenerates into self-centeredness and paranoia, a combination that results in an oscillation between search for appreciation and flight from society. Thus consciousness, particularly self-consciousness, is both effect and cause of social behavior. (More in Sections 11.4 and 11.5.)

What holds for the relation between self-consciousness and social life also holds for language. Self-consciousness and language are likely to be coeval and to have coevolved alongside sociality. This conjecture is suggested in part by our daily experience with silent (internal) speech. It is likely that the primitive system of animal communication, which eventually evolved into articulate language, allowed hominins and Paleolithic men to internalize certain aspects of their social behavior. As they perfected their system of communication, it became more than a means of social intercourse, namely a tool of self-analysis and a carrier of prefabricated chunks of thought that could be summoned and combined almost at will. Eventually our remote ancestors became able to talk to themselves, that is, to internalize conversations, some of which must have referred to their own mental processes. Thus, self-knowledge and language may have evolved through a feed-forward mechanism, and both coevolved with culture.

Note that this hypothesis is at variance with both the hypotheses that the mind is a byproduct of speech (Luria and Vygotsky), and that language is a byproduct of the mind (Chomsky, and Popper and Eccles.) Note also that all of the preceding contradicts both epiphenomenalism and the so-called mirror-theory of knowledge. Consciousness will be examined in some detail in Section 11.2.

In sum, the social psychologists provided the social context of the mind. But some of them were so radical that they showed the person as a lump of passive putty at the mercy of her environment. This might not have happened if they had taken into account that mental life is brain activity, and that the brain is constantly active, even when in isolation from social stimuli. But classical social psychology was brainless, and occasionally even hostile to biopsychology. In particular, the behaviorists claimed that mental processes either do not exist or are just highly complex behavioral processes. This view contradicts the ordinary definition of behavioral process as overt, hence directly observable bodily changes. Moreover, the view in question involves confusing facts with their environmental conditions. Thus, although respiration is impossible in a void, it is not an atmospheric process. By correcting one imbalance, namely the neglect of social stimuli, the behaviorists reinforced another – the neglect of that which controls behavior, namely the brain. Yet all social relations are refracted by the brain, because this organ controls behavior.

The brain-society connection is so strong that it can lead to curbing instinct. For instance, sexual desire can be intensified or extinguished by social intercourse. Freud's conjecture, that incestuous desire is innate, is the basis of the so-called Oedipus complex, but it was never put to the test by any psychoanalysts. Arthur P. Wolf (1995) tested this conjecture by investigating the history of 14,402 Taiwanese married couples during the 1905–1945 period. He found that at that time there were two types of marriage: the "major" one, where the bride was brought up together with her future husband in his home; and the "minor" marriage, where husband and wife met for the first time the day of their marriage. In the former case the children grew up as psychological siblings and developed a strong aversion to the idea of having sex together, as a result of which their marriage was often disastrous. In short, familiarity breeds horror of incest: the Oedipus complex belongs in the dustbin of pop psychology.

Another item found in the same bin is the notion of collective mind. It is often said that organizations, such as businesses, have intentions or purposes. There has also been plenty of talk of collective memories, and even collective unconscious, but no scientific research on such entities. Strictly speaking, all that is nonsense, since mental states are brain states, and brains reside in individual bodies. What is true is that the members of any organization share some beliefs and goals, and that they may engage in collective actions, such as participating in the manufacture of a good, in a street demonstration, or in the worship of a deity. In all these cases, *We = Every one of us*. Hence Le Bon's crowd psychology, according to which the soul of a mass is irrational, was unscientific. But its successor, the psychology of individual behavior in a group, whether amorphous as in a soccer stadium, or organized as in a religious congregation, can be rigorous.

10.3 From Bonding Hormones to Mirror Neurons to Morals

Ironically, neuroscience has recently provided the most compelling argument for the need to integrate individual psychology with social psychology. This is the discovery of mirror neurons in the early 1990s (Rizzolatti and Craighero 2004). These neurons are activated in one's brain when contemplating someone else's behavior. Briefly put, similar (mirror) neurons control similar behaviors: they are copycats. For example, if a person (or a monkey) sees another grasping a thing, the mirror neurons in her motor cortex will fire, as if she were executing the same movements.

The primary role of the human mirror neuron system seems to be to understand the "meaning" (goal) of the actions of others (Rizzolatti and Sinigaglia 2008, 124). This "understanding" is intuitive or pre-conceptual as well as pre-linguistic: it is the *Verstehen* or empathic comprehension extolled by Wilhelm Dilthey and his followers as the superior alternative to the scientific method. But of course the role of such "understanding" is to facilitate the sharing of experiences, learning, and coexistence, not to replace the objective and analytic study of social life. Experience is personal, but its study is expected to be impersonal or, better, objective.

It has been conjectured that mirror neurons make us cooperative and altruistic from birth. Much the same has been claimed for the so-called bonding hormones, oxytocin and prolactin, central to sex and parenting. Yet, about one-third of us are uncooperative. This suggests that bonding hormones and mirror neurons may be necessary to understand others, but not sufficient to feel concern for others. The latter condition seems to emerge only through positive social experiences. (Negative social experiences, such as those lived by children raised in violent inner cities, tend to breed selfishness, the easiest short-term survival strategy.)

The discovery of mirror neurons is bound to have a profound effect on consciousness studies, too many of which have been more literary ("philosophical") than scientific. Indeed, as Rizzolatti has said, it is hard to conceive of an I without an us. This is particularly true of moral behavior, that is, behavior triggered by moral problems, such as whether or not to help a distraught stranger. A number of conflicting views have been proposed to explain moral behavior: That it is instinctive or learned, emotional or calculated, constant or situational, and so on. Unsurprisingly, until very recently only more or less ingenious arguments were offered to prop up these various ethical hypotheses, for it was taken for granted that philosophy, even practical philosophy, cannot be subjected to experimental tests.

As so often happens in other fields, ethical innovation is often due to outsiders: to people who, not being chained to tradition, feel free to challenge it. Indeed, in recent years some psychologists, neuroscientists and economists have been designing and performing experiments on moral behavior, thus continuing the pioneer work of Jean Piaget and some of his pupils. This work has even attracted the attention of a Harvard philosopher (Appiah 2008).

The recent experimental study of fairness is a case in point. Every normal person, in every society, is often torn between self-interest and concern for others: she does her best to balance these two rather natural conations, so as to be fair within bounds. And yet most religions, moral philosophies and economic theories claim that one ought to be either an egoist or an altruist. In particular, standard economic theory asserts that smart people are selfish: that economic "rationality" consists in attempting to maximize one's expected utilities regardless of other people's interests. But this dogma has been subjected to severe criticisms over the past three decades. First, because of the conceptual imprecision of the concept of expected utility (e.g., Blatt 1983; Bunge 1996). Second, because the assumption in question has been experimentally confuted by the so-called behavioral economists (e.g., Gintis et al. 2005; Henrich et al. 2004; Thaler 1992).

These studies have shown that the vast majority of people are strong reciprocators: they not only repay favors but occasionally go out of their way to help total strangers, and run risks in punishing selfish behavior. Obviously, as we grow up we internalize some of the learned rules of social behavior, to the point that in ordinary cases we solve moral problems in an automatic fashion.

At first sight some types of behavior seem impregnable to physiological explanation. For example, although alcohol is a depressant, it can be exhilarating when consumed in moderation in merry company: the social stimuli would seem to obliterate the physiological response. But physiological psychology can account for the effect of social stimuli and their interference with nonsocial ones, whereas classical social psychology can at most describe them.

Indeed, consider two brain systems, *N* and *S*, sensitive to natural and social stimuli respectively, and both innervating a third brain system *C*, in control of either mood or overt behavior of some kind. Furthermore, assume that the activity of this third system is largely determined by the activities of *N* and *S*, and call f_N, f_S and f_C the state vectors of the respective neural systems. Finally, assume that the input f_N into system *C* is weighted by a factor w_N, whereas the input f_S into *C* is weighted by a number w_S. That is, assume that $f_C = w_N f_N + w_S f_S$. Clearly, the net outcome f_C will depend on the relative values of the inputs as well as on the corresponding weights. For example, if $w_N = 0.5$ and $w_N = 0.2$, while the stimuli are $f_N = -10$ and $f_S = 20$, the response will be $f_C = -1$ (inhibition). But if f_S goes from 20 to 30, the outcome of the joint action of the natural and the social stimuli will be $f_C = +1$, that is, an excitation. Admittedly, the preceding sketch has to be fleshed out, or rather brained in, but it sounds like a plausible research project in social neuroscience.

10.4 Evolution: Preliminaries

The immaterial soul could not possibly have evolved along with the body: Thus decreed in 1996 pope John Paul II after having admitted that our bodies had been a result of a long-drawn (though not purely natural) evolution. If, by contrast, mental processes are brain processes, then both the brain and its functions, in particular the mental ones, must have evolved together. Thus hypothesized Darwin in *The Descent of Man* (1871).

The evolution of mentality is the object of study of five budding disciplines: comparative psychology, evolutionary and comparative neuroscience, evolutionary psychology, cognitive archaeology, and evolutionary epistemology. Let us peak at them.

Comparative psychologists try to guess at which stage in evolution the various mental abilities emerged, by identifying them in contemporary animal genera. The underlying guiding hypothesis is that whatever traits are found in all the species included in a modern genus are likely to be ancestral. Two examples must suffice here: the origins of pleasure and of language. Michel Cabanac (1999) conjectured that reptiles were the earliest animals to feel pleasure because contemporary lizards, unlike fish and amphibians, like to be stroked. Hence pleasure is likely to have emerged with reptiles about 400 million years ago.

Likewise, Andrew Bass and colleagues (2008) conjectured that language evolved from the calls that some fish have been using for 400 million years for attracting mates and defending territory. The basis for this hypothesis is the anatomical finding that all vertebrates seem to share a special compartment located in the hindbrain and spinal cord, which "subserves" (effects) social vocalization. So, contrary to intuition, bullfrogs, birds, and Luciano Pavarotti share some important behavioral traits, because they share the vertebrate "body plan," which, in turn, is rooted in a group of common genes.

To be sure, the above hypotheses are speculative. But at least they are philosophically sound, if only because they fit in with evolutionary biology, they elicit experiment, do not detach function from organ, and do not attribute computational abilities to the innumerate. However, biology is necessary but insufficient to account for the emergence of language, because this is symbolic, and symbols, being conventional, are anything but natural: they are social constructions. Hence, any plausible account of the emergence of language will involve social archaeology in addition to evolutionary neuroscience. See Fig. 10.1.

Evolutionary neuroscientists attempt to guess the evolution of the primate brain. For example, its uppermost layer is called the *neocortex* because there is good

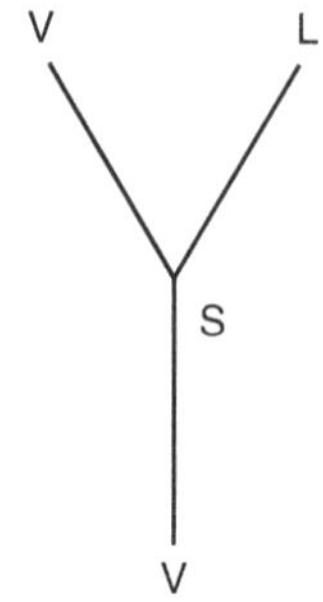

Fig. 10.1 The evolution from vertebrate vocalization *V* to language *L*. The *S* at the fork of the tree denotes the emergence of symbolism

reason to assume that it is the most recent, so that it performs the most sophisticated functions. However, the popular view that the deeper is also the older is not generally true. For instance, we have two visual systems: the ventral and the dorsal, philogenetically newer, which perform somewhat different tasks (Goodale and Milner 2005). Another example is this: Although moral decisions are performed by the frontal lobes, they seem to be motivated by phylogenetically older affective systems, as suggested by the facial expression of disgust, which is the same as that of distaste (Chapman et al. 2009).

Unlike geological strata, the brain subsystems did not just pile up over time: the emergence of every new mental organ is likely to have been accompanied by a reorganization of the entire brain. However, on the whole it is true that going back and down in the brain amounts to receding in evolutionary time. For example, the more abstract and general an idea, the more likely it is to be had in a higher, and therefore newer, cortical layer. Which leads us to evolutionary psychology.

Evolutionary psychology is still embryonic in its scientific phase, and senile in its speculative phase, the one popularized by Barkow, Cosmides, and Tooby (1992). We shall deal with the latter in Section 13.5, as a case of premature wild speculation. As for scientific evolutionary psychology, we shall only recall that it has recently merged with developmental psychology (see Bjorklund and Pellegrini 2002). The reason for this fusion is the same as that for the emergence of evo-devo in general, namely, that evolutionary novelties emerge in the course of individual development. For example, embryology can answer evolutionary questions such as "How did the turtle get its shell?"

Moreover, to generalize something Tulving (2002, 5) said about episodic memory, one may conjecture that whatever brain organ or mental capacity is recently evolved, is likely to be late-developing and early-deteriorating. Examples: intuitive estimates of time, episodic memory (what-where-when?), and moral evaluation. Hence the need to look at phylogeny from an ontogenetic viewpoint, and at ontogeny from a phylogenetic viewpoint.

Indeed, contrary to the older preformationist view, according to which development is rigidly determined by the genetic program in a constant environment, it has been known since the days of Conrad Waddington and Jean Piaget that development is determined by both genome and environment – hence epigenetic rather than genetic. For example, two "identical" twins, that is, individuals with the same genome, may adopt somewhat different lifestyles, as a consequence of which they have acquired different epigenomes, which they may pass on to their offspring. (The difference is that some of the genes have been "silenced" by being coated with foreign molecules.) Consequently, while we cannot be held responsible for our genomes, we are responsible for our epigenomes, as well as for those of our children.

For example, childhood neglect and abuse alter stress responses and increase the risk of suicide; the underlying molecular mechanism is the coating (hence turning off) of a gene involved in glucocorticoid regulation (McGowan et al. 2009). Even the best genetic endowment is powerless in a severely deprived and stressful social environment, which slows down and distorts the development of the child's prefrontal cortex to an extent comparable to an anatomic insult of the same brain region (Kishiyama et al. 2009).

In sum, development is nowadays regarded as *epigenetic* as well as genetic: it is a multi-level and bidirectional process, both bottom-up and top-down: see Fig. 10.2

This is why, far from focusing on the individual organism developing under the tyranny of its genome, conceived of as the unmoved mover insensitive to the environment, nowadays evolutionary developmental neuroscientists and psychologists study the entire developmental system, from gene to cell to whole organism to environment (Gottlieb 1992; Lickleiter and Honeycutt 2003).

I submit that developmental evolutionary psychology has superseded the evolutionary epistemology project sketched in the 1970s by the ethologist Konrad Lorenz (1971), the psychologist Donald Campbell (1974a), and the philosophers Karl Popper (1978) and Gerhard Vollmer (1986). Their correct central idea was that cognition is an adaptation mechanism subject to natural selection. Other than this commonality, the projects of the scholars in question were quite different, as Vollmer (1987) has emphasized. Whereas Popper and Campbell were interested in the evolution (or rather history) of knowledge, Lorenz and Vollmer tackled the problem of the evolution of our cognitive abilities. In particular, they attempted to explain why so many false theories, such as geocentric astronomy, looked attractive in the past – namely because they fit appearances or, as Vollmer puts it, they were quite adequate for the mesocosm. And this brand of evolutionary epistemology suggested that Kant's innate categories and laws of thought are actually products of evolution. But neuroscience and developmental psychology show that we are born mindless (Bunge 1983a, 51–59).

In my opinion the evolutionary psychology project did not come to fruition for the following reasons. First, although it invokes neuroscience, it makes no use of

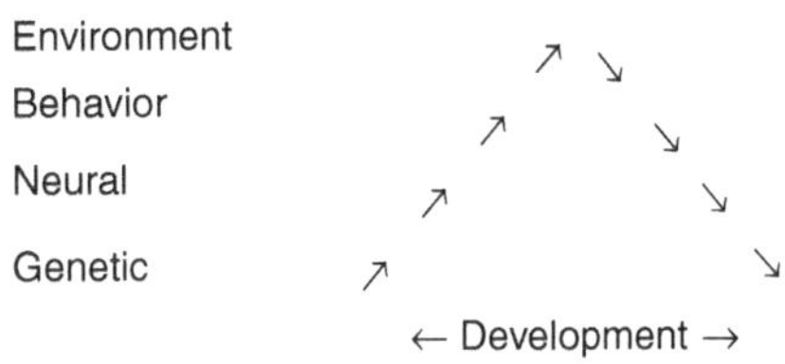

Fig. 10.2 The multi-level process of development. Oversimplified from Gottlieb (1992, 186)

it. Second, and as a consequence, it overlooks emotion despite the intense two-way cortico-limbic traffic that explains not only that we can control emotion, but also that inquiry can be pursued passionately. Third, in the intervening years we have learned that phylogeny cannot be well understood when detached from ontogeny and conversely – hence the need for evolutionary developmental psychology. Fourth, it is likely that the cognitive abilities have coevolved with the manual, artistic, and social abilities – an assumption that cognitive archaeologists take for granted, and which deprives evolutionary epistemology of its independence. Fifth, the evolutionary epistemology project overlooks the social (or cultural) aspect of human evolution. Sixth, evolutionary epistemology is still at the project stage after more than three decades.

Evolutionary epistemology has not explained, for instance, the sudden emergence of cave painting and its lack of subsequent progress; why the ancient Egyptians were far more worried by the afterlife than any other early civilization; why the Chinese have never worshipped any gods, whereas the Indians imagined tens of thousands; why the ancient Peruvians, who were so accomplished in many fields, had no writing; or why the ancient Greeks did not borrow the zero from the Sumerians.

However, evolutionary epistemology has taught us a few important lessons. One of them is that classical epistemology is deficient because it ignores the knowing subject – as Popper (1972) demanded – and consequently overlooks knowledge in the making. Another lesson is that evolutionary novelties allow our children to develop their learning apparatus in a different way from our remote ancestors. While it is true that, contrary to what Kant thought, newborns are mindless and have to start from scratch – since all knowledge is learned, ours have the advantage of being new evolutionary arrivals. For example, their prefrontal cortices have at least two abilities that our hominim ancestor may have lacked. One is that of learning to control emotional impulses. Another advantage is that of being capable of learning how to learn. But all of this and much more occurs in some social context or other. A child reared in a very poor, backward, or violent society won't necessarily have any advantages over her hominim ancestors. So, once again, the naturalist approach must be supplemented with some social research.

In short, if the mind is conceived as a set of brain functions, then it ought to be seen in an evolutionary perspective. But since the human mind develops in a social milieu, human evolution is social as well as biological. This is the subject of the next section.

10.5 Evolution: Biocultural

The reason for dealing at the same time with biological and cultural evolution is, of course, that man is not an animal like others, but the animal that invented culture in the anthropological sense of the word, that is, as the system of economic, political, and symbolic activities that go beyond foraging, mating, breeding, and fighting. Hence genes and culture have been the two sides of the same coin from the moment "modern" humans emerged in Africa around 100,000 years ago (Smail 2008).

As Richerson and Boyd (2005, 194) put it, "[g]enes, by themselves, can't readily adapt to rapidly changing environments. Cultural variants, by themselves, can't do anything without brains and bodies. Genes and culture are tightly coupled but subject to evolutionary forces that tug behavior in different directions." This is why fashionable Evolutionary Psychology, which regards culture as wholly shaped by genes, and genes as impervious to environmental change, is wrong-headed in addition to being an exercise in uncontrolled imagination: see Section 13.4.

A few examples will show the need to fuse the biological and cultural perspectives. (1) The Neolithic Revolution made the emergence of towns possible and, with them, the arrival of new social norms, which in turn changed the rules of sexual selection; for example, wealth and political power conferred a greater Darwinian fitness than physical strength and generosity. (2) Mass migrations transported ideas and habits along with genes (Cavalli-Sforza and Feldman 1981). (3) Crowding in towns facilitated the spread of contagious diseases, and caused inheritable changes in the immune system to the point that they may be the origin of the major genetic differences between ourselves and our ancestors of 10 millennia ago (Keyfitz 1984). (4) In-group solidarity favored the survival of ethnic and cultural groups subject to discrimination. (5) The dairy industry, born only about 5,000 years ago, cannot thrive in places where most people lack the gene involved in the synthesis of lactase, the enzyme that breaks down lactose, the milk sugar. And wherever milk is drunk, a healthy lifestyle is favored, which in turn facilitates the spread of the lactase-related gene. (6) The Industrial Revolution favored brain over brawn, and thus gave the weak but smart a better chance of spreading their genes. (7) The societies that encourage learning support schools and thus favor the spread of the (unknown) genes that facilitate learning. (8) Literacy causes anatomical changes in the brain: it increases white matter in the corpus callosum, and grey matter in the gyri (Carreiras et al. 2009). (9) The increasing importance of intelligence in work and communication, along with progress in informatics and telecommunications, have led to the growth of sedentariness and thus to an increase in the incidence of a number of diseases that are bound to severely distort human demographics and diversity. (10) Current migratory currents are faster and more intense than ever, and they cover the entire planet, favoring miscegenation and two mutually complementary political attitudes: toleration and intolerance of the other, which in turn influence migratory policies. All of the above suggests that human evolution has accelerated since the invention of agriculture about 10,000 years ago (Cochran and Harpending 2009). Incidentally, this view contradicts the dogma of pop Evolutionary Psychology, that human nature has remained unaltered for the past 100 millennia or so.

Let us finally peek at an utterly different but complementary field: cognitive archaeology. The workers in this discipline tackle a formidable inverse problem: that of "inferring" (guessing) the ideas that may have guided the manufacture of the artifacts found in archaeological sites (see Mithen 1996; Renfrew and Zubrov 1994; Trigger 2003a). In principle, it is possible to test the hypothesis that a given ancient artifact was used to accomplish a certain function, and that its manufacture must have involved certain pieces of knowledge. This is done by replicating and using the artifact in question. Experimental archaeology, which does just this, was born more than a century ago.

However, the "conclusions" (conjectures) of archaeological research are conjectural because much of the evidence for them is ambiguous. This stone tool could have been used either for killing or for digging; that vase, for drinking or for offering sacrifice; this building, for worship or for showing off power. The North American mega-fauna became extinct about 11,000 years ago, when the first humans got there. But that was also the time when the last Ice Age ended. So, those animals may have been the victims of climate change, in particular huge floods, rather than of the puny arrowheads of the newcomers. In Australia too the mega-fauna became extinct at the time of the arrival of the first people, about 45,000 years ago. However, there is no independent evidence that those humans were big-game hunters. Rather, remains of charred vegetation suggest that the new arrivals in Australia set prairies on fire, whether deliberately or not, thus depriving the big animals from food. To be sure, further research may resolve these ambiguities. But archaeologists know that all reconstructions of the past are tentative: they do not share the cocksureness of the evolutionary psychologists.

The upshot is that we know very little about the remote past of the mind, or even about its present. And this little we owe to neuroscientists, experimental psychologists, comparative psychologists, and archaeologists, not to evolutionary psychologists. Table 10.1 summarizes and brings together what has been learned about the past of human cognition and culture.

Table 10.1 Stages in the evolution of human cognition and culture (abbreviated from Donald 2001, 260)

Stage	Species/period	Novel forms	Manifest change	Governance
MIMETIC	Early hominids, peaking in *H. erectus* 2 M-0.4 Mya	Action metaphor	Skill, gesture, mime and imitation	Mimetic styles and archetype
MYTHIC	Sapient humans, peaking in *H. sapiens sapiens* 0.5 Mya-present	Language, symbolic representation	Oral traditions, mimetic ritual, narrative thought	Mythic framework of governance
THEORETIC	Modern culture	External symbolic universe	Formalisms, large-scale theoretic artifacts, massive external storage	Institutionalized paradigmatic thought and invention

10.6 What Makes Us Human

The so-called great religions were the first totalitarian ideologies, since they were intent on controlling all aspects of human life. They all erected an impassable wall between humans and other animals. By contrast, Aristotle declared that we are political animals – which allowed people to emphasize either our animality or our political nature. Nearly two millennia later, Descartes took a step backwards when holding that nonhuman animals are automata: that only humans have souls. At the time of Darwin, the famous philologist Max Müller reinforced the Cartesian wall: He claimed that language is the prerogative of humans, and denied that it might have evolved from a protolanguage.

Three generations later, Noam Chomsky repeated this extraordinary contention, and stated that to speak of linguistic evolution is as absurd as speaking of molecular evolution – which did take place and is the subject of a specialized journal. And he added that all humans are born with a "universal grammar," a sort of template for all particular languages, and that "language is the mirror of the mind" rather than having evolved mainly as the most sophisticated means of communication.

On top of dismissing evolutionary biology and sociolinguistics, and of overlooking developmental psychology, Chomsky and his followers adopted psychoneural dualism, and consequently they overlooked primatology and neuroscience – to which they started to pay lip service only in recent years. Wittgenstein's followers chimed in: Max Black, Stuart Hampshire and Norman Malcom declared that it would be senseless to attribute to an animal any concepts at all. They knew, because they had read Descartes, Wittgenstein, and Chomsky.

By contrast, Darwin (1911) had famously suggested that the human-nonhuman difference is "one of degree and not of kind," and fathered animal psychology. During the twentieth century primatologists and neuroscientists kept investigating the similarities as well as the differences among the various primates (see, e.g., Preuss 2007). In particular, Frans de Waal (1998) corrected Aristotle when he found that chimpanzees, our next of kin, are political animals, and Machiavellian ones to boot.

The greatest shock to the anti-evolutionist linguists should have come when some primatologists taught some chimpanzees to use sign language and graphic symbols to interact and express themselves. Thus, the chimpanzee Lana, using a keyboard, produced a number of sentences, such as "Please machine make window open," and "? Beverley move behind room" (Rumbaugh and Gill 1976). Any parent of a two-year old child, and any recent immigrant, would be proud of a similar accomplishment. Not so the anti-evolutionists, who would object that this was no evidence of language. They would claim that what is typical of our language is recursion, or the embedding of sentences into others sentences, as in "Dick knows that Dubya ignores what makes Condi tick." This reaction reminds me of a remark by the mathematician Julio Rey-Pastor: "If a cell were ever synthesized in a laboratory, the vitalists would exclaim: 'Yes, but what about a giraffe?' "

In short, there are three main traditions concerning the problem in hand: spiritualism, naturalism, and sociologism. And we seem to have reached an impasse in the debate over these and other doctrines (see, e.g., Penn et al. 2008). This impasse has two main sources: one is the rigidity or dogmatism of some of the champions of these competing traditions; the other is that, as a matter of fact, there are continuities as well as discontinuities in the evolution of any phylum. For example, many mammals and birds model their surroundings, make hypotheses and try them out, and communicate among themselves. But humans alone possess the additional abilities of reflecting on their own mental processes; of attributing certain events to invented unobservable entities; of transcending the here-and-now; of adopting other people's viewpoints; of designing long-term plans and acting accordingly; of helping others without expecting repayment; of delivering altruistic punishment; of regularly sharing food; of organizing and reforming social systems; and of inventing, breaking, or repealing norms of social behavior. (See further features of the ape-human gap in Adolphs 2009; Gazzaniga 2008; Lorenz 1971; Passingham 2008.)

In addition to disagreeing on what makes us unique, scientists disagree on the source of our peculiarities: is it biological, psychological, or social? In other words, did some of our remote ancestors become human because of a mutation, the emergence of a new mental ability, a radical change in modus vivendi and the attending social organization, or all of the preceding? It is likely that, since humanness is a whole bundle of inter-related characteristics, its origin will eventually be explained by the concerted efforts of students in many disciplines, from genetics and neuroscience to psychology, linguistics, anthropology, archaeology and sociology (Enfield and Levinson 2006).

In particular, when accounting for the emergence of Homo sapiens we should not overlook labor, the factor that Engels (1962) singled out in his 1876 essay on "The part played by labor in the transition from ape to man". This is because labor, much more than mere foraging, requires planning, organization, and instruction. (True, spiders, bees, beavers and bowerbirds make things, but not according to plan and rule: their products are genefacts rather than artifacts.)

Yet, most speculations about the origin of language and other human peculiarities overlook work, perhaps because they take it for granted that hominins were basically sex-crazed gatherers or hunters, more interested in free-loading, cheating, and cheater-detection, than in enjoying life and cooperating to get things done and meet emergencies. (The influence of the ontological and methodological individualism inherent in standard economic theory and the "economic imperialism" it generated is apparent in the hominization literature.)

All animals eat and have sex; many species are gregarious and forge alliances for defense or hunting; several share their food when begged, practice reciprocal altruism, and communicate through calls and signals. Hence humanization cannot have been a matter of subsistence, coexistence, reproduction, and communication alone. Only deliberation, work, and the concomitant social organization, such as cooperation and sexual division of work, tool-making, communication through articulate (syntactical) language, and tutoring of the young, are uniquely human. So are episodic memory, planning, and deliberate social organization. (Note the shift from a single peculiarity, such as tool making or language, to a whole cluster or system of interacting abilities.)

However, all of this called for, and in turn favored, a large increase in brain mass, particularly of the frontal lobes, as well as improved connections among its parts, such as its prefrontal and temporal regions (Calvin and Bickerton 2000; Passingham 2008; Preuss 2007). And of course better brains made it possible to tackle harder problems, think deeper, make better tools, and organize more complex social systems. The improvement in mental operations along with evolution is strongly suggested by comparative neuroscience, which has found that, as the brain size grows, the quantity of white matter fibers – which interconnect different cortical zones – grows much faster than that of grey matter (Allman 1999). That is, the hominization process was not just one of improving adaptation and enrichment of sociality but also, perhaps mainly, one of enrichment of the self. Note, incidentally, that our inclusive materialism shares the idealist view of the all-importance of mind.

Last, but not least, improved brains facilitated cooperation and the concomitant egalitarianism – a moral characteristic of gatherer-hunter communities. Mithen (1999) speculates that better brains facilitated cooperation, in particular coordination, as well as the sharing of information and more effective "mind-reading," which exposed minds to public scrutiny and thus discouraged cheating and plotting. See Table 10.2.

Caution: The inclusion of some items in the list below is currently the object of spirited controversy, both inside and out of the scientific community. For example, episodic memory was regarded as characteristically human until scrub jays were shown to have it; jumping spiders can plan an attack stalking prey; whales too have mirror neurons; chimpanzees are just as adept at politicking as university professors; and the ability to delay gratification has long ceased to be the prerogative of Calvinist traders.

It has been noted that harsh competition for access to resources favors impulsiveness, whereas foraging alone affords the leisure of delayed gratification – and a subordinate status in a rigid social hierarchy imposes self-control (e.g., Genty and Roeder 2006). Behavior of both types has been observed in several species. By

Table 10.2 Some human peculiarities – until further notice

Biological	Psychological	Social
Immaturity at birth	Strong mother-child bond	Children's privileges
Large insula	Disgust	Continence
Large prefrontal cortex	Self-control and planning	Rule-guided action
Lifelong development	Lifelong learning	Teaching
Long lifespan	Foresight	Large social groups
Maximal adaptability	Encyclopaedism	Versatility
Nutrition-dense diet	Sustained physical effort	Labor
Physical weakness	Technical imagination	Reliance on artifacts
Slow maturation	Insecurity in early life	Dependence
Spontaneity	Fantasy	Social plasticity

contrast, only humans point, a gesture that seems to indicate shared intention and attention (Tomasello 2006).

In tackling the problem in hand we must take into account specialist bias. Thus, primatologists tend to emphasize the commonalities of humans and other primates, which dog-lovers tend to extend to their pets; by contrast, linguists, dualist philosophers of mind, and theologians emphasize differences. Evolutionary biologists are the best experts in the matter, yet they are unlikely to claim that their findings and conjectures are final.

At all events, the kind of evolution that brought us to the present stage was biopsychosocial rather than purely biological, psychological, or social. Moreover, the mental aspect of human evolution is affective as well as cognitive. Emphasis on the cognitive aspect of evolution, to the detriment of the emotional and the social aspects, is arbitrary. The understanding of our origins calls for the convergence or merger of all the sciences that study humans.

Closing Remarks

The modern conception of the mind differs in several respects from the traditional concept of the soul. First, minds are material, in the same derivative sense that movements, chemical reactions and social innovations are material, namely because they are changes in concrete things. Second, far from being constant, minds are variable both ontogenetically and phylogenetically. Third, minds develop differently in different social contexts, and are means for social coexistence. All three characteristics suggest that traditional psychology, which ignored societies as well as brains, was badly flawed. We need neuroscientific, developmental, social, and evolutionary psychologies. But we need them to be scientific rather than speculative, and inter-related rather than isolated from one another (see, e.g., Cacioppo et al. 2006; Corballis and Lea 1999).

For example, when studying the development of morality in children, we must remember that the control organ – the prefrontal cortex – being the most recent part of the brain, is also the last to attain full maturity; and we also need to know that morals are culture-bound, and as such subject to historical change, whence moral norms are learned and observed differently in different social groups and in different historical periods. Thus when admiring the bravery and loyalty of the samurai, we must remember that this mercenary killer sells those virtues to a barbarous and cruel feudal lord. And when wondering how to deal with a young offender, instead of remaining satisfied with the verdict that he is simply a bad boy, we must take everything into account: his brain, which may be abnormally immature; his family, which may be dysfunctional or non-existent; his school, which may be the usual authoritarian and boring drudgery; and his neighborhood, which may be infested with drug pushers as well as lacking in sports and cultural centers.

Finally, note that the I-thou-it problem evokes the individualism-holism-systemism trilemma (Bunge 1996, 1998, 2003a). All three perspectives are in evidence in every branch of philosophy although, as usual, ontology is at the root. Ontological individualism holds that reality is a collection of individuals; holism, that it is an indivisible whole, or that the whole is ontologically and epistemologically prior to its parts; and systemism, that reality is a system analyzable into composition, environment, structure, and mechanism.

Psychology shares the individualism-holism-systemism trilemma: the Gestalt school was holistic, whereas behaviorism and its heir, computationalism, is individualist. The previous sections suggest that systemism offers a more suitable framework for the scientific study of mind, since it leads to focusing on the individual-in-her-environment rather than on either the isolated individual or society as a whole. Thus systemism makes room for social psychology as well as for cognitive, affective and social neuroscience, along with both developmental and evolutionary psychology. This strategy has important practical spin-offs, particularly in the design of policies for the control of criminality, alcoholism, drug-addiction, truancy, teen-pregnancy, and welfare-dependency. Indeed, all these cases require focusing on the individual-in-his-milieu rather than on either the isolated individual (the traditional medical and legal approach) or society as a whole (the revolutionary and anti-psychiatry approaches).

Let us now leave society for a while and concentrate on some of the more conventional issues in the philosophy of mind, such as consciousness, free will, and the self.

Chapter 11
Cognition, Consciousness, and Free Will

Cognition is the acquisition of knowledge, or knowledge in the making. This sounds obvious as long as we do not ask what cognition and knowledge are, for we should admit that we don't know much about either. But we are getting to know something about both, particularly since the recent reorientation of the disciplines concerned with them. Indeed, neuroscientists and psychologists have learned that the study of cognition is that of certain brain processes in a social context; they also know that cognition and emotion, though distinguishable, are not separable. Furthermore, contrary to classical epistemology, which focused on the contemporary adult knowing subject, at present the study of cognition includes both its ontogeny and its phylogeny. Thus, what used to be separate disciplines have been converging.

Contrary to the vulgar materialists – whether they be behaviorists, eliminative materialists, or computationalists – we take consciousness and free will for granted, though of course not yet as fully understood. It would have been far easier to declare them nonexistent, or to claim that they are totally unknown, and prophesy that they will never be understood. But the former strategy is escapist, the latter is defeatist, and both are patently wrong. Indeed, whoever denies consciousness cannot have felt its loss when falling asleep or undergoing anesthesia, have never have felt pain or regret, and never questioned their motives for doing something. And whoever denies free will has never taken the initiative or disobeyed orders. Either denial is a case of willful agnosia resulting from an obsolete and paralyzing worldview.

11.1 Cognition and Knowledge

All knowledge is knowledge of something: there is no knowledge in itself. One may know facts and propositions. Consequently ontology should precede epistemology. And yet modern philosophy started rejecting metaphysics. It did so just because the ruling metaphysics around 1600 was obsolete. The price paid for this antimetaphysical turn was subjectivism, outspoken as in Berkeley's case, or shame-faced as in Kant's. Luckily, scientists paid no attention to it until the emergence of quantum mechanics. (Recall Chapter 3.)

We take knowledge of something to be a mental state, and moreover one irreducible to belief. This is because to believe or disbelieve X we must know X to begin

M. Bunge, *Matter and Mind*, Boston Studies in the Philosophy of Science 287,
DOI 10.1007/978-90-481-9225-0_11,

with. Besides, anyone knows many things in which he does not believe; and mystics believe many things that they claim they feel intuitively but do not know. Hence we reject the popular definition of knowledge as justified belief. Nor is truth an ingredient of knowledge, for anyone knows any number of falsities and half-truths.

Finally, we also distinguish knowledge from information, because some pieces of information, such as questions, orders, and absurdities do not constitute knowledge. And also because computers process information but, since they lack minds, they cannot be said to know anything.

Cognitive neuroscience and the associated philosophy of mind regard cognition, or the acquisition of knowledge, as a brain process. Indeed, only the brain can perceive, conceive, plan, evaluate, and know itself – the main kinds of cognition. The preceding sentence is no mere uncertain conjecture: it is a robust result of cognitive neuroscience, as well as of countless experiences and experiments in such diverse fields as cognitive neuroscience, neurosurgery, psychopharmacology, narcotics trade, and politics. Thus, when manufacturing cognitive-enhancing drugs such as amphetamines, the pharmaceutical industry tacitly endorses the principle that cognition is a brain function. Likewise Mussolini: When ordering the jailing of Antonio Gramsci, reportedly he said: "We must keep his brain from functioning for twenty years."

To put it in negative terms: There is no knowledge without a knowing subject, whether *à la* Plato, Hegel, Bolzano, or Popper. But of course one may *pretend* that there is such thing when interested in impersonal matters, such as relevance, reference, sense, and truth. For instance, one can do mathematics without caring who invented, discovered or used the theorems one is studying or using. It is only when checking the truth-value of a new proposition purporting to describe or explain a fact, that one may legitimately ask how it was found or checked it. For instance, a drug company may invest millions of dollars in finding out whether or not a new molecule has the psychotropic effects that its designers conjectured. By contrast, no one will spend a cent in checking Kant's extravagant thesis, that space and time are subjective, for the very planning of even the humblest exploration of the real world presupposes their reality and that of its spatiotemporal features.

Much the same holds for the learning tools, from the humble magnifying lens to the most powerful telescope, and from the pencil to the mainframe: they enjoy no autonomy, but work only as adjuncts to someone's brain. Only live brains can start themselves, ask questions, design research projects, set up automated experiments or calculations, and evaluate results. If all the knowing subjects were removed from laboratories, observatories, libraries, and learned networks, only isolated individuals with rapidly declining cognitive stores would remain. And not even that would remain if humankind were wiped out, because disused learning tools are like fossils. Thus, Popper's (1967) famous thought experiment, wherein knowledge would remain buried in books even after a universal nuclear holocaust, is nonsense: unreadable books are no longer books, just as the fossil remains of dinosaurs are no longer dinosaurs. However, let us move from knowledge back to cognition.

It may be remembered from Section 9.2, that the brain is plastic, that is, some of the connections among the neurons in the brain change in the course of time. We now postulate that cognition is an activity of plastic (or modifiable, uncommitted, or self-organizing) neuronal systems – or psychons, as we may call them. Our second assumption is that some animals have psychons. (Actually, as Eric Kandel has shown, even such primitive animals as sea slugs have plastic neurons, but they are mindless, for they can only learn very primitive motor tasks.) Our third postulate is that all the psychons of an animal are coupled to one another forming supersystems, such as the amygdala and the columns and minicolumns in the cerebral cortex.

Our fourth postulate is that every animal endowed with psychons is capable of acquiring new functions – that is, learning – in the course of its life. (In humans the processes of dendritic sprouting and formation of new synaptic connection stops only at the onset of senility. So, barring senility, we can keep learning new tasks throughout life; and it is well known that learning is self-reinforcing.) Finally, a convention: We call *learned*, as opposed to instinctive or inborn, any neural function involving a psychon that has acquired a regular connectivity, that is, one that is either constant or varies regularly rather than at random. A new connectivity may be formed by chance for the first time – for instance, by the joint firing of two initially independent neurons, as Hebb conjectured. If a connection consolidates, that is, if there is recurrence or recall, whether spontaneously or in response to an environmental demand, we may regard it as established or learned.

11.2 Hebb's Hypothesis

Donald Hebb (1949) revolutionized cognitive psychology when he conjectured that learning an item consists in the formation of a new "cell assembly" or neuronal system. Obviously, for this to happen the cells in question must be uncommitted to begin with; that is, they must not belong to a rigidly wired circuit or network. In other words, their interconnections must be plastic: their synaptic weights must be able to vary in the course of time, either spontaneously or under stimulation. This is the so-called *use-disuse* hypothesis, first proposed by Tanzi, embraced by Cajal, and refined and exploited by Hebb (Cooper 2005). See Fig. 11.1.

The neuroscientific account of learning is roughly this: Learning ability equals plasticity, and learning equals the emergence of new neuron assemblies. Since what learns is a neuronal system, a more accurate reformulation of this hypothesis would be: The learning of a neuronal system is proportional to its plasticity. Assuming

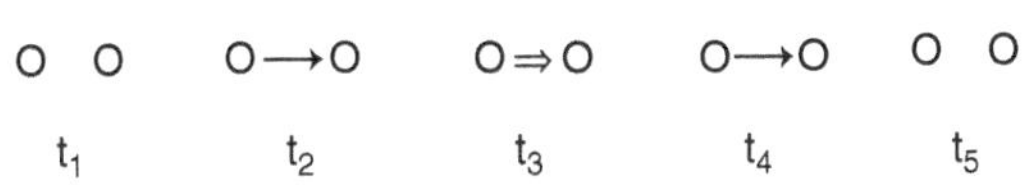

Fig. 11.1 Formation and dismantling of neuronal assemblies. At time t_1 the two neurons are disconnected from one another. At time t_2 one of them synapses onto the other: a system has emerged, that may "do things" (undergo processes) that neither of the single neurons can. By t_3 the connection has been strengthened by use. At t_4 the coupling has weakened by disuse. At t_5 the system has dismantled through prolonged disuse or degeneration, such as excessive dendrites pruning (From Bunge 1985, 25)

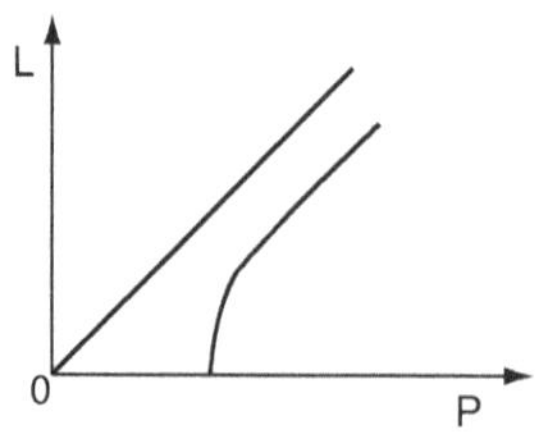

Fig. 11.2 The learning-plasticity relation. Only the first quadrant is shown because the others are psychologically meaningless: neither plasticity nor learning can be negative. In science, math is maid, not mistress

that the two variables have been adequately quantitated, we may abbreviate that hypothesis as $L = kP$, which can be graphed as a straight line in the (L,P) state space of the system in question: see Fig. 11.2. But since in learning rate is just as important as total knowledge acquired, we may consider this more sophisticated conjecture: Learning times learning rate of change is proportional to plasticity times plasticity rate of change, i.e., $L.dL/dt = a\ P.dP/dt$, which entails $L^2 = aP^2 + b$. The graph of this function in the state space (L,P) is a hyperbola. Thus the first graph (linear function) may be regarded as the asymptote of the second. Warning: The preceding is just an exercise in speculative cognitive neuroscience, intended to suggest how to quantitate the merger of the two founding disciplines, and how to build state spaces. Neither of the two hypotheses has so far been subjected to experimental test.

So far we have only considered excitation. The number of possibilities increases dramatically if we add inhibition, which, as Pavlov and von Békesy discovered, is just as basic and pervasive as excitation (Békésy 1967). Behaviorists took no notice of this discovery: they stubbornly clung to their stimulus-response model. And brain-imaging enthusiasts sometimes forget that increased electrical or metabolic activity of a region is an ambiguous indicator: it may indicate either a mental process or the inhibition of another brain "area."

Finally, a few words about the dual or complement of knowledge, namely ignorance. We may distinguish two kinds of it: involuntary and willful. We routinely discard problems and results that do not excite our curiosity. At other times we reluctantly discard or postpone certain interesting problems for lack of time or means, but hope to tackle them at some other time, or else leave them to future generations of investigators. These are cases of the *docta ignorantia* that Nicholas of Cusa discussed in the Renaissance. Finally, there is what may be called *docta stultitia*, the dogmatic claim that certain problems will never be solved. (This case must be distinguished from the unsolvable problems in mathematics: these can be proved to have no solution.)

There are many famous and influential cases of willful ignorance. Suffice it to recall Kant's claim that noumena (things-in-themselves) are unknowable; Comte's decree that the innards of stars will never be known; Emile Dubois-Reymond's dogma that the mind-body problem is insoluble; Noam Chomsky's thesis, that the origin and evolution of language lie beyond the reach of science, just because language cannot possibly have evolved; and the claim of several contemporary philosophers, that consciousness will forever remain mysterious.

We know what happened to these pessimistic prophecies. Physics, chemistry and biology only study things in themselves; only psychology studies things for us, or phenomena, such as seeing red or feeling friendliness. Astrophysics has successfully been investigating the interior of stars for close to a century. Neurolinguistics, primatology, anthropology, and archaeology are currently investigating the origin and evolution of language. And cognitive neuroscience has been investigating the mind-body problem for the last six decades. In sum, the noted prophecies of Kant, Comte, Du Bois-Reymond, Chomsky, and other mysterians, have been falsified.

11.3 Thought, Proposition, Sentence

It is a fact that this is a book. This fact, your seeing it, and your knowing something about books – two further facts – led You to thinking that this is a book. This thought is a brain process. But nothing prevents You from stripping off its neural stuff, and pretending that the given thought can exist in and by itself: this pretense is called a proposition. Finally, this proposition may be formulated as a sentence in thousands of languages. Note that a single fact may be represented by many different thoughts (in different brains under different circumstances); and that these different thoughts are telescoped, as it were, under one and the same proposition, or depersonalized thought. Further, this single proposition may be expressed differently in different languages. The following diagram exhibits the four important ontological categories involved in the previous text: fact, thought, proposition, and sentence.

		Designate		
	Proposition	→	Sentences	
Conceptualizes	↓		↓	*Express*
	Thoughts	→	Fact	
		Represent		

Nominalists, being vulgar materialists (physicalists), claim to shun all talk of concepts and propositions. The few consistent among them identify constructs (concepts and propositions) with their symbols; hence they will say, for instance, that the numeral 7 is (identical with) a prime number, instead of saying that the numeral 7 (a physical object) designates or symbolizes the successor of the number 6. But, as the idealist Frege rightly noted long ago, numerals and other marks have physical and chemical properties, whereas numbers have only conceptual properties.

Nominalists will also identify meaning with reference or denotation, whereas non-nominalists will include sense (or connotation along) with reference. Thus, whereas linguists may admit that Heidegger's *flatus vocis* "Sein is ES selbst" ("Being is IT itself") is an admissible German sentence, no sane person will claim that it makes sense – unless she happens to earn her living teaching such nonsense. Every proposition can be said as a sentence, but the converse is false.

We can get to know fairly well-defined constructs, but not all of them. Indeed, nobody will ever know every single member of an infinite set: it would take forever to count them all. Since this is an impossible task, what mathematicians do is to adopt the systemic approach: they lay down the essential conditions that the given infinity must satisfy. For instance, the natural numbers are nor defined one by one, but all at once through Peano's five postulates. Only one of these characterizes an individual, namely the first postulate: "Zero is a number." (But the predicate "is a number" is defined implicitly, through the postulate system.) The second Peano postulate relates an arbitrary number to its successor. And the fifth postulate, the principle of induction, involves the entire system of natural numbers. Thus the individuals in question, the integers, are defined in a top-down fashion, in violation of the bottom-up prescription of methodological individualism.

All the other number systems are introduced in a similar manner: as systems, rather than element by element (individualism) or as unanalyzable wholes (holism). This holds, in particular, for the system of real numbers, which are non-denumerable, as a consequence of which the overwhelming majority of them remain unspecified and nameless. Only a handful of transcendental numbers, such as π and e, are explicitly defined (by infinite series or products) and thus may be said to be completely known.

Thus, we have the paradoxical situation that we know in outline a whole, but in detail only a tiny minority of the members of this whole. (Political analog: We can know a mass party if we only learn some of its principles and deeds, even if we know only a few of members.) Yet, the huge building of mathematical analysis has been erected on that rather skimpy basis – with, however, the underling logic, a sort of tool for extracting any valid consequence of those few assumptions. Let the methodological individualist, who refuses to countenance systems, despair over this paradox.

What to do with 100^{100} and its relatives, such as its 100th power? Few deny that these numerals symbolize numbers, but no one can claim that they are thinkable. The same holds for the powers of aleph-zero, the numerosity of the set of natural numbers, and the other transfinite numbers. These are neither material nor mental objects. What are they then? I submit that we should distinguish two questions: the ontological and the psychological. From an ontological point of view, those "monsters" belong in the same category as the familiar digits: they are conceptual objects, or constructs. It is only psychologically that 100^{100} is in a different league as 10.

In short, we add the constructs level to the physical and mental strata, and split the category of constructs into intelligible and unintelligible. But we add the following provisos. First, mental objects are brain states or events. Second, Georg Cantor may have understood what we ordinary mortals find unintelligible. Third, constructs do not enjoy autonomy: they are human creations and, moreover, unlike buildings and artifacts, they will leave no ruins behind when the last human being dies.

Let us next glimpse at what is generally regarded as the summit of the mind-body problem.

11.4 Consciousness: The Holy Grail

Consciousness may be regarded as a special case of cognition, namely self-cognition, o awareness of oneself. It is sometimes called "metacognition"; but this is not really a synonym, for one may be conscious of one's own motions, sensations, feelings, and emotions in addition to cognitions. Moreover, we usually succeed in switching consciousness on and off. We do this every time we fall asleep, as well as every time we perform a routine task, that is, one that does not require constant conscious supervision. In the latter case we may be said to engage in tacit cognition, or know-how. Thus, consciousness entails cognition, but not conversely: We often know unconsciously, as Hume, von Hartmann, Tolstoy, Pavlov, Freud, and countless others have noted.

The brain is apt to fluctuate more or less spontaneously between the conscious and the unconscious levels. For example, when attempting to retrieve a forgotten item, we first consciously try out several heuristics, such as searching for semantic relatives of the said item, or attempting to recall the circumstances in which we originally learned it. But this search may clog the memory lane (better: labyrinth) with irrelevant items. So, after repeated unsuccessful trials we give up for the moment, reasonably hoping that the sought item will suddenly resurface spontaneously "out of nowhere" – presumably, from another region of the neocortex. In short, consciousness is very valuable but it should not be overrated; and a satisfactory understanding of it is still elusive, but at least we have a partial functional command of it.

Consciousness used to be the preserve of writers and philosophers, such as Augustine, Jean-Jacques Rousseau, Marcel Proust, James Joyce, Italo Svevo, Robert Musil, Henri Bergson, and Edmund Husserl. The latter, in particular, conceived of his own philosophy, namely phenomenology, as "egology"–the first-person study of the self. But of course such study is unscientific if only because it starts and ends with trivial first-person reports, such as "This tooth hurts." A dentist will certainly examine the tooth indicated by the patient, but he will not treat it without further ado, because he knows that the source of the pain may well be elsewhere. And nobody takes at face value every statement of the form "I am sorry about that", because we are all familiar with hypocrisy. In short, first-person reports are fallible, and there is no first-person (autobiographic) science.

All scientific research seeks what Russell called knowledge by description, by contrast with knowledge by acquaintance. The former may be characterized as knowledge of the third-person kind – such as the anesthetist's report on a patient's fading consciousness. In other words, science is objective. This holds, in particular, for the science of consciousness – to which the phenomenologists, from Husserl to Sartre, have contributed only IOUs and hermetic sentences, such as Husserl's assertion that "consciousness is absolute Being." But phenomenologists and existentialists will not be fazed by the accusation that they have not helped explain consciousness, for their avowed aim is to "understand" (*verstehen*) mind in an intuitive way, not to explain it scientifically.

The science and philosophy of consciousness have exploded in recent years (see, e.g., Blok et al. 1997; Hobson 1999; Laureys and Tononi 2009; Shear 1995; Smith and Jokic 2003.) Regrettably, there is still no consensus on what consciousness is. Worse, some students, starting with the great William James, have denied its existence. Others, such as McGinn (2004), admit the occurrence of conscious states but deny that they will ever be understood, due to our limited "epistemic architecture."

How does he know? And is this a piece of conscious or of tacit knowledge – or rather willful ignorance?

Cognitive neuroscientists have tended to adopt an excessively permissive concept of consciousness. For example, Damasio and Meyer (2009, 6) propose the following working definition: "a momentary creation of neural patterns which describe [map? represent? model?] a relation between the organism, on the one hand, and an object or event, on the other." But presumably the lowly Aplysia, or sea slug, fits this definition, since it can methodically scan and devour all the algae stuck on an irregular rocky patch. As the early ecologists, such as von Uexküll (1921), emphasized nearly a century ago, all animals map their surroundings. Surely human consciousness is more than that: it is the ability to learn new items about the world and oneself.

Philosophers have not fared better. Some of them have described consciousness in cryptic terms, as when Thomas Nagel (1974, 435) famously wrote: "fundamentally an organism has conscious states if and only if there is something it is like to *be* that organism – something it is like *for* the organism." (For an update see Baynes 2008.) One wonders how the predicate "is like to be" might be formalized in the predicate calculus, and how it might translate into other languages. I submit that the Byzantine expression "there is something it is like to be" is at best equivocal, at worst nonsensical, and in any event irrelevant to the problem of consciousness. (For a post mortem of that cryptic phrase, see Lormand 2004.) In particular, it does not suggest criteria that could help anesthetists and surgeons estimate degrees of unconsciousness of surgery patients.

Let us move on past wordplays parading as philosophical depth, and consider a familiar situation. The little girl kicked her dog and walked back to her chair. The dog whined and moved to the corner, where it lied down. After a short while, the girl blushed, shed a few tears, kneeled before the dog, and petted it. The dog got up and licked the girl's hand. So far, we have given only a crude and dry description of two familiar pieces of overt behavior. This story may satisfy the uncurious behaviorist, but it will only whet the appetite of the curious scientist. What caused the behavior of girl and dog? What went on in their heads, which mechanisms activated their overt behavior? We can only guess, as with any other inverse problem of the "Output → Input" type. However, we can make some educated guesses because we are familiar with similar episodes, and because psychologists have made a start in the study of emotion and the awareness of episodes of emotion, feeling, and cognition of self – that is, self-consciousness.

We can guess that, for some reason, the child was at first angry at the dog: in neurophysiological terms, the activity of her orbitofrontal and anterior cingular cortex was enhanced. Subsequently, a region in her prefrontal cortex became active, and sent a signal to some subcortical center. Immediately thereafter, this center signaled to the motor circuitry in her brain, which in turn opened her water gates and dilated the capillaries in her face – hence her blushing. All this happened when the girl "searched her soul" and repented. She then resolved to make amends: Her prefrontal cortex activated her motor strip, which in turn led her to get up, walk towards the dog, and ask it for forgiveness.

What happened inside the dog's skull? Again, we can only guess. Charles Darwin, who pioneered the study of emotions in humans and other animals, might have surmised that the dog was at first frightened, perhaps also angry, and finally became grateful to its young mistress when she caressed him. Is this a case of anthropomorphism? Perhaps, but this is no great sin when dealing with close relatives, particularly relatives that have been largely shaped by humans over a period of about 15,000 years (see Daston and Mitman 2005).

In the above story we met several old acquaintances: behavior, emotion, and reason, along with behaviorism, folk psychology, and cognitive neuroscience. Along the way, the old problem of the nature and even existence of consciousness reared its head. True, the behaviorists, from Watson to Skinner, as well as the philosophers who followed them, from Ryle to Dennett, have claimed that belief in the existence of consciousness is delusional. One may retort that these people were deluded by their own positivist philosophy, according to which there are only phenomena, that is, appearances – in this case overt behaviors. (For good criticisms of the deniers see Donald 2001; Gazzaniga 2008; Searle 1997.)

Given that the literature on consciousness, though bulky, is quite inconclusive, one suspects that the ideas on the subject are confused. It may be that the problem of consciousness is like that of the Holy Grail: Because there is no consensus on what it is, some knights doubt its existence while others wander hither and thither. If this diagnosis is correct, then an analysis like the following is indicated.

11.5 Kinds of Consciousness

The aim of this section is the modest one of pointing out that the word "consciousness" denotes a large variety of mental processes, in particular the following (Bunge and Ardila 1987, 234–5).

1. *Reactivity* or *sensitivity.* When laymen want to make sure that a person is not in a comatose state, they prick her with a pin. But of course all things are sensitive to some physical or chemical agents, so this is actually a test of materiality. To be more precise, we suggest

   Definition 1 Let b denote a thing (living or nonliving), and A an action on b or on a part of b, and originating either outside b or in a part of b. Then b is *A-sensitive* (or *A-responsive*) if b reacts to A (i.e., if A causes or triggers a change in the state of b), either always or with a certain frequency.

2. *Awareness* (or *phenomenal consciousness*). Any animal capable of identifying or discriminating some (internal or external) stimuli, or some of its own actions, can be said to be *aware* of them provided it can do something to control either the sources of stimulation or its own reaction to them – not so if it cannot help responding on cue. For example, the gazelle that approaches a watering hole in sight of a pride of lions, and the rat that accepts an electric shock in exchange for

the chance of eating something or exploring a new environment, can be imputed awareness. Consequently an awareness test or indicator would be he ability to learn new behavior patterns incompatible with inherited or previously learned ones.

Awareness requires neither more nor less than neurosensors of some kind. Hence organisms lacking neurosensors altogether cannot be aware of anything. Not even a sea urchin can be aware of anything, for it lacks sense organs: it is only reactive or sensitive. Actually all animals have this capacity – which is why it is hard to understand why philosophers since Locke have made so much noise about qualia, or phenomenal properties.

In summary, we propose

Definition 2 An animal *b* is aware of (or notices) change *X* (internal or external to *b*) if *b* feels (senses) *X*.

3. *Self-awareness.* An animal can be aware of its surroundings but not always of what it is feeling, perceiving, thinking, or doing. An animal aware of its own feelings or actions may be said to be *self-aware*. When in such a state, it not only moves or feels hungry, but also notices that it is moving or feeling hungry – as suggested by the way it goes about solving the problems it encounters along the way. On the other hand, certain neurological patients are confused as to the origins of some of their own feelings and doings: they are not fully self-aware. (Example: hemineglect, or the failure to recognize half of one's own body, due to either a stroke or anesthesia.) Nor are normal adult humans self-aware all the time: we often manage to temporarily forget hunger or pain, and we perform many actions automatically (unconsciously). Becoming self-aware requires weakening other-awareness and paying mainly attention to oneself, that is, becoming self-absorbed or practicing introspection – or "soul-searching" when confronted with moral problems. (The word "introspection" lost much of its appeal when it was noted that, strictly speaking, we do not watch our own mental processes in the same way that we watch a water stream. But it is still useful.)

   To be self-aware is to be aware of oneself as something different from everything else: it amounts to carving one's own niche in the universe. A self-aware animal notices, however dimly, that it is the subject of its own feelings and doings. Ordinarily self-awareness is so much taken for granted, that we tend to forget that, when absent-minded, we are not aware of ourselves, and that it can be a serious hindrance when performing a non-routine task.

   We condense the above into the following convention:

   Definition 3 An animal *b* is *self-aware* (or has self-awareness) if *b* is aware of some of its inner changes and some of its own actions.

   Note that self-awareness does nor require *thinking* about one's own mental processes. Satisfaction of this additional conditional takes us to the next level.

4. *Consciousness.* An animal aware of what it is perceiving, feeling, or thinking may be said to be *conscious,* even if momentarily oblivious of some of its own

feelings and doings, or not manifestly responding to some external stimuli that normally elicit its reaction.

The goose that rolls an imaginary egg with it beak is not conscious. The early ethologists conjectured that the bird's movements are regulated by a sort of "motor tape" in its nervous system: the animal cannot help moving that way. By contrast, the pigeon that looks attentively at a rotated figure to check whether it is the same as the original, in expectation of a reward, may be said to be conscious: the animal is monitoring and "manipulating" some of its own movements and mental states.

In sum, we stipulate

Definition 4 An animal *b* is *conscious* of mental process *M* if *b* thinks of *M* as occurring in *b*.

According to this convention, an animal can only be conscious of some of its own higher mental processes. Not just feeling, sensing and doing, but also thinking of what it perceives or thinks. (To be sure, thought need not be abstract or even verbalizable; it can be in images, as when we perform a mathematical calculation imagining that we are writing on a blackboard.) An animal conscious of mental process *M* in itself undergoes (either in parallel or in quick succession) *two* different mental processes: *M* (the object mental process or content of its consciousness), and thinking about *M* (i.e., being conscious of *M*). The object of *M* can be a perception (of, e.g., a hot pan), a memory (of, e.g., a tasty morsel), a mathematical formula, or what have you.

Note the difference between consciousness and awareness. Animals of certain species can become aware of certain stimuli, and many are capable of paying attention, but they won't be conscious of anything unless they can think. Conversely, a person lost in daydreaming or in deep, productive thinking, may be unaware of her surroundings. Consequently, the concepts of consciousness and awareness are mutually independent. This being so, they should not be confused. And the hybrid "conscious awareness" should be avoided. Note also that we cannot be conscious of other people's mental processes, because we do not share their brains. But of course we can get to know some of them indirectly, through indicators such as body movements and words. Finally, note that the above definition confutes the claim that recognition in a mirror is a test of consciousness: it only tests for self-recognition.

We come now to the top rung in cognitive abilities:

5. *Self-consciousness.* An animal that is occasionally conscious, that sometimes reflects upon its own perceptions, feelings or thoughts (concurrent or past), and does not attribute them to something or somebody else, can be said to be *self-conscious*. By contrast, an animal that attributes its own perceptions, feelings or thoughts to external objects, fails to be self-conscious; so is someone who "hears voices," imputes his dreams to spirits, or claims to communicate with the dead or with God. Likewise, an individual immersed in a motor or intellectual task who does not pause to reflect upon what she is doing or thinking, who forgets herself, is not self-conscious. She is herself without being conscious

of her self. Anyone who were constantly self-conscious would never get anything done.

Just as self-awareness is one rung higher than awareness, so self-consciousness is one step higher than consciousness. A subject is self-conscious only if she has consciousness of her own perceptions and thoughts as occurring in herself. To put it in epistemological terms, an animal is self-conscious if it knows who and what it is itself – i.e., if it has self-knowledge. Hence the admonition of the Delphi Oracle, *Know thyself!*, amounts to: *Be self-conscious!* This is of course easier said than done, as evidenced by the occurrence of self-deception.

Now, in order for someone to know who and what she is, one must have some recollection of one's past: We are what we have become, and we know what we have learned. Hence the total amnesiac can hardly be said to have a self. On the other hand she need not be able to extrapolate her own life into the future: it may not be able to imagine or plan its next move, except for a very short time, such as the time needed to walk to the next room. Thus, the primate with a frontal lobotomy appears to be self-conscious from moment to moment. As an eminent neuropsychologist put it, in such an animal "the stream of happenings is not segmented and so runs together in a present which is forever, without past or future. The organism becomes completely a monitor at the mercy of its momentary states, instead of an actor on them" (Pribram 1971, 348). The same is likely to happen to animals with a very small prefrontal cortex, like cats – or none, like birds. This calls for the following additional distinctions:

Definition 5 An individual is

(a) *antero-self-conscious* if she recalls correctly some of her recent past;
(b) *pro-self-conscious* if she can imagine (even wrongly) some of her own future; and
(c) *fully self-conscious* if she is both antero- and pro-self-conscious.

Presumably, full self-consciousness is a human prerogative. As Richard Passingham (2008, 30) wrote, "We can rephrase Descartes' Cogito. I am aware of my own thinking, therefore I am human."

Anyone ceases to be fully conscious after suffering a severe concussion. Victims of the barbarous lobotomy operation lost their pro-self-consciousness: they could not even plan a simple meal. H.M., the famous neurological patient studied by Scoville and Brenda Milner, lost his antero-self-consciousness after his operation: he never even remembered having seen the doctors who had examined him the day before. The case of K. C., the patient studied by Endel Tulving, is even more tragic, because he cannot not remember anything about his own life, except for the experiences he has had in the last minute or two; and he has no idea what he is going to do next. His self or inner life is practically restricted to the present. But at least his semantic memory (of "whats") is nearly intact, which suggests that semantic and episodic memories have different locations in the brain.

If psychoneural dualists can explain such deficits in purely psychological terms, they have not published their findings. By contrast, the biological approach to both conscious and nonconscious processes has vindicated the materialist hypothesis, that consciousness is a brain process (Place 1956). And this hypothesis is not only philosophically significant: it also underlies the search for brain indicators (or measures) of consciousness in addition to self-reports and behavioral indicators of conscious states (see Seth et al. 2008).

11.6 The Neuroscientific Approach

Being conscious of a mental process in oneself is to be in a certain mental state. Now, according to cognitive neuroscience, being in a mental state is the same as a brain being in a certain state (or rather undergoing processes of a certain kind).

Hence consciousness, which in classical (brainless) psychology is conceived of as an entity, is better regarded as a collection of brain states. For this reason we stipulate

Definition 6 The *consciousness* of animal *b* is the set of all the states of the brain of *b* in which *b* is conscious of some feeling, perception, or thought in *b* itself.

Most cognitive neuroscientists assume that any mental process, such as the recognition of a face, the identification of a sound, the recall of an episode, or the completion of a figure, mobilizes a large number of neurons grouped into specialized assemblies or modules. Imaging studies have repeatedly shown that this number increases with the complexity of the task and the level of consciousness of the operation.

However, it is also well known that practice reduces the required level of consciousness: think of near-automatic driving in a highway, performing routine calculations, or correcting printer's proofs. A computer scientist might say that, with practice, software gradually hardens into hardware. (The great polymath Euler has been credited with saying "My pencil knows more than I.") The level of consciousness rises only in the face of a difficulty: otherwise we proceed automatically, or nearly so, as when reading an emotionally neutral text on a familiar subject.

A heightened level of consciousness is only required by unexpected novelties and by the deliberate effort to formulate the new strategies called for by non-routine problems, such as the search for new information sources, new problems, new conjectures, or new methods. All such tasks call for sharply focused attention and strong motivation in addition to the cooperation of a number of cortical modules such as the anterior cingulate cortex. (Incidentally, this seems to be the organ of cognitive conflict-resolution, as when one is asked to read the word *red* displayed in green: see Botvinick et al. 1999.)

Dehaene and Naccache (2001, 14) following a suggestion of Dehaene, Kerszberg and Changeux (1998), propose that consciousness emerges under attention: "top-down attentional amplification is the mechanism by which modular processes can be temporarily mobilized and made available to the global workspace [the site of working memory], and therefore to consciousness." However, human subjects can perform certain discriminations, such as telling male faces from female ones, without paying attention (see Tononi and Koch 2008).

Tononi and coworkers, by contrast, hypothesize that consciousness is either cortical integration or cortical information capacity, particularly in the posterior parietal area (Alkire et al. 2008). But integration calls for neuronal synchronization, as Sherrington had suggested, and Wolf Singer (2009) and others have confirmed in recent years. This is an obvious condition for any complex thing: Synchronization is necessary for integration, which in turn is necessary for a system to act as a unit in some respect. Thus the unity of the mind, the "binding" of the processes occurring concurrently in different subsystems of the brain, is just the functional counterpart of the unity of the brain. He and Raichle (2009) conjecture that the slow cortical potential (of frequencies between 1 and 4 Hz) is the messenger that communicates distant regions of the neocortex.

Under either hypothesis (attention or integration), the dimming and final temporary loss of consciousness caused by alcohol, sleeping pills and anesthesia is identical with the more or less pronounced weakening of the cohesion or systemicity of the cerebral cortex. Thus the anesthetist's job is to loosen and monitor such integration – and to make sure that integration is eventually restored.

The subjective (or phenomenological) unity of consciousness may thus be (tentatively) explained in terms of the interconnection of a number of specialized neuronal modules (or circuits) in the global workspace. The modules in this supersystem are not necessarily fixed and adjacent. They are plastic and may be itinerant, as it has been suggested earlier. Consequently "the contours of the workspace fluctuate as different brain circuits are temporarily mobilized, then demobilized" (Dehaene and Naccache, loc. cit.).

These hypotheses explain not only consciousness but also unconscious processes: these occur in neuron assemblies that remain isolated from the "global workspace", as is the case with subliminal perception (in particular blindsight) and the automatic detection of typos, and syntactic flaws. More on this anon.

11.7 The Dual Role of Consciousness

Except for a few philosophers, such as Dennett (1991) and McGinn (2004), nearly everyone takes it for granted that consciousness has two roles: the monitoring of mental processes, and the control of mental and motor activities. Cognitive neuroscience can model roughly as follows this dual role of consciousness (Bunge 1980a, 176–8; Bunge and Ardila 1987, 145).

Consider a neural system made up of two subsystems, *N* and its consciousness *C*. Suppose further that these two units are linked by bonds of two kinds: *N* stimulates or inhibits C, which in turn either activates or inhibits *N*. See Fig. 11.3.

The subject is aware of activity in *N*, or in a muscle innervated by *N*, just in case *N* stimulates *C*, or *C* acts upon *N*, either by stimulating or inhibiting *N*. In the first case, *C* passively monitors *N*: the dashboard metaphor. In the second case, *C* exerts a causal action on *N*: the steering-wheel metaphor. Normally we are conscious

C

↑↓

N

Fig. 11.3 The dual role of consciousness: dashboard ($N \rightarrow C$) and steering wheel ($C \rightarrow N$)

of several events at the same time, as in multi-tasking. Hence we assume that the consciousness of an animal over a given period is the activity of a system of a number of such units in C while connected with a second neural system N.

Admittedly, the above schema is bound to raise many a neuroscientist's eyebrow. For example, the study of involuntary (unconscious) motor behavior has shown that, if the subject becomes conscious of it, this happens only after he has performed the movement – a fact that would confirm the Lange-James hypothesis. But this does not warrant the conclusion that "conscious experiences are consequences of brain activity, rather than causes" (Haggard 2005).

This dualist interpretation is all the more bizarre for coming from an avowed anti-dualist. What, if not a brain activity, can consciousness be according to monism? Besides, experimental findings on involuntary motor behavior do not necessarily carry over to voluntary behavior, which is often conscious though not necessarily self-conscious. For example, after I have consciously decided to drive myself home, I will myself to walk to the garage, start the engine, and drive into the street; but once I join the traffic flow, I cease to pay attention to the driving, and resume thinking about some conceptual problem.

Could a machine be conscious? Yes, up to a point. Indeed, there are plenty of machines equipped with both measuring devices and controls, which may be regarded as analogs of self-cognition and action respectively. But machines lack the spontaneity and freedom that characterize human consciousness: they work in accordance with the program specified by their inventors and human controls, without understanding what is going on, and without the possibility of deliberately altering the program. For example, unmanned aircraft "sense" their momentary state, search and destroy targets, and can even adapt their trajectories to unforeseen obstacles. But they have no choice of mission: they can only do what they have been designed to do. In short, it is possible to construct self-aware machines, but not fully conscious ones.

The strength of the C-N connection is likely to vary in the course of time, either spontaneously or as a result of interactions with other parts of the nervous system. This explains the fact that, as William James pointed out, consciousness comes in degrees – something we experience daily when slowly falling asleep and waking up. If the two systems become totally disconnected, the corresponding conscious experience is interrupted. This explains the momentary loss of consciousness in deep sleep or as a result of a concussion. The hypothesis also helps explain blindsight, unconscious learning, tacit memory, the loss of episodic memory, and the like. All of these would be disconnection syndromes, hence basically similar to the aphasias, agnosias, and apraxias, as well as to some amnesias, classically described by Normand Geschwind in the 1960s. But disconnection also occurs when learned behavior, initially conscious, becomes unconscious with practice.

The above model is at variance with the view that consciousness is the activity of a single supervisor or "central executive." This view is inconsistent with an extensive body of neurological data about patients who have sustained lesions in any of the brain centers, yet have retained the ability to perform mental operations requiring not only a number of "faculties", but also mental effort and concentration, i.e., consciousness. However, there is little doubt that self-consciousness is located in the prefrontal cortex – which in humans occupies no less than 29% of the total neocortex. It is also doubtless that ours is only a coarse model. But it is one that can be perfected, as more is learned about the prefrontal cortex.

At this early stage of the scientific investigation of consciousness, we need not commit ourselves to any of the hypotheses we have just discussed. There are several alternative conjectures. However, to qualify as scientific, any hypothesis about conscious states must construe them as states of certain neuronal systems: there is no more brainless mind than there are faceless smiles. Besides, a scientific theory is much more precise than any of the extant models, which are of the hand-waving kind.

Might the quantum theory help craft a full-fledged theory of consciousness, as Roger Penrose and others have claimed? The orthodox or Copenhagen interpretation of that theory suggests an intimate quantum-consciousness connection, for it states that consciousness causes the collapse or projection of the state function. But this is just an obiter dictum, since the quantum theory does not contain any variables denoting mental properties. As we saw in Chapter 3, the said collapse is nowadays conceived of as the decoherence resulting from the interaction between a quanton and its macrophysical environment, which need no be manned by an experimentalist: state-function collapse is a purely physical process.

Just as quantum physics does not contain psychological variables, so cognitive neuroscience has no need for quantum mechanics, because its anatomical units, neurons, are strongly entangled with their environment. Hence even if any of them were to be in a coherent superposition of states at a given instant, it would promptly decohere ("collapse") due to the strong interactions with the surrounding tissue; consequently all of its properties would become sharp and, in particular, free from the "uncertainty" or bluntness peculiar to quantons (recall Section 3.2).

The same holds, a fortiori, for any of the multi-neuron systems capable of experiencing mental processes. Therefore the project of constructing a "quantum cognitive science," as proposed by Quentin Smith (2003), is at best future music, and at worst wrong-headed fantasy and consequently a waste of time. In any event, anyone seriously interested in merging cognitive science with quantum mechanics should stop writing symbols that look like state (or wave) functions but are not such, because they are not mathematically well-defined. In fact, they are not solutions of Schrödinger equations for neural nodules capable of performing mental functions, but just scribbles. So far as I know, nobody has done this for a neuron, or even for a water droplet. And yet even Stephen Hawking has indulged in claiming that a certain letter stands for the state function of the universe. As the German say, paper is tolerant.

Biopsychology is not only finding more and more about consciousness: it can also dispel dualist myths, such as that of zombies. The zombie, a character of Haitian folklore and contemporary philosophy of mind, is said to be a living dead: someone who is just like a live person with a normal brain, but behaves like an automaton or as a Golem, i.e., it has no consciousness. Saul Kripke and other philosophers have held that just imagining the existence of zombies refutes psychoneural monism. This wild fantasy has prompted a small academic industry: many papers and several web sites devoted to zombies (see, e.g., Chalmers 1996). Let us see how a cognitive neuroscientist might react.

If zombies had normal brains, they would also have normal mental properties, because one cannot have the same base properties without the same emergent properties. And if, on the contrary, zombies had abnormal brains, then they would behave in such an extraordinary way, that they would be promptly spotted and disqualified as human beings. Indeed, they would not respond to questions or commands; they would be unable to solve problems requiring imagination; they would be unable to take the initiative and make plans; they would commit plenty of antisocial actions without remorse; they would feel no pain; and they would not suffer from mental disorders; they would not even dream or hallucinate that they are zombies. In short, humans without consciousness would not behave like human beings. So, far from falsifying psychoneural monism, the zombie fantasy illustrates it.

Let us finally write a few words about unconscious processes. Nearly all of classical psychology dealt with conscious processes. Pavlov earned the Nobel Prize for inaugurating the experimental study of automatic behavior, both unconditioned and learned. Freud, his near contemporary, got the Goethe Prize for crafting amusing but outrageous stories about "the" unconscious – the reification of conscious states. Whereas Freud saw hidden goals even in emotional facts such as the child-parent bond, today's cognitive neuroscientists are uncovering the neural mechanisms of unconscious as well as of deliberate behaviors. They even study the automatic regulation of behavior and emotion, as when one keeps walking towards a goal, or abstains from shouting angrily in public, without deliberation.

Pavlov's work is relevant to a tricky neurological problem: that of ascertaining whether the victim of a brain accident, who has been knocked unconscious, is in an irreversible vegetative state, and so not worth being kept on a life-support system. The most plausible test is the so-called trace conditioning of the eyeblink response, which consists in learning to associate a tone with an air puff in the eye. The ability to learn this simple trick is regarded as evidence of minimally conscious state, hence a reliable predictor of recovery (Bekinschtein et al. 2009).

11.8 The Self

Selfhood is usually identified with consciousness. That is, a person is defined as a being capable of being in conscious states. I submit that this definition is far too restrictive, for it deprives babies, senile individuals, and a fortiori non-human animals, of personhood. Such exclusion does not agree with the common practice of attributing different personality traits to babies and senile individuals, as well as to apes, monkeys, dogs, cats, and other animals.

Anyone who has handled animals knows that some members of any given higher-vertebrate species of are more curious, smart, teachable, gregarious, playful, affectionate, or gluttonous than others: they have different personalities, hence they are persons to begin with, though of course not necessarily human persons. I therefore suggest this alternative definition: A person is an animal with a mind, i.e., capable of having mental processes. Not all the time, of course: We do not surrender our selves and cease to be persons when drunk, in deep sleep, or under general anesthesia. To be sure persons, just like all other concrete things, are always in a state of flux, but they lose their selves only when brain-dead: when they are no longer able to have a mental life.

The further problem of identifying the animal species with mental capacities is best left to ethologists and animal psychologists. There is no consensus on this matter, partly because of the religious and idealist underrating of nonhuman animals, and partly because of the difficulty in attributing mentality on the strength of purely behavioral indicators. However, there is a tendency to attribute minds to all higher vertebrates, that is, mammals and birds.

Most philosophers of mind are likely to regard all of the above as either wrong or irrelevant to the philosophical problem of personhood, which they deem to be very hard or even insoluble. But this problem is trivial to a psychoneural monist, for she takes the whole person, body and its functions, as a whole, and so does not wonder what the "link" between mind a body may be, anymore than a cardiologist wonders about the "link" between a heart and its contractions.

There is one outstanding exception, though: the neuroscientist Rodolfo Llinás (2001, 128) believes that the self is unreal: that "it is a very important and useful construct, a complicated eigen (self) vector. It exists only as a calculated entity." This extravagant statement makes no sense, for there are no eigenvectors in themselves: every eigenvector is attached to some operator or other – and no such mathematical object is even mentioned in the text. Besides, the thesis that selves are abstract evaporates as soon as one moves from the first-person mode to the third-person one. For example, if Rodolfo tells me "I am now going to insert a microelectrode into this neuron," I may translate this sentence into "Rodolfo is about to insert a microelectrode into that neuron." I will be talking about Rodolfo him*self* doing something by him*self*, not about his hands alone, nor just about his brain, let alone about someone else.

Nor does the materialist agonize over whether she is a body, a mind, or a body-mind composite. She knows that she is an animal endowed with a brain capable of experiencing mental events. She also knows that she will lose her self when her brain goes. This is why she knows that she will miss her own funeral. For once, the expression "mortal remains", used in religious obituaries to denote cadavers, is correct. But religions are wrong in overrating natural death, since it is just the termination of life. However, let us turn to a more cheerful and less trivial subject.

What is not trivial, but the subject of ongoing empirical research, is the problem of finding out the stages of the development of personality and, in particular at what age children become self-conscious and start attributing minds to other beings. (Preliminary result: at about 5.) Given the limited linguistic ability of young children, this investigation demands unusual ingenuity in experimental design. (Unlike the scientific method, which is universal, scientific techniques are topic-specific. Once again, ontology precedes epistemology: recall Section 1.5.)

As we grow up, we acquire knowledge of ourselves and others, and these two pieces of knowledge are involved in the way we deal with others. And unless we exert some self-control, we are bound to get into trouble. Self-control is normally learned through experience and self-reflection. But it won't if the organ of social regulation, namely the orbitofrontal cortex, is badly damaged (e.g., Beer 2006). This has been known for over a century, when the case of Phineas Gage was given considerable publicity: he suffered a radical deterioration in social behavior after a piece of iron injured precisely that part of his brain. This episode might have given birth to social cognitive neuroscience, had it not been for the prevailing philosophical preconception at the time, that the natural and the social sciences are separated by an unbridgeable chasm.

However, drives and self-control are not the only determinants of social behavior: social bonds are just as important, and occasionally they overpower self-control, as in the case of the behavior of members of closely knit groups such as criminal gangs. In fact, the data (e.g., Wright et al. 1999) appear to support the systemic hypothesis that criminal behavior results from a combination of low self-control or morality (individualism) with social bonds and circumstances (holism). The following diagram telescopes our systemic view of human social behavior.

Social bonds
↓↑ ↘
Self-control → Social behavior
↓↑ ↗
Drives

To conclude this section, the biosocial approach to mind suggests that the philosophical problem of the self has been artificially inflated by the separation of mind from both its brain and the latter's social environment. Science does not study "the self and its brain", as the popular book by Popper and Eccles (1977) has it. Rather, it explores "the socially-embedded brain and its self." One reason for this reversal is that mental events are brain events, just as digestion is a process in the guts, and walking the specific function of legs. Another reason is that mindless brains precede minding brains both in development and in evolution: We are born with uneducated brains that, in suitable environments, actualize or acquire mental faculties that differ from one animal species to the next. This mind-building process is called "learning", and it goes on under dual control: that of oneself and one's immediate social surrounding.

11.9 Free Will

I do something of my own free will if I can choose to either do or not do it, and if this decision is not forced upon me by circumstances. For example, I write this sentence because I want to, not because anyone ordered me to. The same holds for many other actions and inactions in the course of the day. I experience free will every time I am given a choice among alternatives, and a fortiori when I invent a new option. Moreover, I assume that the same holds for all adult human beings, even those under duress, for they can drag their feet, cheat, or rebel – albeit at risk. It is well known that all codes of behavior presuppose free will. For example, usually soldiers are pardoned for committing war crimes if they acted under strict orders.

Furthermore, I venture to suggest that other higher vertebrates too enjoy a measure of free will, as when wild animals tackle new problems in novel ways, or domestic animals escape or disobey a command. In any event, as Nicholas Rescher (2009, xii) put it, "the emergence of free will is one of evolution's crowning glories." Yes, evolution has been in part a liberation process: one of decreasing dependence on the environment (thanks to improvements in homeostasis) and increasing empowerment (ability to do), thanks to improvements of the brain and its social uses.

And yet free will has generally been regarded as illusory, not only by philosophical naturalists like Spinoza and d'Holbach, but also by scientists like Laplace and Skinner. The most popular argument against free will is the one from classical or Laplacean determinism: the state of a person at any time is an effect of all the causes in the person and her environment, indeed the whole universe, at previous times. Accordingly, a person would lack initiative, so that she could never be in control of herself, whence she cannot be held responsible for her own actions. Hence benefactors and criminals would be equivalent, because both would be products of their environments. Consequently we should neither reward good actions nor punish wicked ones.

In other words, the naturalist who denies free will is likely to argue, in a behaviorist vein, that the brain is the slave of its environment, so it cannot help making the decisions it makes: all of these would be stimulus-bound. This is what classical neuroscience and behaviorist (or stimulus-response) psychology used to teach. But we have known for a while that this view is wrong: that the brain is active all the time, even during sleep, and that most of its activities are spontaneous or self-generated rather than responses to external stimuli. We conjecture this because the brain applies only a tiny percentage of its energy budget to its transactions with the external world (recall Section 9.2).

We also know that self-control, which is a necessary condition for free will, is a learnable function of the prefrontal cortex, the philogenetically newest area of the brain. So much so, that people with serious damage to that brain region lack free will: they are swayed by the stimuli impinging on them. Thus, the removal or disconnection of significant portions of nervous tissue in the prefrontal cortex causes "environmental dependency syndrome" – an irreversible disorder. However, self-determination is not the prerogative of the prefrontal cortex: it is a property of matter on all levels: recall the cases of inertia, self-organization, and spontaneous light emission and radioactivity (Chapters 2 and 3).

In conclusion, negative freedom (or freedom *from* all constraints) can never be total, because we are constantly bombarded by our natural and social environment. But positive freedom, or freedom *to* (*potestas agendi*), is possible, and it is compatible with determinism. The reason is that, to exercise it, we use our volition areas (in the frontal lobes and the parietal cortex). For example, the political prisoner who is given the option of revealing the names of his comrades or being shot, is certainly deprived of negative liberty. But he can exert his free will to implement his positive liberty: better dead than betraying the cause. Lobotomy patients cannot enjoy such freedom because their main volition center has been surgically cut off from the rest of the brain. In conclusion, there is freedom *to* because causation is for real: there would be no positive freedom in a world ruled by chance (Bunge 1959a). This view is called Compatibilism.

By contrast, according to idealism, the soul is immaterial and therefore out of all causal streams, so it may start itself and even guide the body. Emergent materialism concurs, except that it replaces "soul" with "brain." Under this version of materialism, the person retains the initiative and moral responsibility that idealism traditionally attributed her. Moreover, the free will thesis ceases to be a dogma, to become an empirically testable hypothesis. Indeed, a subject's decisions can be altered by tampering with her brain in various ways, such as subjecting her to stressful situations or electrically stimulating her frontal lobes or parietal cortex (e.g., Desmurget et al. 2009). And the assumption has the additional advantage that it does not involve telekinesis, or the mental power to move matter without energy expenditure.

Hebb (1980) was one of the rare experimental psychologists who thought that the problem of free will could be attacked scientifically, because free action originates in spontaneous (or non-stimulus-bound) brain activity. His own experiments on sensory deprivation had shown that, contrary to the stimulus-response dogma, the living brain is constantly active. But, of course, freedom is never complete, because there are self-imposed moral constraints, and we are all subject to natural and social constraints and pressures. Among the latter are the limitations that we freely consent to in exchange for the right to participate in social networks that protect our interests. Furthermore, positive freedom will lead to inefficient or even evil action unless accompanied by the knowledge and the moral conscience required to wanting to do the right thing.

Since Hebb's days, free choice has been studied in the laboratory, not only in humans but also in monkeys. For example, Bijan Pesaran and his coworkers (2008) studied the brain activity of monkeys making free choices, as well as instructed choices. They trained monkeys to reach visual targets for rewards of juice in two different ways. In one of the tasks the animals were instructed to search in a fixed sequence; in the other they had choose the sequence. Neuron firing in two brain areas was recorded: the prefrontal cortex and the parietal cortex. It was found that both areas are active during both tasks, but that coherence between them was far stronger during free search. So, this difference in intensity may be regarded as a physiological indicator of the degree of free will.

Let us now glimpse at the ethics of free will. A central problem in this field is that of responsibility and the corresponding blame. A person may be said to be primarily *responsible* for X only if she enjoys free will, knows how to handle X, and is in charge of X. If any of these three conditions is missing, then the person in question cannot be blamed when something goes wrong with X during her watch. This is why it is foolish to entrust young children, the mentally handicapped, and the unskilled with heavy responsibilities. The fact that everyone knows this, is evidence for the conjecture that free will is usually taken for granted in real life – even by psychologists and philosophers who deny it on paper.

Is free will limited to humans? Two famous mathematicians have recently published what they call "the strong free will theorem". This theorem "asserts, roughly, that if indeed we humans have free will, then elementary particles already have their own small share of this valuable commodity" (Conway and Kochen 2009).

The reader might think that the said mathematicians have found a bridge between the science of the presumptive seat of free will, namely our prefrontal cortex, on the one hand, and the quantum theory on the other. Nothing of the sort: Conway and Kochen focus on a well-known feature of quantum matter: that some of its properties are not fully determined by its immediate past and its surrounding. This has been known for about a century. Suffice it to recall radioactivity, the spin measurement, and the countless scattering experiments, where incident particles directed at a target in the same direction and with the same speed will end up in different places. (In the latter case physicists calculate the probability that the incident particle will be deflected into a given solid angle.)

A popular example is that of an electron moving in a magnetic field, and which may end up, with equal probabilities, with its spin either parallel or antiparallel to the external field. The authors claim that this constitutes evidence that the particle is *free to choose* either direction. (Actually they analyze a more complex measurement on a spin 1 particle, but the basic idea is the same.) The vast majority of physicists describe such results in strictly physical terms; and students might be flunked if they were to indulge in anthropomorphism and claim, as Conway and Kochen do, that "the universe has by definition taken a free decision." In fact, *by definition* free will is "the ability to act at one's own discretion" – and only persons in a conscious state have this ability. Moreover, free will does not come free, but involves energy expenditure. Hence, if Conway and Kochen took particle free will seriously, they would tell us which part of the Hamiltonian (energy operator) of their particle corresponds to its free will.

Which might be the sources of the extravagant claim of Conway and Kochen? I surmise that it has two sources: sloppy language and the Copenhagen interpretation of the quantum theory. The first culprit made its appearance when the authors attributed quantum indeterminacy to the free will of fundamental particles. It also occurs in their warning (op. cit.: 228) that they will use the words "property", "event", and "information" almost interchangeably, although they designate radically different ontological categories. Actually even the title of their paper, "The strong free will theorem", is careless: what does "strong" qualify, free will or theorem?

It is also obvious that Conway and Kochen adopt uncritically the Copenhagen interpretation. In fact, they focus exclusively on experimental situations, as if the quantum theory held only in the laboratory, while in fact it is used to explain such empirically inaccessible processes as the nuclear reactions occurring in the stars. The upholders of the Copenhagen interpretation overlook this fact and, following the Berkeley-Hume-Kant-Mach-Vienna Circle line, claim that, as Leo Rosenfeld once said, "the experimenter conjures up the quantum phenomenon." While the experimenter is indeed free to try to measure what he wants (or what his research grant can afford), to be successful he must bow to the laws of nature. (So, Spinoza was right for this very particular case: The experimenter's freedom consists in his knowledge of the relevant natural laws. If he does not know them, the best he can do is to play the trial and error game.)

To seriously check the claim that the experimenter's free will implies the quanton's, one would need two items absent from the paper in question: (a) a theory of free will and a bridge formula between it and the quantum theory; and (b) experimenters willing to be locked inside a magnetic resonance imaging apparatus, with their brains wired to a measurement device. Regrettably, our authors offer no suggestions about either item. Nor do they say how quantons manage in the absence of experimenters, particularly before the emergence of humans.

However, there is a simpler and cheaper refutation of the Conway-Kochen theorem, namely this. Free will is the ability to do what one wants, and both volition and decision are functions of the human (or higher vertebrate) prefrontal cortex. As such, these functions are way beyond the possibility of brainless things, such as elementary particles, atoms, or even individual neurons.

Why react so vehemently to the extravagance of two mathematicians? Not so much to protect physics from esotericism, as to argue for the reality of human free will. Indeed, the Conway-Kochen "theorem" is a conditional of the form "If *H*, then *P*". Now, since *P* is obviously false, then *H* too is false. That is, if we admit the said "theorem," we must conclude that we lack free will. And I, for one, am not prepared to surrender this human prerogative. Six decades ago, when I asked the chief of the Argentine political police for identity documents, he told me that I would get them if I signed an authorization to search my house. When I refused to do this, he asked me why, and I replied: "Because I want to keep my free will." Let the reader imagine the sequence.

In concluding, an obvious political point: Free will is necessary but not sufficient for the enjoyment of civil liberties. These can be exercised under a liberal social order, where "liberal" is meant in the broad sense, not in the narrow economic sense intended by the neo-liberals, who are only interested in free enterprise. A fair social order will make it possible for nearly everyone to enjoy her free will, with Mill's condition that it won't restrict another person's liberty. But such nearly universal empowerment is unattainable in a very divided society, where power – whether economic, political or cultural – is monopolized by a minority (Bunge 2009). So, the problem of free will is as important in politics as it is in ethics and in theology.

11.10 Explanation by Causes and by Reasons

A favorite tenet of the neo-Kantian philosophy of the social is that, whereas the natural sciences explain facts in terms of causes, social facts can only be explained by reasons. The reason for this divergence would be that, unlike natural facts, social facts would be eminently cultural, moral, or spiritual – whence the qualifiers "cultural", "moral", and "spiritual" for the social studies (see, e.g., Dilthey 1959; Putnam 1978).

There are at least two problems with this thesis. A first problem is that it omits feelings and emotions, as when one says that the shares of a company fell because investors feared that the company would be the victim of a hostile take-over, or because they heard an alarming rumor that turned to be faflse. A second objection to the thesis in question is that it focuses on social actions resulting from deliberate decisions, and consequently it overlooks the fact that many macrosocial events are the result of events that are beyond our control, such as natural disasters and plagues. Outstanding examples are the volcanic eruption that destroyed Pompeii, the Black Death, the storm that sunk the Spanish Armada and thus spared Elizabethan England, and the sand storms that are currently threatening Beijing.

Nor are natural disasters the only socially effective causes without attendant reasons. Most of the flaws of our industrial societies have not been willed by anyone. For example, no one wants inflation, unemployment, or business cycles to happen: they are inherent in capitalism and can only be mitigated by strict regulations and strong social programs like those instituted by the New Deal.

A third objection to the cause/reason dichotomy is that, from a neuroscientific viewpoint, proposing a reason amounts to triggering a causal process in the prefrontal cortex. Consider the following familiar situation. A man is caught stealing. Asked to explain his action, the thief responds: I lost my job because of the current recession; and, since I have the duty to feed my family, I am forced to steal. This chain of reasons may in turn be translated into the following causal chain: Environmental event → Deliberation and decision processes in the prefrontal cortex → Motor cortex → Act of stealing.

Thus, from the point of view of cognitive neuroscience, reasons for acting are efficient causes. Hence the reason/cause dichotomy results from psycho-neural dualism, which is ontologically shallow. Of course, for analytic purposes, and particularly to evaluate the merits and demerits of alternative reasons for taking action, one may distinguish reasons from causes. But distinction need not entail a separation, for the simple reason that there are neither disembodied reasons nor actor-free actions. We may only pretend that there are.

The preceding consideration should impact action theory, a branch of practical philosophy that has remained at a rather primitive stage because it has been approached from a linguistic angle, whereas what is needed in this discipline is *res, non verba* (deeds, not words).

Closing Remarks

Epistemology has traditionally been defined as the philosophical study of knowledge. In turn, knowledge is usually understood as the product or final state of cognition, or knowledge in the making. And cognition can be studied either psychologically or neuroscientifically. The pioneering studies of Jean Piaget on the way young children acquire certain key concepts are in the best classical tradition, that which overlooks the brain.

The neuroscientific approach to cognition should attract those who, like Quine, have been demanding the naturalization of epistemology. But of course it is easier, and therefore more popular, to demand the naturalization of epistemology than to work on this project and investigate cognition as a process in a brain immersed in a social surrounding.

This fondness for promissory notes, along with the persistence of psychoneural dualism, helps to explain the backwardness of the cognitive neuroscience of child development. But this very same imperfection is a powerful stimulus for research – as Rita Levi-Montalcini wrote in her scientific autobiography, *In praise of imperfection*. Which is one of the marvels of scientific research: that it can transmute scarcity into plenty.

Chapter 12
Brain and Computer: The Hardware/Software Dualism

The early-modern philosophers, particularly Descartes, regarded the mind as the reasoning engine: they overlooked motivation and emotion. The contemporary version of this ultra-rationalist view is computationalism, the thesis that all mental processes are computations.

From a philosophical viewpoint, computationalism comes in two varieties: materialist and idealist. The former asserts that *brains* are computers (e.g., Churchland and Sejnowski 1993; McCulloch 1965). By contrast, idealist computationalism holds that the *mind* is either a computer or a collection of computer programs, and in either case detachable from the anatomical "hardware" (Putnam 1960; Pylyshyn 1984). This is often called the "multiple realizability" thesis.

In either version, computationalism is the newest phase of the information-processing psychology that displaced behaviorism in the 1960s, and that characterizes brainless cognitive science, still the dominant school in psychology. Thus, Computerese is just a dialect of Informationese.

From a historical viewpoint, computationalism is a sophisticated version of behaviorism, for it only interpolates the computer program between stimulus and response, and does not regard novel programs as brain creations. See Fig. 12.1.

The root of computationalism is of course the actual similarity between brains and computers, and correspondingly between natural and artificial intelligence. The two are indeed similar because the artifacts in question have been *designed* to perform analogs of certain brain functions. And the computationalist program is an example of the strategy of treating similars as identicals. The history of science and technology shows that this strategy fails about as often as it succeeds. The same history also shows that success often blinds to failure. Let us glimpse at the brain-computer analogy from a philosophical viewpoint.

Computationalism raises at least three problems of interest to philosophers: the capabilities and limitations of computers; the respects in which computers

Stimulus → Black box → Response	Stimulus → Program → Readout
(a) Classical behaviorism	(b) Computationalism

Fig. 12.1 Classical (**a**) and contemporary (**b**) behaviorism

M. Bunge, *Matter and Mind*, Boston Studies in the Philosophy of Science 287,
DOI 10.1007/978-90-481-9225-0_12, © Springer Science+Business Media B.V. 2010

successfully mimic brains; and the degree to which theoretical computer science can double as theoretical psychology. Let us tackle these three problems.

12.1 Do Computers Reason?

Everyone admits that, whereas people can feel and reason, computers cannot feel anything. But most people believe that computers, or rather computer programs, are the greatest thinkers ever: the most accurate, fastest, and the only infallible reasoners. There is some truth to this, though not much. First of all, most mental processes are not computations: they are desires, hopes, fears, pictures, and other non-propositional items (Mellor 1991). Second, computers are severely limited even in the propositional realm. For example, they are incapable of detecting problems or inventing any programs or, a fortiori, the ideas involved in them (Bunge 1956a). No algorithm, no computation.

John Searle's (1980) ingenious and well-known Chinese Room argument should persuade anyone that computers are purely syntactic engines: that meaning (reference cum sense) escapes them by design. Gregory Chaitin (2006, 7), the well-known IBM computer scientist, put it this way: the computer-programming languages "are formalisms for computing and calculating, not for reasoning, not for proving theorems, and most emphatically not for inventing new mathematical concepts nor for making new mathematical discoveries."

However, a computer enthusiast might raise the following objection, as Hilary Putnam once did at a meeting: While present-day computers are indeed limited, nothing guarantees that this limitation won't be overcome in the future, because the mind is nothing but a set of computer programs. Let us see. An original invention is a process in two phases: design and realization. The design is a representation or description of the desired artifact, and the realization is the latter's construction. In the case of a conceptual artifact, such as a theory or a computer program, the two phases coincide: the artifact emerges as it is being described. And of course, by definition of "original," an original design is one that has never been described before – that is, one that is so far unknown. Thus the task of inventing a radically new idea, such as a novel computer program, is just as hard as the task of trying to find the Holy Grail: all one knows is the use or function it would perform if it were available.

More precisely, the inventor faces a hard inverse problem: Given the desired output, design an artifact capable of delivering it. And this is of course an underdetermined and therefore ill-posed problem, one requiring creativity far beyond the reach of anything working to rule, the way the known machines are supposed to work. Indeed, algorithms can handle only direct problems, such as calculating the next state of a Turing machine given its next-state function, current state, and input. A well-posed direct problem if soluble, has a single solution. By contrast, an inverse problem is open ended: it has either multiple solutions or none; and, if soluble at all, it calls for ad hoc tricks, such as additional special assumptions (see Bunge 2006).

Familiar examples of inverse problems are: finding the epicenter of an earthquake, the dispersion center of a biospecies, and the source of an epidemics; spotting

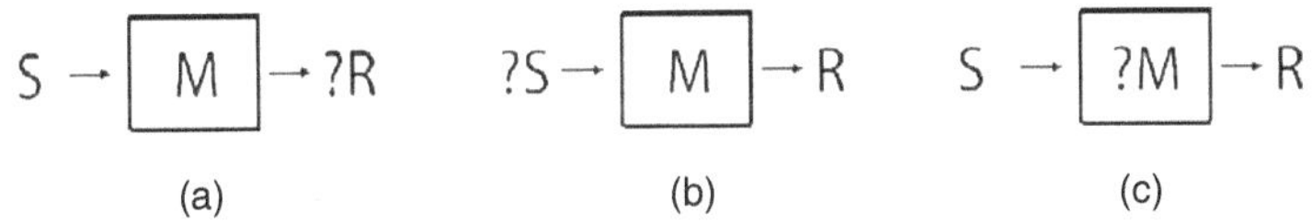

Fig. 12.2 (**a**) Direct problem: Given the mechanism *M* and the input(s) *S*, find the output(s) *R*. (**b**) Inverse problem: Given the mechanism and the output(s), find the input(s). (**c**) Hard inverse problem (identification): Given the input(s) and the output(s), find a mechanism

a place from which a given thing may become visible; identifying the cause(s) of a medical syndrome; and guessing the intentions behind a piece of human behavior. The most famous historical example of a direct problems is: Given a mechanical system and its equation(s) of motion, compute its orbit(s). The corresponding inverse problem is Newton's: Given the orbit(s) of a mechanical system, find its equation(s) of motion. Whereas the direct problem may demand some talent, solving its inverse counterpart required *the* Newton. See Fig. 12.2.

It is an article of scientific and technological faith that, in principle, all well posed direct problems, however hard, are soluble. No such faith is possible with regard to inverse problems, because they are in principle ill posed: their statement is incomplete, and consequently they demand extra guesses. And, because guessing cannot be programmed, inverse problems are beyond the ken of computer programs. Now, the most interesting and demanding problems in science, engineering, social technology, the humanities, and ordinary life are of the inverse kind.

Just think of the problem of inventing a theory to account for a mass of empirical data; of designing an artifact to accomplish a novel task; of designing a government program aiming at solving a given bundle of social problems; or of identifying the author of an unsigned text (or painting, or musical score, or scientific paper, or technological design, or artifact). In other words, the most intriguing and difficult problems, in all fields, are inverse and therefore beyond the reach of computer technology. Shorter: the most important tasks are not programmable. Or, to put it cynically: whatever is programmable is trivial even if laborious.

Evidently, once a task has been programmed, it may require expertise but not original brainpower. For example, there is an algorithm for discovering regularities of particular mechanical systems from data about their movements (Schmidt and Lipson 2009). But the resulting equations do not involve any high-level variables, such as mass, momentum, and energy; and they do not cover all the possible mechanical systems, the way theoretical mechanics does, so that they are useless for tackling new problems. Science cannot be automated, but many time-consuming scientific procedures, such as sequencing genomes, and detecting rare and novel particles in a collider, can be automated once the corresponding algorithms have been invented.

Still, it is often claimed that all the limitations of present-day computers might be overcome in the future. Of course prudence counsels never to say that something cannot be done. But the problem in hand is logical, not empirical. Indeed, the expressions "programmed spontaneity," "programmed creativity," and "automated programmation" are oxymorons, for it is in principle impossible to program an utterly unknown task, or to invent an idea of which one has no clue.

To use Cattell's (1987) well-known distinction, one may admit that computers are capable of "crystallized reasoning," while denying that they can engage in "fluid reasoning," which is the ability of tackling new problems. Indeed, no computer can be self-activated: it can only be activated by a computer program, and every such program is a detailed and precise prescription for attaining a goal of a known kind with means of a known type. One might as well order a monkey to design a piece of software.

To indulge in a technicality, a computation is what a Turing machine does. And this is for the machine to undergo a finite sequence of configurations following one another in accordance to precise rules. And one of these rules stipulates that a computer will stay in the same state unless it admits a stimulus. In other words, computers lack spontaneity or free will: they are at the mercy of their users. In particular, they lack the plasticity, freedom and willpower required to decide to learn new items through submitting voluntarily to a demanding apprenticeship.

Thus, contrary to a computer that has been badly damaged, the brain of a person who has suffered a severe insult, such as a stroke, a bullet wound, or the surgical removal of a substantial part, can recover the lost function to some extent: it can learn to walk on prosthetic legs, it can relearn to speak and calculate, and so on. Such self-repair ability is beyond any machine.

Of course, one may imagine robot surgeons, nurses and teachers capable of diagnosing and repairing damaged machines. But they would not be self-repairing artifacts, and they would be other-programmed rather an self-programmed. This is where Kant's distinction between the autonomous and the heteronomous agent comes in.

The enormous power of digital computers has reinforced mechanistic materialism, whereas their limitations have encouraged idealists. In mathematics, this conflict translates as the debate between those who believe that all of mathematics can be captured by formal systems, and those who emphasize the creative and therefore unpredictable aspect of mathematical work. Actually there is no need to take sides on this issue, because each of the contenders is partially right: In mathematics there is invention as well as discovery (or postulation and proof), and consequently the right attitude "straddles the mechanist and anti-mechanist viewpoints" (Feferman 2009). And such compromise does not involve restricting materialism: it only entails admitting that the Turing machine is not a true model of the human brain if only because, unlike the machine, the brain is spontaneous and creative.

12.2 The Computer Metaphor

One often reads in the psychological literature that animals "compute" every bit of experience, whether it be seeing or sweating, jumping or feeling thirsty, choosing or deciding, discovering or inventing, and so on. But surely this is just a metaphor, since computations proper are operations on symbols, whereas mental operations are processes in neural systems, which are made of living cells that communicate through physical or chemical signals, not symbols. These are artifacts and, moreover, the earliest symbols were invented no more than about 30,000 years ago, when our species had been in existence for about 100,000 years. It is therefore shocking to read that "there must be an addressable read/write mechanism in brains that encodes information received into symbols" (Gallistel and King 2009). Only the Holy Ghost could compete with the mystery of the miracle of the symbols embedded in brains before having been invented.

The further claim that neuroscience tries "to discover the algorithms used in the brain" (Sejnowski et al. 1988, 1300) is no less puzzling, since an algorithm is, by definition, a rule for calculating something, such as the logarithm of a number, whereas the brain "uses" biological laws. Likewise, planets do not compute their own orbits, but "use" (or "come with") Kepler's laws, which are natural regularities, not rules for processing information. Consequently, the search for logarithms in the untutored brain is similar to the search for wood in the cell on the strength of the traditional Chinese doctrine of the five elements, one of which would be wood. Since they are artifacts, algorithms cannot be discovered: they must be invented. And invention is anything but a rule-directed activity: it is an art. In particular, there can be no algorithms for designing algorithms, just as there are can be no rules for creating original works of art. The reason is obvious: By definition, any rule-directed task is routine rather than original.

Compare computational neuroscience with computational physics or chemistry. A computational physicist or chemist designs and uses computer programs to compute functions, or solve equations, that are known or hoped to represent physical traits, such as atomic or molecular energy levels. He is a theoretical (or rather mathematical) physicist or chemist who uses more advanced tools than paper and pencil. By contrast, a computational neuroscientist models brain "structures" (organs, regions) as computers. Hence he looks for algorithms embodied in neural networks: he shares the Pythagorean belief that the world, or at least the brain, is made up of mathematical objects. For example, he may claim that the untutored brain performs Bayesian inferences: that it arbitrarily assigns initial probability values ("priors"), and computes posterior probabilities using Bayes' theorem – where "probability" is equated with degree of belief. To me, this sounds like holding that planets compute the prior subjective probability of meteorite impact, and regulate their angular velocity so as to avoid the catastrophe. I suggest instead that the brain computes only when it does.

More precisely, I submit that a Computational X-science can research either of two possible research projects:

(A) To design computer programs and simulations to solve mathematical problems arising in Theoretical (or Mathematical) X-science, which in turn attempts to model X-facts; and
(B) To discover the algorithms lurking in X-facts.

I further submit that, while project A is being successfully carried out in physics and chemistry, project B is highly problematic, not to say wrong-headed.

Furthermore, the computer metaphor is incomplete, since we are seldom told how such "computations" are carried out – for instance, what are the algorithms for feeling surprise or fear, for falling in love or down the stairs, for asking questions or criticizing answers. Such imprecision is characteristic of an immature science. It is similar to the molecular biologist who assured us that DNA molecules "specify" proteins (or "instruct" about their synthesis), instead of exhibiting the corresponding chemical reactions.

Computationalism is popular not only among computer fans, but also among the cognitive psychologists and philosophers of mind reluctant to learning some neuroscience. By contrast, computer designers and robotics engineers have learned long ago that, to design better AI machines, they have to learn more about natural intelligence. After all, one cannot imitate X unless one knows something about X. And one of the things we do know about natural intelligence is that it involves memory; and that human memory, unlike computer memory, is reconstructive – impoverishing or enriching, always distorting – rather than replicating.

Philosophers seem to love the computer metaphor for two reasons. First, because it looks like a simple, rational, and modern solution to the age-old mind-body problem. Second, because computationalism is ontologically ambiguous: it attracts the (superficial) idealist because it assures him that matter is unimportant; and it also satisfies the materialist because it reassures him that mental processes are material. (In fact, computationalism is hylomorphist, whence it might satisfy a non-Christian Aristotelian.)

But of course the consistent idealist won't be fooled, if only because this neo-mechanist view of the mind makes no room for spontaneity, free will, or free creations – as Einstein characterized the theoretical concepts. Nor will the emergentist materialist be fooled by computationalism, because live brains can invent qualitatively new tricks, whereas the goal of a programmed process is fixed in advance by the user. There can be no such things as programmed spontaneity and ruled creativity, for the simple reason that these expressions are self-contradictory.

Of course computers are similar to human brains in some respect, but they are dissimilar in most. The most obvious difference between them is that brains, unlike computers, and machines in general, lack initiative, whereas brains are constantly active even in the absence of external stimulation (recall Section 9.3). This point is best appreciated by recalling that the Turing machine, which is the basic blueprint for the digital computer, produces no outputs unless it receives some inputs. Indeed, one of the postulates of the theory of Turing machines states that, if a Turing machine in a given state receives the null input, it remains in that state. In obvious symbols: For every s in the input alphabet, $M(s,0) = s$, where M is the transition or next-state function, and 0 designates the null input. By contrast, the transition function for a live brain would satisfy something like this: For every state s there is another state t, such that $t \neq s$, and $M(s,0) = t$.

12.3 Criticism

Computationalism, the view that all mental processes are operations upon symbols carried out by the brain in accordance with precise rules (algorithms), is open to the following objections.

1. The vast majority of the "computations" in question are just glorified hand-waving, since they are seldom specified. Hence to state that a bird, or an acrobat, "computed" a certain bodily movement, is no more informative than saying that it or she performed it.
2. Unlike computers, people and other animals are not designed: they are products of evolution and experience. The hardware-software distinction does not apply to people, because mental processes cannot be detached from the brain where they happen, except by abstraction. Human "software" grows with learning: it is subject to both development and evolution. And minds do not evolve by themselves: only minding brains evolve. Hence computationalism implies, mistakenly, the irrelevance of both developmental and evolutionary biology to psychology.
3. Computer programs, or algorithms, are not living things or processes but artifacts. Like machines, they are designed and repaired, ordered and sold, lost and stolen. Hence algorithms satisfy not only certain laws, but also certain technical norms (or conventions), some of which are smart but others less so – which is why they are continually being upgraded.
4. Only routine computations proper are algorithmic. All other mental processes, from feeling love, fear or hatred to guessing, inventing and criticizing, are non-algorithmic. In particular, there are no known rules for hitting on good ideas. Nor do we know the "neural computations" that allegedly guide computations proper.
5. Information-processing psychologists use freely the words "information" and "computation", but they make no use of the corresponding technical concepts, that are elucidated in Shannon's theory of communication and in metamathematics, respectively. Hence their discourse remains on an intuitive and metaphorical level. Actually, information-processing psychology is just classical psychology rewritten in Informationese.
6. Because computationalism is exclusively interested in rational (and particularly algorithmic) processes, it cuts the strong links between cognition, on the one hand, and motivation, emotion, and sociality on the other. Hence it can explain neither curiosity nor learning, nor the fact that social circumstances now stimulate, now inhibit learning. By contrast, the biological approach to cognition exhibits its strong interactions with emotion and social context (e.g. Phelps 2006).
7. Unlike computers, which are natural conservatives designed to obey, humans have the ability to innovate, disobey and cheat. In particular, they can craft,

discuss, criticize, and put into practice rules of conduct, some of which are moral principles, and others technical or legal norms. And, although any livable rules of conduct are informed by knowledge, they are not epistemic items. Moreover, such norms are motivated and enacted by social emotions, such as empathy, sympathy, pity, shame, pride, trust, and mistrust, which are beyond the ken of machines. In short, unlike people, computers have no moral sense. Further abilities beyond the ken of any conceivable machines, for not being programmable, are the ability to be self-conscious and self-critical; to distinguish the essential from the secondary; to use clues, cut corners, suggest alternatives, and understand metaphors; initiative (or spontaneity), creativity, common sense, curiosity, intuition, and systemic (as opposed to analytic or sectoral) thinking; and to handle continuity and actual infinity (except as symbols).

8. Computationalism ignores the fact that, unlike computers, brains are social and learn through interaction, adaptation, cooperation, and conflict.
9. Because they are social, human brains are n-th order intensional: they can get to know that someone else knows that someone else knows something, and so on. (First-order intensionality is exemplified by self-consciousness; and being able to form a "theory of mind," i.e., to figure out what other people are thinking, counts as second-order intensionality. Note the difference between intension and intention: the former is related to meaning, whereas intention is related to goal.)
10. Computationalism violates the very first principle of the philosophy of technology. This is the thesis that made things, unlike found ones, "embody" ideas, whence they constitute an ontological level of their own: that of things made rather than found. (See further limitations of computers in Bunge 1956a, 1985.)

In short, computationalism does not fare better than Cartesian dualism. The reason is that it, too, refuses to take the brain seriously. Both are functionalist: they adopt the stuff/function dichotomy. Hence neither of them can explain anything, since explanation consists in exhibiting mechanism. Basically, brainless psychology is no better than heartless cardiology. (More on the modes and limitations of fuctionalism in Mahner and Bunge 2001.)

12.4 Software is Rather Hard

A computer won't do anything unless a user feeds it a program. That is, an artificial information-processing unit is a system with three components: (a) a user with a brain trained to operate computers and, in particular, to encode and decode messages – that is, to pair symbols to meanings; (b) a computer together with its energy source; and (c) a computer program. In short, $IPU = U + H + P$. The removal of any of these three components disables the information-processing system IPU.

A computer program specifies a sequence of machine states. It is said to be a piece of *software* because it can be replaced with another program, and inserted into the same piece of hardware, to produce a different information-processing unit.

Still, the hardware/software nomenclature is misleading, because software is just as "hard" or material as hardware. True, unlike an ordinary piece of matter, a piece of software has a semantic content – or rather it elicits one. More precisely, a piece of software, when inserted into a computer, evokes meanings in its user's brain. Hence, the "content" of a piece of software is very different from the content of a bottle: it can only be poured into a trained brain – and only metaphorically at that. That is, software belongs in the semiotic level of reality, along with bank bills, sentences, and diagrams, neither of which fulfils its function without a brain able to understand its meaning. (Recall Section 4.1.)

The hardware/software distinction made in computer technology is useful but superficial. It is useful because one may use different pieces of software to activate the same machine. And also because software design is conceptually and socially different from the design of machine architecture, to the point that most software engineers, unlike their hardware counterparts, do not hold engineering degrees. But the distinction is ontologically shallow because every piece of software is materialized in a disk: the distinction in question does not involve a detachment.

The hard/soft distinction is even less justifiable in the case of animals, in particular humans. This is because human "software," the mind, is only a collection of brain functions, and these, unlike pieces of computer software, are undetachable from the hardware or brain. Furthermore, whereas brain "software" grows and decays with experience and reflection, a piece of computer software emerges fully developed and stays unaltered in a disk. Indeed, it has been known since the work of Sir Frederick Bartlett in 1932 that human memory is constructive as well as destructive, to the point that one may "recall" events that never happened (see Schacter 2001; Tulving 2002). Computers may fail but can be repaired, whereas brains may become enriched up to a certain point, after which they decay beyond repair.

12.5 Machine vs. Man?

It is widely believed that computers can replace chess players and even mathematicians. Deep Blue, the famous IBM machine, is said to have defeated the world chess champion Gary Kasparov. But this is mistaken, because the machine was programmed by Dr Feng-siung Hsu, aka CrazyBird; moreover, the program was upgraded after every move. What actually defeated Kasparov was not the machine but the system composed of the machine and Dr Hsu.

The claim that computers can do mathematics is equally mistaken: They only process physical (electromagnetic) correlates of mathematical concepts (Bunge 1956a). Moreover, such correlations are conventional. In particular, machines process numerals, not the numbers these designate. (Unlike numbers, which have conceptual properties, numerals have only physical and chemical properties.) Because they handle names, not concepts, one might say that computers are nominalist machines.

True, some computer programs prove theorems from given axioms and definitions. But these premises have got to be first invented by live mathematicians. It is also true that such machines are able to choose among alternative strategies – but someone has to invent such strategies, and equally live programmers have to feed them into the machines. In general, it is one thing to follow a program, and another to write a thoroughly new one. Again, it is one thing to select the most suitable means for attaining a given goal, and quite another to conceive of a goal and of a whole family of means to reach it.

Besides, and this is essential, all a computer program can do is to help solve a problem: Machines cannot discover new problems because there are no algorithms for inventing or discovering them. And such problems, to be tackled with the help of a computer, have got to be computational (programmable). But the vast majority of interesting mathematical problems are not of this kind. Indeed, consider this small sample:

- Formalize a hypothesis or theory stated in ordinary language.
- Given the first few terms of an infinite series, guess the general term.
- Invent methods for adding infinite series or products, or computing integrals.
- Given a continuous function, find the differential equation(s) that it satisfies.
- Find the premises (axioms, definitions and lemmas) that entail a given proposition. (That is, given B, find an A such that $A \vdash B$.)
- Discover the abstract theory underlying two different theories.
- Invent new algorithms.
- Given a set of empirical data, invent a theory (other than an inductive generalization) accounting for them.

In sum, machines cannot replace brains: at most they can assist us in performing some routine tasks. In particular, scientific research is not being fully automated, since only routine operations are rule-directed: finding problems, inventing ideas, and evaluating cannot be programmed.

But there is no doubt that the spread of the information and computing technology has altered the manner in which we live, think, work, and interact. Just look around and you will see plenty of people talking to a cell phone or texting – too busy to talk to themselves, examine their own lives, and plan. But this happens mainly in the public spaces in cities of the industrialized world: it is not typical of the Third World, where the vast majority of people live. Communication has always been coextensive with humanness, since humans are essentially sociable. But we do not grow, eat, wear, or clobber bits. So, it is not true that ours is the information society.

No doubt, life would become extremely difficult in the industrialized nations if all the communication channels were suddenly shut off. But it would not stop: We would keep feeding babies, using energy sources, growing vegetables, milking cows, transporting goods, moving around, studying, playing, and so on. And, as a compensation for the good things that the information channels deliver, we would be spared all the cultural garbage and political poison they spread. In short, contemporary society differs only in degree from its ancestor half a century ago. And the differences in social interactions are two-faced: the weak bonds (with strangers) have become stronger, while the strong ones (with relatives and friends) have been weakened, because warm face-to-face interactions are being replaced with cold screen-to-screen interactions.

Closing Remarks

Computer worshippers believe that the future of psychology belongs to AI. This is like saying that the future of human anatomy and physiology belongs to robotics. Since the goal of AI and robotics is to imitate us in some regards, they can advance only insofar as they learn about humans from the sciences of man. In general, to imitate anything, start by learning about he genuine article.

Much the same holds for the current vogue of the "its from bits" recipe in popular physics: it is wrong to try and reduce the natural to the artificial, and in particular to attempt to base physics upon engineering, rather than the other way round, and this for two reasons. First, because machines and engineering are artificial items built by brains, which belong in both nature and culture. Second, because the basic sets of digital computer science are denumerable, whereas reality is continuous in most respects, which is why continuous functions and differential equations have been so successful in physics and engineering.

In sum, computers are useful as long as they are regarded as aids to brains, not as substitutes for them.

Chapter 13
Knowledge: Genuine and Bogus

It has been said that humans can fake anything, even altruism and love, as well as science and philosophy. Some theories and practices can be recognized at first sight as being bogus and unrecyclable. This is the case of palmistry, homeopathy, and creation "science." Others demand plowing through esoteric texts. This is the case of phenomenology, which its founder, Edmund Husserl (1960), launched as a "rigorous science," yet at the same time as "the polar opposite" of science proper. Phenomenologists claim to be able to know anything instantaneously, apodictically, and exclusively through introspection – by pretending that the external world does not exist, instead of assuming it and exploring it. And, since they reject the scientific method, they do not feel obliged to offer any evidence for their claims. But, since phenomenologists have never produced anything other than phenomenology itself, why should we feel compelled to believe the claim that theirs is a rigorous science?

Assessing other theories and practices is harder, because they contain fragments of mathematics or of science. For example, alchemy has been the paradigmatic pseudoscience since Robert Boyle's time. But to their credit, in their mad pursuit of cheap gold from "base" metals, the alchemists designed many apparatuses and procedures that later proved indispensable to chemical research. Astrology was a wrong personality theory; but the art of casting horoscopes required precise astronomical knowledge – and fed a number of scientists. The Pythagorean brotherhood produced a mixture of science and pseudoscience: it created at once mathematics, theoretical physics, and mystical nonsense. Aristotle was the top logician of antiquity and the earliest marine biologist, but his physics and astronomy were wide of the mark.

Kant made important contributions to ethics, but invented a subjectivist physics and an aprioristic psychology that may have paved the way to the absurd *Naturphilosophie* of Hegel and Schelling, as well as to "interpretive" (*verstehende*) social studies. Marx mixed sound economics, brilliant historical insights, and sound social criticism with Hegelian nonsense and apocalyptic prophesying. The neoclassical economists pioneered the use of mathematics in social science but never bothered to put their hypotheses to the test. Freud speculated about emotions, sexuality, and unconscious processes – all of them neglected by the scientific psychology of his time – without concern for their empirical validation. (To this day there are no psychoanalytic laboratories.) And the founders of the quantum theory

M. Bunge, *Matter and Mind*, Boston Studies in the Philosophy of Science 287,
DOI 10.1007/978-90-481-9225-0_13, © Springer Science+Business Media B.V. 2010

misinterpreted it in the light of Berkeley's subjectivism, overlooking the fact that physicists are very late arrivals in the universe.

In short, sometimes intellectual gold comes mixed with muck – whence the need to design and use a sifting device. This is our task in the present chapter: to construct a sort of litmus test for ideas and procedures advertised as scientific. Such test should help not only to protect us from intellectual swindle but also to evaluate research projects.

13.1 Science and Pseudoscience

We shall be concerned only with the sciences and pseudosciences that claim to deal with facts, whether natural or social. Hence, we shall not deal with mathematics except tangentially, and only as a tool for the exploration of the real world. Obviously, this world can be explored either scientifically or non-scientifically. In either case such exploration, like any other deliberate activity, involves a certain approach. The latter may be construed as a set of general assumptions, together with some antecedent knowledge of the items to be explored, a goal, and a means.

In a way, the general assumptions, the extant knowledge of the facts to be explored, and the goal dictate jointly the means or method to be employed. Thus, if what is to be explored is the mind, if the latter is conceived of as an immaterial entity, and if the goal is to understand mental processes in any old way, then the cheapest means is to engage in free speculation. Given an idealistic assumption about the nature of mind, it would be preposterous to try and catch it by exploring the brain. If, on the other hand, mental processes are assumed to be brain processes, and if the aim is to understand the mechanisms underlying mental phenomena, then the scientific method, particularly in its experimental version, is mandatory. (This is the philosophical rationale of cognitive, affective and social neuroscience.) That is, whether or not someone studies the brain in order to understand the mind, depends critically upon her more or less tacit philosophy of mind.

In general, one starts research by picking a domain *D* of facts, then makes (or takes for granted) some general assumptions (*G*) about them, collects a body *B* of background knowledge about the *D*s, picks an aim (*A*), whether conceptual or practical, and in the light of the preceding determines the proper method (*M*) to study the *D*s. Hence, an arbitrary *research project* may be sketched as the ordered quintuple = <*D, G, B, A, M*>. The only function of this list is to keep track of the essentials in framing the following definitions.

A *scientific* investigation of a domain *D* of facts assumes that these are material, lawful and open to scrutiny, as opposed to immaterial (in particular supernatural), lawless or inscrutable; and the investigation builds on a body *B* of previous scientific findings, and it is done with the main aims (*A*) of describing and explaining the facts in question with the help of the scientific method (*M*). In turn, the latter may be summarily described as the sequence:

Choice of background knowledge – statement of problem(s) – tentative solution (e.g., hypothesis or experimental technique) – run of empirical tests (observations, measurements, or experiments) – evaluation of test results – eventual correction of any of the preceding steps, and new problems posed by the finding.

Contrary to widespread belief, the scientific method does not exclude speculation: it only disciplines imagination. For example, it is not enough to produce an ingenious mathematical model of some domain of facts, the way mathematical economists do. Consistency, sophistication and formal beauty are never enough in scientific research, the end product of which is expected to match reality – i.e., to be true to some degree. Pseudoscientists are not to be blamed for exerting their imagination but for letting it loose. The place of unbridled speculation is art, not science.

The scientific method presupposes that everything can in principle be debated, and that every scientific debate must be logically valid (even if no logical principles or rules are explicitly invoked). This method also involves two key semantic ideas: those of meaning and truth. Nonsense cannot be investigated, hence it cannot even be pronounced false. (Think of calculating or measuring times of flight using Heidegger's definition of time as "the maturation of temporality.") Furthermore, the scientific method cannot be practiced consistently in a moral vacuum. Indeed, it involves the ethos of basic science, which Robert K. Merton (1973) characterized as universalism, disinterestedness, organized skepticism, and epistemic communism – the sharing of methods and findings with the scientific community.

Finally, there are another four distinguishing features of any authentic science: changeability, compatibility with the bulk of the antecedent knowledge, partial overlap with at least one other science, and control by the scientific community. The first condition flows from the fact that there is no "live" science without research, and that research is likely to enrich or correct the fund of knowledge. In sum, science is eminently changeable. By contrast, the pseudosciences and ideologies are either stagnant (like parapsychology), or they change under pressure from power groups or as a result of disputes among factions (as has been the case with theology, Marxism, and psychoanalysis).

The second condition, partial intersection with tradition, can be restated thus. To be worthy of the attention of a scientific community, a new idea must be neither obvious nor so outlandish that it clashes with the bulk (though not the totality) of the antecedent knowledge. Compatibility with the latter is necessary not only to weed out groundless speculation but also to understand the new idea, as well as to check it. Indeed, the worth of a novel hypothesis or of a proposed experimental design is gauged partly by the extent to which it fits in with reasonably well-established bits of knowledge. (For example, telekinesis is out of the question if only because it violates the principle of conservation of energy.) Typically, the principles of a pseudoscience can be learned in a few days, whereas those of a genuine science may occupy a lifetime, if only because of the bulkiness of the body of background knowledge it is based upon.

The third condition, that of either using or feeding other research fields, follows from the fact that the classification of the factual sciences is somewhat artificial. For example, where does the study of memory fall: in psychology, neuroscience, or both? And which discipline investigates the distribution of wealth: sociology, economics, or both? Because of such partial overlaps and interactions, the set of all the sciences constitutes a system. By contrast, the pseudosciences are typically solitary.

The fourth condition, summarized as control by the scientific community, can be spelled out as follows. Investigators do not work in a social vacuum, but experience the stimuli and inhibitions of fellow workers (mostly personally unknown to them). They borrow problems and findings, and ask for criticisms; and, if they have anything interesting to say, they get both solicited and unsolicited advice. Such interplay of cooperation with competition is a mechanism for the generation of problems and the control and diffusion of results: it makes scientific research a self-doubting, self-correcting and self-perpetuating enterprise. This renders the actual attainment of truth less peculiar to science than the ability and willingness to detect error and correct it. (After all, everyday knowledge is full of well-attested trivialities that have not resulted from scientific research.)

So far, the distinguishing features of genuine factual science, whether natural, social, or biosocial. Let us now take a quick look at the kind of philosophy that favors the advancement of science.

13.2 Philosophical Matrix of Scientific Progress

I submit that the progress of science depends on conditions of three types: psychological, such as curiosity; social, such as research freedom and social support; and philosophical, such as realism. The first two conditions have been studied by several scholars, particularly Merton (1973). By contrast, the philosophical conditions have hardly been studied because of the dogma, common to both idealism and positivism, that science and philosophy are mutually disjoint; and yet they are no less important. I submit they are the ones summarized in Fig. 13.1.

Perhaps "humanism" and "scientism" are the only words in the diagram below in need of clarification. Humanism is understood here as a secular worldview together with a moral philosophy emphasizing the basic human rights and duties, as well

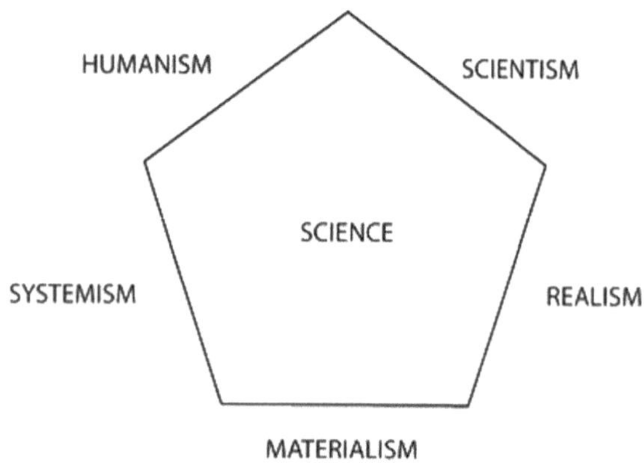

Fig. 13.1 The matrix of scientific progress. Imagine the prospects of scientific progress if scientism were replaced with irrationalism, realism with subjectivism, materialism with idealism, systemism with either holism or individualism, humanism with mercenariness, and the center with superstition

as the desirability of solving social issues in a peaceful and rational way. Thus a humanist scientist will abstain from using his skills to harm the innocent, and will put humankind before King and Country: he will side with Professor Einstein rather than with Dr. Strangelove. By contrast, a non-humanist scientist may engage in mercenary pursuits, and thus end up doing pseudoscience. Clear examples are the anthropologists who attempted to justify the racial prejudices of colonialists and Nazis, and the economists who could not care less for the social injustice and environmental destructiveness inherent in raw capitalism.

The word *scientism* too calls for elucidation because it is sometimes employed in a derogatory sense, and at other times confused with either positivism or naturalism. As understood in the present work, scientism is the research strategy that may be compressed into the following argument:

> Everything knowable is best studied scientifically.
> Mind, society and morals are knowable.
> ∴ Mind, society and morals are best studied scientifically.

This conclusion coincides with the definition of *scientisme* in Lalande's classical *Vocabulaire* (1938, II: 740): "the idea that the spirit and the methods of science should be extended to all walks of intellectual and moral life without exception." Condorcet (1782) was the first to formulate scientism explicitly. Positivists and Marxists adhered to it even when they failed to comply with it. The unity of science movement, which flourished between the 1930s and ca. 1960, was a blend of scientism and positivism. Ironically, the last fascicle of the *International Encyclopedia of Unified Science*, which started publication in 1938, was Thomas Kuhn's *Structure of Scientific Revolutions* (1962), the first constructivist-relativist salvo in the contemporary science wars. True, Kuhn later distanced himself somewhat from the social constructivists, and regretted writing his "purple passages". But in 1979 he wrote an unrepentant preface to the English translation of Ludwik Fleck's *Genesis and Development of a Scientific Fact*, of 1935. And, when asked by a *Scientific American* reporter whether the universe changes every time a scientific revolution occurs, Kuhn answered emphatically: "Of course!"

Scientism opposes the reactionary dogma, first stated by Kant and adopted by the so-called interpretive (or *verstehend*) approach to the study of mind and society, that these are beyond the reach of science, so that they should remain the property of the anti-empirical and innumerate. In particular, philosophical (or "humanistic") psychology and anthropology are the prescientific counterparts of the corresponding sciences: they belong in the same league as the *Naturphilosophie* of Hegel and Schelling.

Alfred Schütz, a prominent member of the *Verstehen* school and a phenomenological sociologist, conducted no empirical research. Worse, he discouraged scientists from tackling big social problems, such as unemployment, nationalism, and war. By contrast, Claude Lévi-Strauss did field work on Amazonian Indians, and Clifford Geertz studied village life in Bali. But, because both of them focused on the symbolic and ludic aspects of societies, while disdaining their so-called material basis, they were unable to explain how those peoples coped with shortages, natural disasters, land-owners, and central governments. This is why the writings of Lévi-Strauss and Geertz are closer to literature than to science – an evaluation that they might have agreed, since they were no friends of scientism.

Warning: Like anything else, scientism can be faked. The positivists unwittingly faked scientism for two centuries. Indeed, they proclaimed their love of science while attempting to force it into the phenomenalist straitjacket, which precludes conjecturing imperceptible entities and properties, such as atoms, genes, minds, and social forces. The confusion of positivism with scientism was so common in the second half of the nineteenth century, that it spilled over into politics. Indeed, positivism was so prestigious that, between ca. 1870 and 1920, the Argentine progressives swore by Comte and Spencer. These were the very same heroes of the self-styled scientific party headed by the dictator Porfirio Díaz, which froze Mexican society from 1876 to 1910.

Unsurprisingly, after that period the Argentine conservatives, as well as the Mexican progressives, replaced positivism with idealism. These changes in philosophical climate affected the social studies and the science and education policies, and thus the training and recruitment of scientists. But it did not affect the content of research in natural science: I can only see myself through an idealist microscope or telescope. However, let us go back to our philosophical pentagon.

The said pentagon is regular only in exceptional cases, such as those of Galileo, Newton, Euler, Berzelius, Ampère, Marx, Darwin, Bernard, Boltzmann, Ramón y Cajal, Einstein, and Keynes. (Actually, Newton and Marx are borderline cases. Newton, because he expected the Deity to replenish once in a while the momentum dissipated in friction; and Marx because he did not always distinguish between social science and political ideology.) We should add to the above the vast majority of modern chemists. Most of them have been materialists and realists, partly because chemists are more attracted to laboratory work than to theoretical speculation; and partly because, since the time of Dalton (1808), chemists have cultivated atomism – the most tangible fruit of ancient materialism.

In most cases the sides of the philosophical pentagon are unequal: Sometimes it is short on materialism, at other times short on realism, and in still other cases short on humanism – as in the cases of the great scientists who sell out to the powers that be. For example, until a century ago there were vitalist biologists; even nowadays some students of society claim to profess Neo-Kantian subjectivism (the *Verstehen* school); most economists worship standard economic theory, or at least pay lip service to it even after its repeated failures; and most nineteenth century physicists opposed the atomic theories because they shared the positivist bias against positing unobservable entities.

Paradoxically, most of the founders of the modern (quantum) atomic theory, particularly Bohr, Heisenberg, Born, and Pauli, followed the same positivist line. Indeed, they denied the autonomous (objective) existence of the very entities they were successfully describing. (Recall Chapter 3.) But of course they did not practice the subjectivism they preached. Indeed, their equations did not contain any psychological variables. This is why the positivist varnish can be scraped off quantum mechanics, leaving the neutral mathematical formalism, which can then be reinterpreted in a strictly objectivist fashion (Bunge 1967b).

(For example, according to the Copenhagen school, the eigenvalues of an operator representing a dynamical variable, such as the angular momentum, are values that an *observer* will *find* when *measuring* the said property. But this is false, for what ones reads in a meter are values of an indicator functionally related to a property, as in the case of the angle of a pressure gauge (see, e.g., Bunge 2006a). Hence the realistic reinterpretation of the said postulate: The eigenvalues of an operator representing a dynamical property are the values that this property may take. A second famous example is this. According to Born's orthodox interpretation, $|\psi|^2\Delta v$ is the probability of *finding* the particle in question inside the volume Δv. By contrast, on de Broglie's realist interpretation, the same quantity is the probability that the thing in question be *present* in Δv. The realist prefers the second interpretation for two reasons: because ψ does not involve any variables representing human actions, and because the two probabilities, of presence and of finding, are bound to be different, since finding a quanton depends on the search method as well as on the state of the quanton.)

In the case of quantum physics, a bad philosophy did not succeed in dashing the atomistic program of explaining the visible through the invisible. Also Newton unwittingly suffered from a parallel inconsistency: although his *Principia* contains the earliest hypothetic-deductive system in factual science, in the same work Newton included a philosophical defense of inductivism. The great investigator does not succumb to a bad philosophy; but the latter, which he may incur in his popular writings, may mislead philosophers and frustrate young vocations.

However, good philosophy does not guarantee good science. For example, the psychoneural identity hypothesis, adopted by medical psychology since the times of Hippocrates and Galen, did not help physicians to perform experiments or to craft models of brain functions, because physicians, with few exceptions, have been empiricists rather than scientific investigators.

Similar events happened in social science. For example, Marx was an important original social scientist despite claiming to use the dialectical "method" (ontology) that he had learnt from Hegel, one of the most confused and obnoxious philosophers in history. And a few Marxists made original contributions to anthropology, archaeology, and historiography thanks to their materialist, systemic, and scientistic approach (see, e.g., Barraclough 1979; Trigger 2006).

By contrast, no Marxist philosophers have gone beyond embroidering Marx's nonsensical dialectical ontology, because their task was to catechize, not to investigate. They kept doing Marxist scholasticism even after the precipitous fall of Soviet communism (see, e.g., Panasiuk and Nowak 1998). Likewise, although Max Weber's early work was marred by the neo-Kantian variant of subjectivism, his mature work was free from it despite his paying lip service to that antiscientific philosophy (Bunge 2007a).

In sum, a philosophy may be characterized as *progressive* if it promotes the advancement of knowledge, and *regressive* if it blocks it.

13.3 Pseudoscience

So far, the distinguishing features of genuine factual science, whether natural, social, or biosocial. Let us now list the distinguishing traits of pseudoscience. (More in Bunge 1998; Frazier 2009; Gardner 1983; Kurtz 2001; Mahner 2007c; Park 2008; Wolpert 1992.) A *pseudoscientific* treatment of a domain of facts violates some of the conditions listed in Section 13.1 above, while at the same time calling itself scientific. It may be inconsistent or it may involve unclear ideas. Or it may assume the reality of utterly unlikely items, such as alien abduction or telekinesis, selfish genes, innate ideas, brain-independent minds, memes, or self-regulating markets. The said treatment may postulate that the facts in question are immaterial, inscrutable, or both. It fails to build on previous scientific findings. It may perform deeply flawed empirical operations, such as inkblot tests, or it may fail to include control groups. It may fake test results, or it may dispense with empirical tests altogether.

Besides, the pseudosciences do not evolve or, if they do, their changes do not result from research. Thus, Ernest Jones (1961, 235) informs us that Freud's foundational work on the interpretation of dreams, first published in 1900, was reissued eight times in Freud's lifetime: "No fundamental change was ever made, nor was one necessary." Much the same can be said of neoclassical microeconomics, which has remained stagnant since its birth in the 1870s except for mathematical embellishments, as Milton Friedman (1991) triumphantly declared.

The pseudosciences are characteristically isolated from other disciplines, although occasionally they interbreed with sister bogus sciences, as witnessed by psychoanalytic astrology. And, far from welcoming criticism, they attempt to fix belief. Their aim is not to search for truth but to persuade: they posit arrivals without departures and without travel. Whereas science is full of problems, and every one of its important findings poses further problems, pseudoscientists are characterized by certainty. In other words, whereas science begets more science, pseudoscience is barren because it generates no new problems. In sum, the main trouble with pseudoscience is that its research is either deeply flawed or non-existent. This is why, contrary to scientific research, pseudoscientific speculation has not delivered a single law of nature or of society.

However, some disciplines and practices are hard to diagnose because they make intensive use of sound mathematics. Bayesian statistics and inductive logic are cases in point, for they involve standard probability theory, but assign probabilities to propositions, not to facts (states of affair and events), and they do so arbitrarily rather than in accordance with impersonal and explicit rules. Moreover, they do not even tell us what it means to say that a scientific conjecture is probable rather than true in the light of certain data, or plausible in the light of antecedent knowledge. Bayesian statisticians only tell us that "probability measures the confidence that a particular individual has in the truth of a particular proposition, for example, the proposition that it will rain tomorrow" (Savage 1954, 3). But since mere opinion does not qualify as science, Bayesianism is pseudoscientific (Bunge 2008b). For the same reason, inductive logic is a piece of pseudoexact philosophy. Ironically, Popper too assigned probabilities to propositions while rightly attacking inductive logic.

13.4 Immaterialism in the Study of Matter

Berkeley (1710) started the modern anti-science movement by rejecting the very notion of matter. Being a radical empiricist, he reasoned that knowledge starts with sensation, and that anything beyond sensation is unwarranted speculation. This included assuming that some sensations are caused by external objects, as well as attributing them primary (or subject-independent) properties, such as that of satisfying Newton's laws of motion. In short, Berkeley subordinated metaphysics to a radical empiricist epistemology: Antirealism entailed immaterialism.

Hume, Kant, Comte, Mach, and the logical positivists followed suit: they too adopted phenomenalism and rejected the concept of matter. But only Mach (1942), an eminent experimental physicist and psychologist, attempted to *prove* that there is no such thing as matter, and therefore materialism is false. He proceeded as follows. Mach focused on Newtonian mechanics, in particular the second law of motion, namely Force = Mass × Acceleration. Moreover, he defined "material" with "having mass". (He did not even acknowledge the existence of electrodynamics.) And he, Mach endeavored to prove that the concept of mass is redundant.

To achieve this goal, Mach analyzed the very special case of two bodies joined by a spring. In this case Newton's second law reads: $F_1 = m_1 a_1$, $F_2 = m_2 a_2$, and $F_1 = -F_2$. These equations entail $m_1/m_2 = -a_2/a_1$. This formula says that the ratio of masses equals negative the reciprocal of the ratio of accelerations. So far, so good. But now Mach takes the plunge: He posits that the given consequence of Newton's law *defines* the concept of relative mass: he conflates law and convention. And, since an explicit definition of the form "Definiendum = Definiens" renders the definiendum redundant, Mach concludes that the concept of matter is unnecessary, whence materialism is false.

That was not Mach's only logical *faux pas*: he also declared that the formula "$F = m.a$" defines "force". And, since he believed that the right-hand side of this formula is a product of accelerations, he declared that the concept force is just as redundant as that of mass. So, Mach crowed that he has slain the "stuff and force" dragon of the mechanistic materialism of Büchner, Vogt, and Moleschott. In short, Mach confused laws with definitions, and inverted the correct logical relation "Dynamics entails kinematics." He thus sacrificed Newtonian mechanics on the idealist altar. (More in Bunge 1966.) Regrettably, many physics textbooks adopted Mach's flawed version of classical mechanics Their authors did not realize that, in Newtonian mechanics, the concepts of acceleration, mass, and force are primitive (undefinable), and mutually independent, as shown by logicians (see Suppes 1957).

Another attempt to dematerialize physics was made when Einstein advanced his theory of gravitation: it was claimed that this effected the geometrization of physics. This claim is false: what the theory does is to represent the gravitational field by a manifold described by the so-called geometric tensor G. But the latter is in turn determined by the stress-energy tensor T, which describes the distribution of particles and fields in space-time, in accordance to the formula "$G = kT$". Since this formula can be read in either direction, we must conclude that matter and space shape each other. Hence, contrary to what Misner, Thorne, and Wheeler (1973) once claimed, Einstein's theory of gravitation does not realize William Clifford's 1870 dream of a space-time theory of matter.

Wheeler, a coworker of Niels Bohr's as well as Richard Feynman's thesis supervisor, was the most imaginative and persistent of all the distinguished physicists intent on slaying the materialist dragon. In 1960 he believed that the building material of the physical world is empty curved space-time; in particular, he wished "to build mass out of pure geometry." But his theory, geometrodynamics, did not work, for it made no room for spin ½ particles, such as electrons, protons, neutrons, and neutrinos. So, in 1970 Wheeler announced that propositions are the building blocks of the universe, so that physics is a "manifestation of logic". This did not work either. So, in 1990 Wheeler declared that the bit, or unity of information, is the basic or elementary component (see Barrow et al. 2004). In all three cases, physical entities, possessing energy, would be composed of nonphysical units, thus violating the principle of conservation of energy. Worse, none of the three Quixotic attempts mentioned above has solved any physical problem, and neither has succeeded in even bruising the said dragon.

In short, physics remains the basic science of matter. And the claim that mathematics may secrete physics is a Pythagorean fantasy, and one that violates Leibniz's dichotomy between truths of reason and truths of fact.

13.5 Exploring the Unconscious: Fact and Fancy, Science and Business

Although Sigmund Freud is usually credited with having discovered the unconscious, the truth is that people have been talking since time immemorial about doing certain mental operations "without thinking." In the middle of the Enlightenment, Leonhard Euler said "My pencil knows more than I." One century later Hermann von Helmholtz, another great polymath, wrote about unconscious inferences. The self-taught philosopher Edward von Hartmann published his monumental best-seller, *Das Unbewusste*, when Freud was 13 years old. In his classic *Principles of Psychology* (1890), William James wrote about "the fortunate lapse of consciousness" that allows us to leave a warm bed to go to work. And of course Pavlov got the Nobel prize in 1904 for having shown experimentally (not just claimed) that animals can learn to perform certain actions in an automatic (unconscious) fashion. What Freud did was to claim that most of our mental life is unconscious, that "the unconscious never lies," and that dreams reveal our subconscious desires – provided of course they are suitably "interpreted" by analyst. He did not propose interpretation rules; he overlooked the fact that many dreams are unrealistic or even irrational; and he never put his conjectures to the experimental test: he relied on the gullibility of his readers.

While no one disputes the Pavlovian ("Freudian") thesis that many, perhaps most, of our actions are unconscious, no serious scientists hold dreams in awe, although a few do study dreaming in the laboratory; and most of Freud's fantasies, in particular the Oedipal one, have been discredited. Eric Kandel (2006, 363) is likely to be the only eminent scientist to have held that the emergence of psychoanalysis was a revolution. The serious historians of psychiatry, such as Shorter (1997), know that it was a *counter*revolution, for it replaced experimental psychology, that had been officially born in 1879, with wild (and often silly) speculation. In fact, while there have been thousands of profitable psychoanalytic offices around the world, no psychoanalytic laboratories have ever been set up since Freud started his school in 1900. In this regard psychoanalysis is even less scientific than parapsychology.

The scientific study of unconscious mental processes began only in mid-nineteenth century with observations on split-brain and blindsight patients. Since then, the various brain-imaging techniques, such as PET scanning and functional MRI, have made it possible to ascertain whether someone feels or knows something even though he or she is not aware that he feels or knows it. Moreover, these techniques make it possible to localize such mental processes in a non-invasive way. An example is the paper by Morris, Öhman, and Dolan (1998) – which, unsurprisingly, does not cite any psychoanalytic studies. Let us peek at it.

The amygdala is the tiny brain organ that feels such basic, strong, and ancient emotions as fear and anger. When this system gets damaged, a person's emotional and social life is severely stunted. The activity of the amygdala can be monitored by a PET scanner: This device allows the experimenter to detect a subject's emotions, and even to locate them in either side of the amygdala. However, such neural activity may not reach the conscious level. In this case, only a brain scanner can help.

For example, if a normal human subject is briefly shown an angry face as a target stimulus, and immediately thereafter an expressionless mask, he will report seeing the latter but not the former. Yet, the scanner tells a different story. It tells us that, if the angry face has been associated with an aversive stimulus, such as a burst of high-intensity white noise, the amygdala is activated by the target even though the subject does not recall having seen it. In short, the amygdala "knows" something that the organ of consciousness (whichever and wherever it may be) does not. Psychoanalysts could use this very method to measure the intensity of a male's hatred for his father. But they don't, because they don't believe in the brain: their psychology is idealistic, hence brainless.

The number of examples of pseudoscience can be multiplied at will. Astrology, alchemy, parapsychology, characterology, creation "science," "intelligent design," Christian "science," dowsing, homeopathy, and memetics are generally regarded as pseudoscientific. (See, e.g., Kurtz 1985; Randi 1982; and *The Skeptical Inquirer*.) On the other hand, it is less widely accepted that psychoanalysis, widely regarded as the science of the unconscious, is a fake science too. Let us check whether it meets the conditions that, according to Section 13.1, characterize the mature sciences.

To begin with, psychoanalysis violates the ontology and the methodology of all genuine science. Indeed, it holds that the soul ("mind" in the standard English translation of Freud's works) is immaterial, yet can act upon the body, as shown by psychosomatic effects. However, psychoanalysis does not advance any mechanisms whereby an immaterial entity can alter the state of a material one: it just states that this is the case. Moreover, this statement is dogmatic, since psychoanalysts, unlike psychologists, do not perform any empirical tests. Freud himself had emphatically dissociated psychoanalysis from both experimental psychology and neuroscience. So much so, that the course of studies at the Faculty of Psychology that he sketched did not include any courses in either discipline.

To mark the first centenary of the publication of Freud's *Interpretation of Dreams*, the *International Journal of Psychoanalysis* published a paper by six New York analysts (Vaughan et al. 2000) who purported to report on the first experimental test ever of psychoanalysis in the course of one century. Actually this was no experiment at all, since it involved no control group. Hence those authors had no right to conclude that the observed improvements were due to the treatment: they could just as well have been spontaneous. Thus, psychoanalysts make no use of the scientific method because they do not know what it is. After all, they were not trained as scientists but only, at best, as medical practitioners.

The French psychoanalyst Jacques Lacan – a "postmodern" cult figure – admitted this and held that psychoanalysis, far from being a science, is a purely rhetorical practice: "*l'art du bavardage*." Finally, since psychoanalysts claim that their views are both true and effective, without having submitted them to either experimental tests or rigorous clinical trials, they can hardly be said to proceed with the intellectual honesty that scientists are expected to abide by (even if they occasionally lapse). In sum, psychoanalysis does not qualify as a science. Contrary to widespread belief, psychoanalysis is not even a failed science, if only because it makes no use of the scientific method and ignores counterexamples. It is just quack clinical psychology.

13.6 Speculative Evolutionary Psychology

The moment Darwin's theory of evolution emerged it became clear that, as Theodosius Dobzhansky famously put it, "nothing makes sense in biology except in the light of evolution". Regrettably, it has not always been clear what evolution itself is. Indeed, Darwin's theory was distorted from the start. It was distorted by Herbert Spencer, who though that evolution consists in the survival of the fittest and illustrated the universal progression from the simple to the complex; by the social Darwinists, who claimed that it confirmed the ancient myth that social standing is in the blood; by Richard Dawkins, who held that evolution is essentially a sequence of self-replicating and self-serving genes, so that the very existence of organisms is paradoxical – an idea that renders biology redundant; and by Daniel Dennett, who claims that evolution is guided by algorithms, i.e., computation rules – alas, unspecified ones. The latest bastard of evolutionary biology is the fashionable Evolutionary Biology (see Barkow et al. 1992; Buss 2004; Pinker 2003).

The worthy goal of this discipline is to trace the origin and evolution of such mental abilities as speech and moral judgment, as well as such social attitudes as jealousy, altruism, dominance, incest-avoidance, and violence, And the torch used in this exploration is the hypothesis that mind and behavior are ruled by genes, which in turn are insensitive to changes at the organismic and social levels. That is, there would be no intermediaries between molecule and mind.

One might object that, since, primitive man is no longer around, and since neither brains nor ideas fossilize, evolutionary psychology is untestable, hence confined to concocting entertaining tales. But the evolutionary psychologists have an ingenious answer: We are still primitive, our minds were formed during "the Pleistocene," and humans stopped evolving long ago. Actually, they tell us, we are essentially living fossils. All we have to do to explain present mental and behavioral traits is to figure out the problems that our remote ancestors, the fabled hunter-gatherers, might have faced.

Everyone assumes that the circumstances of our remote ancestors were quite different from ours. But the evolutionary psychologists assure us that human nature, which is defined by a set of "domain-specific" (special purpose) algorithms "designed" to cope with the problems faced by our remote ancestors, has not changed significantly over the past 100,000 years or so. Suffice it to imagine how we would feel, think, and act if we lived in "the Pleistocene." This is how the evolutionary psychologists have made up dozens of fascinating stories – mostly about sex and dominance – that purport to account for nearly every mental and social trait. They also purport to explain our shortcomings. For instance, we find it difficult to imagine tiny invisible particles moving at very high speeds because our minds were designed to cope with visible slow-moving things. Obviously, this conjecture fails to explain why modern physicists, chemists, and biologists are much more proficient at dealing with microphysical entities than at throwing javelins or fashioning stone axes.

Let us glimpse at just five of the basic assumptions of evolutionary psychology. The first one is borrowed from pop genetics, pop ethology, pop anthropology, and pop psychology, particularly psychoanalysis. This is the triple hypothesis that (a) sex is the prime mover of all animal behavior; (b) all males desire to spread their seed (or their genes) as widely as possible, while (c) all females prefer the strongest (or richest) partners because they wish to produce the best offspring. These assumptions presupposs in turn that (a) all drives are ultimately rooted in sex; (b) all animals always seek sex *consciously* rather than instinctively; and (c) they do so because they *know* that copulation leads eventually to pregnancy. But it is well known that (a) fear, hunger and thirst trump sex; (b) most animal behavior is instinctive, and (c) the coitus-pregnancy connection, though presumably suspected by people very long ago, was confirmed experimentally only in the eighteenth century by Lazzaro Spallanzani working with frogs. In sum, only Freud and the Pope share the obsession with sex that characterizes evolutionary psychologists.

Another basic assumption of evolutionary psychology is that the mind is a computer that runs on innate algorithms. We rejected this non-biological view of mind in Chapter 12, for it overlooks (a) the huge differences between artifacts, such as computers and algorithms, and natural items, such as brains and the laws of nature inherent in them; (b) spontaneity, initiative, and creativity – the ability to invent original ideas and actions that do not respond to environmental stimuli; and (c) sociality. Besides, the thesis of inborn knowledge is inconsistent with developmental neuroscience and psychology, as well as with the experience of parents and teachers: We are born ignorant of nearly everything, though of course endowed with the organ of learning anything.

A third basic assumption of evolutionary psychology is that the mind is composed of hundreds or thousands of mutually independent modules, or microcomputers, every one of which performs a given task, such as detecting cheaters or identifying potential mates. We know from cognitive, affective and social neuroscience that the brain is indeed composed of specialized subsystems, such as the ones for perceiving faces or sounds. But we also know that these modules are not mutually independent. For example, perception depends not only upon the current stimulus but also upon memory and expectation. Thus, if we are waiting for someone at a street corner, we are likely to misidentify many a passerby as the expected individual.

A fourth basic hypothesis of evolutionary psychology, and the one distinctive of it, is that our minds were "designed" by natural selection to cope with "the Pleistocene environment," that started nearly 2 million years ago and ended some 50,000 years ago ago. There are several problems with this claim. One of them is that it denies the existence of the modern mind, characterized by rationality and abstraction. Another problem is that it does not explain the fact that most of us manage far better in the urban jungle than in the wilderness. A third problem is that it is in principle impossible to know which were the precise survival problems that our remote ancestors faced, hence the mental abilities (or modules) that were selected. We are born learners, not learned; adaptable, not adapted, let alone to any of the long-gone and largely unknown Pleistocene environments – which, incidentally, are likely to have undergone huge changes every few thousand years. Last, but not least, it is not true that our genome has remained unchanged over the past 50 millennia. For example, the ability to digest milk after infancy emerged in some peoples only about five millennia ago, with the domestication of goats and cows. (More on the genetic changes brought about by the invention of agriculture about 10,000 years ago in Cochran and Harpending 2009.)

Finally, another postulate of evolutionary psychology (as well as of standard economic theory) is that we are all basically selfish: that altruism is nothing but smart egoism. In other words, it is assumed that we are weak reciprocators rather than strong ones: that we do something for others only because we expect them to eventually pay us back. This hypothesis overlooks bonding and the social emotions, particularly empathy. Ethologists have always known that gregarious animals "invest" in cooperation leading to long-term bonds, without which there would be no animal societies. And more recent observation and experiment have also shown that humans and other vertebrates engage in genuinely altruistic actions (e.g., Gintis et al. 2005); and that they do so not only after careful deliberation, but often instinctively and out of empathy (e.g., Preston and de Waal 2002). As de Waal (2008, 292) puts it, the beauty of the empathy-altruism connection is that "the mechanism works so well because it gives individuals an emotional stake in the welfare of others."

Which findings do evolutionary psychologists boast about? Here is a sample of favorites (Buss 2004). (1) Typically, men invest less than women in parental care because they are uncertain about paternity – not because they spend more time working away from home. (2) Women prefer strong or rich men, individuals able to support them and their offspring, as proved by the fact that, in singles bars, females tend to make advances to prosperous-looking males – not because the ones who look for such men frequent such bars. (3) Depression has not been eliminated by natural selection because it is good for you: it induces you to lower your bar to a more realistic height. (4) Social dominance has nothing to do with economic and political power: it has purely biological roots, and an exclusively sexual motivation. (5) All social features are encoded in the human genome, whence social revolutions have at best ephemeral results (Barkow 2006). (6) Cultural activities are only "mating strategies." Thus, Plato, Thomas Aquinas, Leonardo, Michelangelo, Descartes, Spinoza, Newton, and other famous bachelors presumably worked only to seduce women. Likewise the inventors of writing, computing, chronicling, philosophizing, and the like, must have been just as sex-crazed as monkeys and apes.

In short, the main hypotheses of evolutionary psychology are either untestable or testable but implausible, hence unscientific in either case. (See further criticisms in Buller 2005; Cochran and Harpending 2009; Gould 1997a; Lickleiter and Honeycutt 2003; Lloyd 1999; Smail 2008.) Still, the project of constructing a scientific evolutionary psychology is interesting. Whether it is viable, is yet to be seen.

So much for pseudoscience. The subject of its underlying philosophy is intriguing and vast, yet largely unexplored. (See, however, Bunge 2006c; Flew 1987.) Just think of the many pockets of pseudoscience ensconced in the "hard" sciences, such as the anthropic principle in cosmology; the "its from bits" myth and the attempt to craft a theory of everything; information talk in biochemistry; the "it's all in the genes" dogma in biology; human sociobiology; equilibrium economics; and game-theoretic models in economics and political science. Analyze an egregious error in science, and you are likely to find a philosophical bug.

13.7 Borderline Minefields: Proto and Semi

Every attempt at classifying any collection of items outside mathematics is likely to meet borderline cases. The main reasons for such vagueness are either that the classification criteria themselves are imprecise, or that the item in question possesses only some of the features necessary to place it in the box in question. For instance, we still do not know how to tell with certainty certain hominins from humans.

Whatever the reason, in the case of science we find plenty of disciplines, theories, or procedures that, far from falling clearly either in science or outside it, may be characterized as protoscientific, semi-scientific, or as failed science. Let us take a quick look at these cases.

A *protoscience*, or emerging science, is obviously a science *in statu nascendi*. If it survives at all, such field may eventually develop either into a mature science, a semi-science, or a pseudoscience. In other words, at the time a discipline is said to be a protoscience, it is too early to pronounce it scientific or nonscientific. Examples: physics before Galileo and Huygens, chemistry before Lavoisier, and medicine before Virchov and Bernard. All of these disciplines eventually matured to become fully scientific. (Medicine and engineering can be scientific even though they are technologies rather than sciences.)

A *semi-science* is a discipline that started out as a science, and is usually called a science, yet does not fully qualify as such. I submit that cosmology, psychology, economics, and politology are still semi-sciences despite their advanced ages. Indeed, cosmology is rife with speculations that contradict solid principles of physics. There are still psychologists who deny that the mind is what the brain does, or who write about neural systems "subserving" or "mediating" mental functions. And of course many of the so-called Nobel prizes in economics are often awarded to inventors of mathematical models that have no resemblance to economic reality – if only because they ignore production and politics – or to designers of economic policies that harm the poor.

So much for glaring cases of pseudoscience. In some cases it is hard to know whether something is scientific, semi-scientific, or pseudoscientific. For instance, the vast majority of nineteenth-century physicists regarded atomistics as a pseudoscience because it had produced only indirect evidence for the atomic hypothesis. Worse, since there was no detailed theory of individual atoms, atomistics was only weakly testable, namely through the predictions of statistical mechanics and chemistry. But the theory became scientifically respectable almost overnight as a consequence of Einstein's theory of Brownian motion and Perrin's experimental confirmation of it. Only die-hard positivists, like Ernst Mach and Pierre Duhem, opposed atomism to the last.

Another case of philosophical interest is string theory. It looks scientific because of its impressive mathematical apparatus, and because it tackles an open problem that is both important and hard: that of constructing a quantum theory of gravitation. For this reason, and because it has generated great mathematics, it is still attracting some of the brightest young brains. But the theory postulates that the physical space has ten dimensions rather than three, just to secure the mathematical consistency of the theory. Lisa Randall, a prominent string theorist, speculated that physical space had initially all the extra dimensions, but lost them as it aged. This maneuver certainly saves the theory from refutation, but it sounds like an excuse, particularly since no shrinking mechanism was advanced. Since the extra dimensions are unobservable, and since the theory has resisted experimental confirmation since its inception in 1968, it looks like science fiction, or at least like failed science, as one of its erstwhile enthusiasts has admitted (Smolin 2006).

The case of phrenology, the "science of skull bumps," is instructive. It revived Galen's testable materialist hypothesis, that all mental functions are precisely localizable brain processes. But, instead of putting this exciting hypothesis to the experimental test, the phrenologists sold it successfully at fairs and other places of entertainment: They went around palpating people's skulls, and claiming to locate such the centers of imagination, altruism, philoprogenitivity, and so on. The emergence of modern neuroscience at the hands of Ramón y Cajal finished off phrenology. What is emerging now is a symbiosis of localization with coordination, as we saw in Section 9.3.

The discredit of phrenology casts doubt not only on radical localizationism but also on the scientific attempts to map the mind onto the brain. In particular, the brain-imaging devices invented over the past three decades were first greeted with skepticism because the very attempt to localize mental processes sounded like phrenology. But these new tools have proved very fruitful and, far from confirming the phrenological hypothesis (one module per function), it has given rise to many new insights, among them the view that all the subsystems of the brain are interconnected, and that such connections are the stronger, the more difficult the task. If a tool or a theory leads to important findings, it cannot be pseudoscientific, because one of the marks of pseudoscience is that it is built around some old superstition.

A useful procedure for evaluating the scientific character of any new approach is to assess separately its body H of substantive hypotheses and its method M. In this way we get a 2×2 matrix

$$\begin{matrix} HM & H\overline{M} \\ \overline{H}M & \overline{HM} \end{matrix}$$

I submit that, at present, cognitive neuroscience is in the NW corner (right H and M), whereas two centuries ago phrenology was in the NE corner (right H, wrong M). Currently the Gestalt school, behaviorism, and cognitive psychology (in particular computationalism) are in the SW corner (wrong H, right M), while psychoanalysis, parapsychology and contemporary evolutionary psychology dwell in the SE one (wrong H and M). This is the pseudoscientific corner. The non-diagonal items should be regarded as semisciences: they may eventually develop either into full-fledged sciences or into pseudosciences – or they may quietly disappear for lack of new findings.

A final word of caution. Most of us are suspicious of radically new theories or tools, and this for either of two reasons: because of intellectual inertia, or because it is necessary to grill every newcomer to make sure that it is not an impostor. But one must avoid confusing the two reasons. Rigorous inquisitive types like novelty, but only as long as it does not threaten to discredit or dismantle the entire system of knowledge.

13.8 The Pseudoscience-Politics Connection

Pseudoscience is always dangerous because it pollutes culture and, when it concerns health, the economy, or the polity, bogus science puts life, liberty or peace at risk. But of course pseudoscience is supremely dangerous when it enjoys the support of governments, organized religions, or large corporations. A handful of examples should suffice to make this point.

From the Enlightenment on, most progressives have held that genome is not destiny: that we can learn not only to think but also to feel and act, both directly through imitation and learning, and indirectly through social reform. By contrast, conservatives and reactionaries of all stripes have adopted nativism, the view that we are born with all the traits that emerge in the course of life. Thus, the Hindu sacred scriptures consecrated the caste system; the Bible held that the Jews were chosen by Yahveh; Aristotle, that "barbarians" are inferior to Hellenes; the European colonists, that the conquered peoples were savages only good to be enslaved or exterminated – and so on. The nativism-conservatism connection was considerably weakened by the Enlightenment and the subsequent spread of leftists ideologies, but it resurfaced once in a while, with particular virulence as social Darwinism and, more recently, under the wing of Evolutionary Psychology. Let us recall the most recent revival of "scientific" nativism.

Steven Pinker (2003), a Harvard professor and the most popular psychologist of the day, devotes an entire chapter of one of his influential books to the political issues surrounding the environmentalism/nativism dilemma. He states that "the new sciences of human nature," from genetics to evolutionary psychology, vindicate what he calls the Tragic Vision. This is none other than the individualism and pessimism of orthodox economics and conservative political philosophy, from Hobbes to Burke to Schopenhauer to Nietzsche to Hayek to Thatcher to Reagan.

Pinker cites (op. cit.: 294), in particular, the following "discoveries" of those "new sciences": "the primacy of family ties" – despite the fact that in most cases the members of business firms, political cliques, laboratories, regiments, and sports teams – are not genetically related; "the limited scope of communal sharing in human groups"–although all primitive societies and many modern enterprises are cooperatives; "the universality of dominance and violence across human societies" – even though the murder rate has been declining in all civilized societies over the past century, and not even the most divided societies are basically tyrannical and violent; and "the universality of ethnocentrism and other forms of group-against-group hostility across societies" – as if such undeniable strife were not balanced by cooperation, law-abidance, and material interests.

But that is not all: to persuade us that basically we are all selfish nasty brutes, Pinker (loc. cit.) completes the previous list with the following: "the partial heritability of intelligence, conscientiousness, and antisocial tendencies" – although all such capacities can be either enhanced or repressed by education and informal social control; "the prevalence of defense mechanisms, self-serving biases, and cognitive dissonance reduction" – which, though real, are surely less severe in welfare societies than in "liberal" ones; "the biases of the human moral sense," including nepotism and conformity – true enough, but let us not overlook the facts that altruism and disconformity occur along with selfishness, and that political progress has often involved moral progress. The preceding is a clear example of failed radical reductionism, in this case of social science to genetics and psychology. Besides, Pinker's list of accomplishments of the "new science of human nature" reads like the preamble to a New Right Manifesto rather than a summary of scientific findings. Commitment to a regressive political ideology is a reliable indicator of the pseudoscientific nature of a discipline.

Much the same holds for the self-styled evolutionary psychologists, whom Pinker admires: they too assert confidently that social inequality is in our genes, and consequently social revolutions are bound to fail. Thus Barkow (2006, 37), one of the founders: "social stratification is a reflex of the evolutionary fact that people do want more for their children than they want for the children of others." But surely class barriers, by definition, slow down social mobility or even prevent it altogether.

Hence only a classless society, or at least one with porous social barriers, allows for personal advancement. Note that this is a purely logical argument. What needs empirical evidence is the assumption that ambition for one's offspring is inborn and therefore universal. But human genetics has not provided evidence for this piece of pop genetics.

Far from having biological roots, social stratification has a strong impact on the quality and length of life: People on top live better and longer than their subordinates. The psycho-neuro-endocrine-immune mechanism is roughly this: Subordination → Stress → Release of cortisol → Rise in blood pressure and blood sugar levels → Higher morbidity. This is why life is better and longer in Japan and the Nordic countries than in the more inegalitarian countries, such as the United States and the United Kingdom (see, e.g., Wilkinson and Pickett 2009).

Besides, social archaeologists have found that social stratification emerged only about five millennia ago, along with civilization. As Trigger (2003b, 44) informs us in his monumental treatise, "[a]nthropologists apply the term 'early civilization' to the earliest and simplest form of societies in which the basic principles governing social relations was not kinship but a hierarchy of social divisions that cut horizontally across societies and were unequal in power, wealth, and social prestige." However, let us move on to other pseudoscience specimens.

Eugenics, once promoted by many bona fide scientists and progressive public intellectuals, was invoked by American legislators to introduce and pass bills that restricted the immigration of people of "inferior races," and led to the institutionalization of thousands of children regarded as mentally feeble. The racial policies of the colonial powers and the Nazis were justified by the same "science," and led to the enslavement or murder of millions of Amerindians, Indians, Blacks, Slavs, Jews, and Gypsies.

The worldwide economic crisis that started in 2008 is a more topical example of the disastrous social consequences that result from social policies inspired by wrong economic and political philosophies. Indeed, there is consensus that this crisis is to be blamed on the laissez-faire policies pursued by the American and British governments since the days of Ronald Reagan and Margaret Thatcher. Now, laissez-faire is not a stray ideological slogan: it is the logical consequence of two dogmas that have been held uncritically, despite sea changes in economic reality, since Adam Smith (1776) advanced them in his great work. These are the principles (a) that the only goal of economic activity is the pursuit of private profit; and (b) that the free (unregulated) market is self-regulating – that it is always in equilibrium or near it, whence any intervention is bound to hurt it.

In turn, the above hypotheses rest on three unexamined philosophical doctrines: an individualist ontology, an unscientific epistemology, and an individualist ethics. Individualism is the thesis that there are only individuals: that collective entities, such as business concerns and nations, are figments of the imagination. This thesis is wrong: what is fictitious is the isolated individual. As we have argued elsewhere (e.g., Bunge 2003a), everything in the real world is either a system or a component of one. In particular, an individual's actions can only be understood in her social context. One may start analysis at either the micro or the macrolevel, but no analysis will be satisfactory if it misses either. The methodological lesson is that any satisfactory explanation of a social fact will involve what I have called Boudon-Coleman diagrams (Bunge 1996). Here is a recent example:

Macro-level	Easy credit	→	Real-estate bubble			Crisis
		↘	↗	↘		↗
Micro-level		Rise in housing demand			Default	

Boudon-Coleman diagrams go against the grain of radical methodological individualism, which enjoins staying always at the microlevel. This methodological viewpoint cannot remain neutral in the epistemological controversy between realism (or objectivism) and subjectivism: if consistent, it starts from individual cognitive experience rather than from knowledge, which is learned in society and checked by scientific communities (Merton's "organized skepticism"). Hence the methodological individualist should be either a radical subjectivist (like Berkeley, Kant, Fichte, or Husserl), or a radical empiricist (like Hume, Comte, Mill, or Carnap). Popper's combination of radical methodological individualism with epistemological realism does not work.

Just as holism comes with an ethics of duty, such as Confucius' and Kant's, individualism is wedded to the egoist devise "Everyone for himself." Systemism, by contrast, suggests a humanistic ethics where rights and duties are on a par. In such a moral philosophy every right implies a duty, and conversely. For example, my right to earn a living implies my duty to help others survive; and my duty to pay taxes implies my right to have a say in the way they are spent. I submit that ordinary folks abide by such moral philosophy, whereas orthodox economists and conservative politicians preach deontology to the masses while advising egoism to their clients.

All of the advanced economies are ruled by economic policies of some kind. In turn, such policies are designed on the strength of economic theories and moral principles, and they are pushed or executed by political parties and governments:

POLITICS
↓
ECONOMIC POLICIES
↗ ↖
ECONOMIC THEORIES MORAL PHILOSOPHIES

The orthodox economist will object to the inclusion of politics and morals among the determinants of economic policies: He will claim that these are purely technical rules belonging in the operating manual of the macroeconomic machinery. But of course this claim is mistaken at best and disingenuous at worst, for every political economy is bound to advance some interests while harming others. For example, free trade favors the strong while arresting the development of the weak; and the welfare state improves the lot of the poor through taxing the wealthy. In short, all public policy is morally committed. As the great socioeconomist Gunnar Myrdal enjoined long ago: Declare your values! Not doing so may help condone either pseudoscience or mercenary science – on which more anon.

13.9 Mercenary Science

The fact that a great many scientists signed Faustian pacts with the war devil throughout the twentieth century has given science a bad name, and has discouraged many able youngsters from pursuing a scientific career. We shall therefore presently glimpse at mercenary science – the kind that erases the humanist side of the philosophical pentagon in Fig. 13.1 in Section 13.2.

Whereas basic science is disinterested, the goal of applied science projects is to obtain results of possible practical interest; and of course technologists design, repair or maintain artifacts of all kinds, from machines to social organizations. For instance, basic chemistry and biology provide the background knowledge for pharmacology; the latter analyzes or synthesizes compounds with possible therapeutic use; and industrial pharmacology, along with medicine, designs and tests drugs and therapies. The usual flow of knowledge between the three domains is this:

$$\textit{Basic science} \rightarrow \textit{Applied science} \rightarrow \textit{Technology}$$

Besides, there is mercenary science, or science for hire regardless of moral consequences, or even knowing that its results will be used for evil purposes. Mercenary science is exceptional because the vast majority of scientists are not interested in practical applications; moreover, they are incompetent to "translate" science into technology. Still, in a number cases patriotism, ideological zeal, greed, or the wish or rub shoulders with the powerful (the so-called Kissinger complex) have prevailed, and some great scientists reinvented themselves as occasional Mephistophelian technologists: recall phosgene, the nerve gas, the hydrogen bomb, napalm, Agent Orange, the antipersonnel and cluster bombs, and the terminator gene (see Cornwell 2003; Schwartz et al. 1972).

Some of the scientist who engaged in mercenary science were not anonymous company or government employees, but eminent physicists or chemists, some of them Nobel laureates. On the other hand, the scholars who worked for either the CIA or the KGB while pretending devotion to freedom or peace, or advised the Pentagon on how best to kill Vietnamese peasants, hardly count as mercenary scientists, because theirs was pseudoscience (see, e.g., Lang 1981; Stonor Saunders 2000) The same holds for the Nobel laureates in economics who, to the delight of the oil industry, denied the reality of global warming, or even the finiteness of the oil deposits.

Mercenary politologists are the most sinister but not the only violators of the ethical code of science. The food scientists who seek the best combination of fat, sugar and salt to entice us to overeat cookies or baby food, and get hooked on them, are not far behind. They are in the company of the chemists employed by Big Tobacco, who manipulated nicotine to make cigarette smoking ever more addictive. Add the psychologists who help design deceptive advertising, and you get a picture of the huge contingent of natural, social and biosocial scientists paid to use science against the people. The only efficient defense against this mercenary army is more and better science education.

13.10 Philosophy: Genuine and Bogus, Proscience and Antiscience

Pseudophilosophy is nonsense parading as deep philosophy. It may have existed since Lao-tzu, but it was not taken seriously until about 1800, when the Romantics challenged the Enlightenment. By giving up rationality, they generated a lot of pseudophilosophy: Recall the lunatic pronouncements of Hegel, Fichte, Schelling, and their British followers (see, e.g., Stove 1991). True, Hegel tackled a number of important problems, so his work cannot be dismissed lightly. However, his work, when understandable at all, was usually wrong in the light of the most advanced science of his own time. Worse, it enshrined the equivocation that depth must be obscure.

Our characterization of scientific research in the previous chapter involved philosophical ideas of five kinds: logical, semantical, ontological, epistemological (in particular methodological), and ethical. In particular, it involved the notions of logical consequence and logical consistency; the semantic notions of meaning and truth; the ontological concepts of real fact and law (objective pattern); the epistemological concepts of knowledge and test; and the moral principles of intellectual honesty and public service. Why such philosophical commitments of science? Let us see.

Scientific research has philosophical underpinnings because it is, in a nutshell, the *honest* search for *true knowledge* about the *real world*, particularly its *laws*, with the help of both *theoretical* and *empirical* means – in particular the *scientific method* – and because every body of *scientific knowledge* is expected to be *logically consistent*, as well as a subject of *rational discussion* in the bosom of a community of investigators. All the expressions in italics occur in (metascientific) discourses about any factual (empirical) science. And the discipline in charge of elucidating and systematizing the corresponding concepts is philosophy. Indeed, philosophy is the study of the most fundamental and cross-disciplinary concepts and principles. Hence, philosophers are expected to be generalists rather than specialists. And some of us often assume the ungrateful task of passing judgment on the credentials of some pseudoscientific or ideological beliefs.

Now, different philosophical schools treat the above philosophical components of science differently or not at all. Let us recall briefly only four influential contemporary examples: Marxism, existentialism, logical positivism, and Popperianism.

Marxism introduced some revolutionary ideas in social science, particularly the materialist conception of history and the centrality of social conflict. However, Marxian materialism is narrowly economist: it underrates the roles of politics and culture (in particular ideology). Moreover, following Hegel, Marxism confuses logic with ontology, hence it is diffident of formal logic; its materialist ontology is marred by the Romantic obscurities of dialectics, such as the principle of the unity of opposites; its epistemology is naive realism (the "reflection theory of knowledge"), which makes no room for the symbolic nature of pure mathematics and theoretical physics; it glorifies social wholes at the expense of individuals and their legitimate aspirations; it exaggerates the impact of society on cognition; and it adopts utilitarian ethics, which has no use for disinterested inquiry, let alone altruism.

No wonder that, while in power, dialectical materialist philosophers opposed some of the most revolutionary scientific developments of their time: mathematical logic, relativity theory, quantum mechanics, quantum chemistry, genetics, the synthetic theory of evolution, and post-Pavlovian neuropsychology (see, e.g., Graham 1981). But the worse sin of dialectical materialism is that, because it overrates conflict, it underrates cooperation, and it is actually a philosophy of war, even though its followers proclaim their love of peace.

And yet, the serious flaws of Marxism are beauty spots by comparison with the absurdities of existentialism. This hermetic doctrine rejects logic and, in general, rationality; and it is centered in an extremely sketchy, nearly unintelligible, and even ridiculous ontology. Suffice it to recall Heidegger's characterization of being and time, the central subjects of the book that made him famous overnight: "Being is IT itself," and "Time is the maturation of temporality." Another two famous aphorisms by the same author are "The world worlds," and "The word is the dwelling of Being." All this is just nonsense, hence not even false. But it may sound profound to the naïve; and it attracts the lazy, for it discourages rational discussion.

Furthermore, existentialism has no use for semantics, epistemology, or ethics. No wonder that it has had no impact on science – except indirectly, and negatively, through its debasement of reason and support of Nazism. No wonder, too, that it has not produced an intelligible philosophy of science, let alone a stimulating one. Actually existentialism is a prime example of a pseudophilosophy.

By contrast, logical positivism uses modern logic, defends scientism, and criticizes oibscurantism. But it has no defensible semantics beyond the operationist thesis that meaning equals testability – a category mistake. It has no ontology beyond phenomenalism ("there are only appearances"). Its epistemology overrates induction and misunderstands or underrates scientific theory, which it regards as a mere data compendium. And it has no ethics beyond Hume's emotivism. Unsurprisingly, logical positivists have misinterpreted relativistic and quantum physics in terms of laboratory operations instead of as representing objectively existing physical entities that exist in the absence of observers (see, e.g., Bunge 1973a). Still, logical positivism is scientistic, and therefore far superior to the anti-science stance of postmodernism. But, for better or for worse, logical positivism is dead except in some science textbooks.

Karl Popper was instrumental in killing logical positivism. He praised rationality and the pursuit of knowledge. But he rejected the very attempt to elucidate the concepts of meaning and interpretration, without which it is impossible to use mathematics in science. He had no ontology beyond individualism (or atomism, or nominalism), which renders social engineering impossible – in spite of which he approved of it. Popper valued theory to the point of regarding observation, measurement and experiment as only ways of testing hypotheses. He overrated criticism, underrated discovery and induction, had no use for positive evidence, and had no ethics beyond the Buddha's, Epicurus', and Hippocrates' injunction not to harm. Because of his abuse of the word "no", Popper's philosophy may be called *logical negativism* (Bunge 1999). Yet, Popper has the merits of having defended rationality and a realist interpretation of physical theories, and of having deflated inductivism and shown the impossibility of inductive logic.

But Popper first underrated, and later admitted but misinterpreted evolutionary biology as consisting exclusively of culling misfits; he opposed the psychoneural monism inherent in biological psychology; he rejected the materialist conception of history adopted by the most advanced historical school – that of the *Annales*; and he defended neoclassical microeconomics, which – as will be argued below – is pseudoscientific in being conceptually fuzzy and immune to empirical falsification.

In short, none of the four above-mentioned schools matches the philosophy inherent in science. I submit that any philosophy capable of understanding, using and promoting scientific research has the following characteristics.

1. *Logical* Internal consistency and abidance by the rules of deductive inference; acceptance of analogy and induction as heuristic means, but no claim to a priori validations of analogical or inductive arguments.
2. *Semantical* A realist theory of meaning as intended reference (denotation) – and as different from extension or truth domain – together with sense or connotation. And a realist view of factual truth as the matching of a proposition with the facts it refers to.
3. *Ontological* Materialism: All real things are material (possess energy) and fit some laws (causal, probabilistic, or mixed). Mental processes are brain processes, and ideas in themselves, however true or useful, are fictions. Dynamicism: All material things are in flux. Systemism: Every thing is either a system or a (potential or actual) component of a system. Emergentism: Every system has (systemic or emergent) properties that its components lack.
4. *Epistemological* Scientific *realism*: It is possible to get to know reality, at least partially and gradually; and scientific theories are expected to represent, however imperfectly, parts or features of the real world. Moderate *skepticism*: Scientific knowledge is perfectible as well as fallible. However, some findings – e.g., that there are atoms and fields, that there are no disembodied ideas, and that science pays – are firm acquisitions. Moderate *empiricism*: All factual hypotheses must be empirically testable, however indirectly; and both positive and negative evidence are valuable indicators of truth value. Moderate *rationalism*: Knowledge advances through educated guessing and reasoning combined with experience. *Scientism*: Whatever is knowable and worth knowing is best known scientifically.

Ethical Secular humanism: The supreme moral norm is "Pursue the welfare (biological, mental, and social) of self and others". This maxim directs that scientific research should satisfy either curiosity or need, and abstain from doing unjustifiable harm. Epistemic socialism: Scientific work, however artisanal, is social, in that it is now stimulated, now inhibited, by fellow workers and by the ruling social order; and that the (provisional) umpire is not some institutional authority but the community of experts. Every such community prospers with the achievements of its members, and it facilitates the detection and correction of error. (Warning: This is a far cry from the Marxian claim that ideas are exuded and killed by society; it is also at variance with the constructivist-relativist view that "scientific facts" are local social constructions, that is, mere community-bound or tribal conventions.)

I submit that the above philosophical principles are tacitly met by the mature or "hard" sciences (physics, chemistry, biology, and history); that the immature or "soft" sciences (psychology and the non-historical social sciences) satisfy only some of them; and that the pseudosciences violate most or all of them. In short, I repeat that great science is nurtured by sound philosophy, as suggested by Fig. 13.1. By contrast, pseudoscience grows in a nest built with bad elements, some of them characteristic of postmodern pseudophilosophy: see Fig. 13.2.

Moreover, the pseudosciences are akin to religion, to the point that some of them serve as surrogates for it. The reason is that they share philosophical idealism and a non-humanistic ethics. Indeed, both pseudoscience and religion postulate immaterial entities, paranormal cognitive abilities, and a heteronomous ethics. Let us spell this out.

Every religion has a philosophical kernel, and the philosophies inherent in the various religions share the following idealist principles. *Idealist ontology*: There are autonomous spiritual entities, such as souls and deities, and they satisfy no scientific laws. *Idealist epistemology*: Some people possess cognitive abilities that fall outside the purview of experimental psychology: divine inspiration, inborn insight, or the capacity to sense spiritual beings or prophesy events without the help of science. *Heteronomous ethics*: All people are subject to inscrutable and omnipotent superhuman powers; and nobody is obliged to justify their beliefs by means of rational argument or scientific experiment.

All three philosophical components common to both religion and pseudoscience are at variance with the philosophy inherent in science. Hence, the theses that science is one more ideology, and that science cannot conflict with religion because

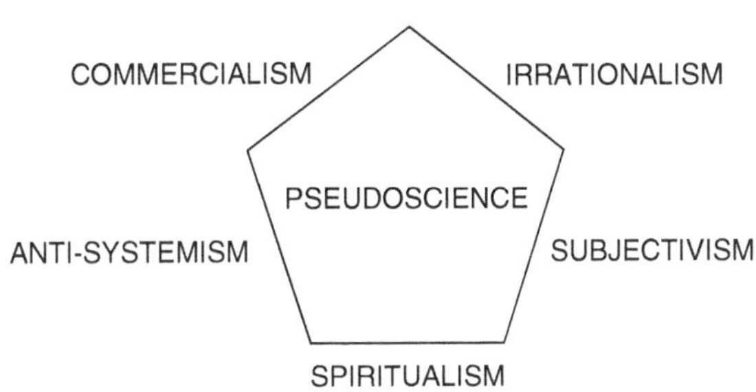

Fig. 13.2 The ideological matrix of pseudoscience

they address different problems in different but mutually compatible ways, are false. (More on religion and science in Bunge 2009; Mahner and Bunge 1996.)

Closing Remarks

The serpent was blamed for coaching Eve, on the ground that knowledge is evil. As societies became secularized, we learned that knowledge is intrinsically good, if not always worth acquiring. We also learned that it may be misused, and also that it may be faked – just like nearly everything else. Since scientific knowledge is in the public domain, to misuse it, sell it for private profit, or fake it, amounts to stealing from the commons.

Medical quackery is a case in point. It is far easier to identify and pillory than any other kind of bogus knowledge, because it openly contradicts biology, and it is practiced only on faith, without any laboratory of clinical evidence showing that it works better than placebos. Yet, alternative medicines continue to flourish under the protection of the free market ideology. Some of them, such as phenomenological medicine, narrative medicine, existentialist psychiatry, and transcultural psychiatry, are even taught in prestigious medical schools along with scientific medicine. If this tendency were to prevail, sick people would have to resort to veterinary clinics.

One can safely bet that alternative medicines will continue to thrive wherever the practice of medicine is regarded as a business rather than a social service, and wherever pseudophilosophy is held in awe just because of the tendency to mistake hermeticism for depth. So, it is up to the sanitary authorities, medical associations, serious scholars, and journalists, to alert the public and teach that the only alternative to dangerous medical quackery is better medicine.

In academia much bogus knowledge is tolerated in the name of academic freedom – which is like allowing for the sale of contaminated food in the name of free enterprise. I submit that such tolerance is suicidal: that the serious students must be protected against the "anything goes" crowd. The philosophies of matter and mind ought to provide such protection. To provide it they must be sandwiched between logic, on the one hand, and up to date science on the other. It is a waste of time, as well as misleading, to either defend or attack materialism without knowing what matter is; likewise, it is not productive to either defend or attack psychoneural monism without knowing something about biological psychology.

In sum, it is not enough to produce genuine knowledge: We should also examine and denounce bogus knowledge, because it diverts and deceives. However, we should never damn new ideas as pseudoscientific just because they are outlandish. Remember that throughout the nineteenth century nearly all physicists and philosophers condemned atomistics as pseudoscientific. Let us be neither narrow-minded nor utterly lacking in philosophical filters.

Part III
Appendices

Chapter 14
Appendix A: Objects

In ordinary language, the word "object" denotes a material thing that can be seen and touched. By contrast, in modern philosophy "object" (*objectum*, *Gegenstand*) stands for whatever can be thought about: it applies to concrete things and abstract ones, arbitrary assemblages and structured wholes, electrons and nations, stones and ghosts, individuals and sets, properties and events, facts and fictions, and so on.

The concept of an object is thus the most general of all philosophical concepts. In fact, this concept is so general that it is used in all the branches of philosophy in all languages – though not always consistently. For instance, someone might say that the subjects of this chapter are objects, whereas its object or goal is to elucidate "object".

Yet, to my knowledge there is no generally accepted theory of objects. True, mereology, or the calculus of individuals, was expected to fill that gap. Regrettably, this theory is extremely complicated, uses an awkward notation, and does not accomplish much because, following the nominalist program, it eschews properties. As for the theories of objects proposed by Meinong and Routley, they are only moderately well known, possibly because they include impossible objects on a par with possible ones. The goal of this paper is to formulate a general theory of objects free from those flaws. However, the reader with no taste for symbolism is invited to skip this chapter.

14.1 Individuals and Properties

We shall presently propose an axiomatic theory of individuals of any kind. The first section presupposes only the classical predicate calculus with identity, a bit of set-theoretic notation, and another of semi-group theory; the balance of the chapter also uses the concept of a mathematical function. The specific primitive (undefined) concepts are those of individual and property. Like all primitives, these are elucidated by the postulates where they occur.

The set of all individuals will be called $\mathbb{X}$, and that of all properties (of any n-arity) $\mathbb{P}$. As usual, the formula "Px", where P is in $\mathbb{P}$ and x is in $\mathbb{X}$, will be read "x is a P", or "individual x has property P". Similarly, the formula "Rxy", where x and

M. Bunge, *Matter and Mind*, Boston Studies in the Philosophy of Science 287,
DOI 10.1007/978-90-481-9225-0_14, © Springer Science+Business Media B.V. 2010

y are in $\mathbb{X}$, and R is in $\mathbb{P}$, is to be read "x is R-related to y", or "R relates individuals x and y".

Definition 1 Every object is either an individual or a property:

$$\Omega = \mathbb{X} \cup \mathbb{P}.$$

Hence, our axioms for objects will refer to individuals, properties, or both. We start with

Axiom 1 No object is both an individual and a property:

$$\mathbb{X} \cap \mathbb{P} = \emptyset.$$

Axiom 2 All individuals have at least one property:

$$\forall x \exists P (x \in \mathbb{X})(P \in \mathbb{P}) Px.$$

The dual of this postulate says that there are no properties in themselves:

Axiom 3 Every property is possessed by at least one individual:

$$\forall P \exists x (P \in \mathbb{P} \ \& \ x \in \mathbb{X}) Px.$$

This postulate, first suggested by Aristotle when criticizing Plato's theory of forms, is tacitly adopted by all the sciences. But it is negated by the functionalist philosophy of mind, which postulates mental functions that are not brain functions.

Axiom 4 Every individual is related to at least one other individual:

$$\forall x \exists y (x, y \in \mathbb{X})(R \in \mathbb{P}x)[\neg(x = y) \& Rxy].$$

Axiom 5 Every property of an individual is related to at least one other property of the same:

$$\forall x \forall P (x \in \mathbb{X})(P, Q \in \mathbb{P})\{Px \Leftrightarrow \exists Q[\neg(Q = P) \& Qx]\}.$$

In other words, properties come in clusters rather than independently from one another. The reason for this is that properties satisfy laws, the great majority of which relate two or more properties.

Finally, we introduce the following relative-existence predicate.

Definition 1 Let C be non-empty subset of some set X, and χ_C the characteristic function of C, that is, the function $\chi_C \colon X \to \{0,1\}$ such that $\chi_C(x) = 1$ if and only if

x is in C, and $\chi_C(x) = 0$ otherwise. The relative (or contextual) existence predicate is the statement-valued function

$$E_C : C \to \text{The set of statements containing} E_C$$

such that "$E_C(x)$" is true if and only if $\chi_C(x) = 1$.

The formula "$E_C(x)$" is interpreted as "x exists in C", and it is equivalent to $\chi_C(x) = 1$.

Notice that the above existence predicate is unrelated to the "existential" quantifier, which I prefer to call "particularizer". I submit that "$\exists xPx$" only says that *some* individuals have the property P. Their existence must be assumed or denied separately. For instance "Some postulated entities exist in the real world" can be symbolized as "$\exists xE_W Px$", where W stands for the collection of real things.

Whoever makes an ontological commitment uses tacitly E, not $\exists$. Of course, it might be objected that it is far simpler to state that x is in C, or $x \in C$. But this objection misses the point, which was to introduce an existence *predicate* E_C, and to demarcate it from the particularizer $\exists$. Only a logical imperialist could fail to draw this distinction, which, though unnecessary in formal matters, is essential in factual ones, where no fiat can guarantee existence.

So far we have not distinguished between concrete objects, such as numerals, and ideal objects, such as numbers. We proceed to introducing this distinction.

14.2 Material Objects

Consider the algebraic system

$$S =< \mathbb{M}, \oplus, \mathbf{0}, \mathbf{1} >,$$

where $\mathbb{M}$ is a subset of the set $\mathbb{X}$ of all individuals, whereas $\mathbf{0}$ and $\mathbf{1}$ are distinguished elements of $\mathbb{M}$. We assume that any two members x and y of $\mathbb{M}$ conjoin (or concatenate) to form a third member $x \oplus y$ of $\mathbb{M}$, called the physical (or mereological) sum of x and y.

A concatenate need not be a system; that is, no bonds need be involved: an arbitrary assemblage of things counts as an object. In this regard, this technical concept of an object does violence to common sense (Koslicki 2008). But we have another word to denote complex objects endowed with a structure that gives it unity and cohesion: *system*.

In any event, we need the more general concept of an object because there are simples, that is, objects devoid of structure. And also because some objects, such as properties, events, and constructs, are not things. However, let us go back to our concatenates or mereological sums.

We formalize the above intuitive notion of concatenation by introducing

Axiom 6 S is a monoid, that is, $\oplus$ is a binary, commutative and associative operation in $\mathbb{M}$: For any x, y and z in $\mathbb{M}$,

$$x \oplus y = y \oplus x, (x \oplus y) \oplus z = x \oplus (y \oplus z).$$

This assumption allows us to introduce the part-whole relation $\angle$ by means of

Definition 2 For any x and y in $\mathbb{M}$, x is a *part* of y if there exists a third individual z that, if juxtaposed with x, results in the whole y:

$$x \angle y = \exists z\, E_{\mathbb{M}} z[z \in S \& (x \oplus z = y].$$

In the simplest case, part and whole coincide: this is the case of the basic or elementary constituents of the universe, such as electrons and photons. In general x is *elementary* $= x$ has no parts. This definition is preferable to the identification of elementarity with simplicity because, according to the quantum theory, electrons and photons are far more complex than the point particles of classical mechanics.

Now we define the two distinguished elements of S: the null individual and the universe. The former is that individual that makes no difference to any individual:

Definition 3 The null individual $\mathbf{0}$ is the individual that is a part of every individual: For all x in $\mathbb{M}$,

$$\mathbf{0} \oplus x = x.$$

From this definition and the earlier one it follows that a thing is part of itself. In other words, the part-whole relation $\angle$ is reflexive.

At the other extreme, the universe is the maximal individual:

Definition 4 The universe is the individual such that every individual is a part of it:

$$\mathbf{1} = (\iota x)\forall y(y \in \mathbb{M} \& y \angle x),$$

where ι designates the definite descriptor.

I submit that our concept $\oplus$ of physical (or mereological) sum is tacitly used in all the factual sciences. For example, the physical law that the electric charge of a composite thing equals the sum of the charges of its components can be symbolized as $Q(p_1 \oplus p_2 \oplus \ldots \oplus p_n) = Q(p_1) + Q(p_2) + \ldots Q(p_n)$. This law is completely general: it places no restriction on the kind of thing. The same holds for more familiar examples of concatenation, such as combinations of countries or letters.

14.3 Emergence and Levels

Now for the fashionable, albeit somewhat unclear and even controversial, concept of emergence. Actually we will introduce two different concepts of emergence: those of a bulk or systemic property, and of a radically new trait or feature arisen in the course of a process, such as morphogenesis, bioevolution, or history. The former is this: An emergent property of an individual is one that no part of it possesses. More precisely, we posit

Definition 5 For any individual x in $\mathbb{M}$, P_B is a bulk property of x if no part y of x possesses P_B:

$$\forall x[P_B x =_{\text{df}} Px \& \neg \exists y(y \angle x \& P_B y)].$$

To define the second or diachronic concept of emergence we need that of descent (or that of precursor, which is the dual or converse of the former). Here we cannot avail ourselves of the concept of time, for we have not defined it. A possible axiomatic definition of the atemporal relation D of descent is this: For any members x, y, and z of $\mathbb{M}$,

D1 Irreflexivity: $\neg Dxx$.
D2 Asymmetry: $Dxy \Rightarrow \neg Dyx$.
D3 Transitivity: $Dxy \& Dyz \Rightarrow Dxz$.

This concept allows us to introduce

Definition 6 For any individual x in $\mathbb{M}$, P_N is a radically new property of x if x descends from a precursor y that lacks P_N:

$$P_N x = \exists y(Dxy \& \neg P_N y).$$

The two concepts of emergence that have just been elucidated are distant relatives of that of supervenience, and they occur with increasing frequency in the scientific and technological literature. Thus, one says that life is an emergent property of cells because these are alive, whereas their components are not. And of course chemists assume that all molecules have atomic precursors, and evolutionary biologists have conjectured that cells emerged from prebiotic things.

Finally, we introduce the concept of a level of being. A set of objects can be said to constitute a level of reality if all of them possess bulk properties that their parts lack. For example, the elements of the social level are systems composed of gregarious organisms. One may say metaphorically that the social level emerged from the life level, or that the latter precedes the former, or $B < S$. Thus the part-whole relation $\angle$, and in particular the component-system one, induces the levels relation $<$.

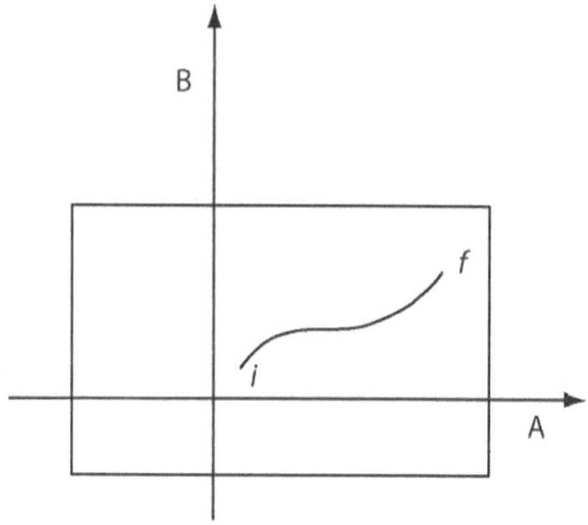

Fig. 14.1 State space for a thing with two salient variables, *A* and *B*. The curve represents a *process* undergone by the thing, and every point in the curve represents a *state* of it; in particular, the points *i* and *f* represent the initial and final states. The box contains all the possible states of the thing, i.e., those allowed by the laws involving at once *A* and *B*

(Since the defining relation $\angle$ is a strict partial ordering, the defined relation< must be of the same type.) To put it formally,

Definition 7 Every object on a given level is composed of objects in the next lower level:

$$L_n < L_{n+1} =_{\mathrm{df}} \forall x \exists y [x \in L_{n+1} \& y \in L_n \Rightarrow y \angle x]$$

We use this definition to posit

Axiom 7 The levels order< is a strict partial ordering of the set *L* of all the levels of being.

14.4 State and Process

Let us now introduce the concept of a state space, which occurs in all of the factual (or empirical) sciences and engineering. This will require using some formal resources that outstrip the predicate calculus – which suggests that logic is insufficient to do metaphysics.

For the sake of simplicity we shall define state spaces for things that are adequately described by non-quantum theories. The simplest such state space is the one with only two elements, such as the "on" and "off" states of a switch. A state space for a system composed of *n* independent switches will have 2^n elements. If a neuron could be in one of two states, as McCulloch and Pitts assumed in the earliest mathematical model of the human brain, its state space would consist of 2^{10} raised to the 11th power – which is much, yet as nothing compared to the non-denumerable infinity of the state space of a simple mechanical system, such as a linear oscillator, characterized by only two dynamical variables. See Fig. 14.1.

Let us next describe a state space for a classical thing with *n* salient properties.

Call $P_1, P_2, \ldots, P_n$ the properties of interest of the members of a subset K of $\mathbb{M}$, and F_i the function or operator representing P_i, with $1 \leq i \leq n$. Let us agree on

Definition 8

(i) The function $\mathbb{F} = <F_1, F_2, \ldots, F_n>$ from $\mathbb{M}$ to the set of complex numbers, is to be called the state function for the individuals in K;
(ii) The space S_K spanned by $\mathbb{F}$ is the state space of the Ks;
(iii) The particular value of $\mathbb{F}$ for an individual x in $\mathbb{M}$ is called the state of x;
(iv) K is the kind or species characterized by $\mathbb{F}$.

Any state space S_K is an n-dimensional abstract space. A point in it may be regarded as the tip of the arrow $\mathbb{F}$. And a pair of points in a state space represents a point event, or change in the state of an individual of the kind in question. In other words, an event can be represented as an element of the Cartesian product of S_K by itself.

Note that, according to the preceding, there are no states in themselves: Every state is a state that an individual can be in. Hence the concept of an individual logically precedes that of a state, so that an ontology according to which the bricks of the world are states, o else changes of state (events), is inconsistent with the above a well as with the way states and events are treated in science. (For example, there can be no chemical reaction without reactants, anymore than there can be smiles without faces – except in the case of Lewis Carroll's Cheshire cat.)

The previous criticism applies to Whitehead's process metaphysics – which Russell shared for a while – as well as to David Mermin's (2009) claim that states are unreal because only events would be observable. Since by definition events are changes of state, these must be real for their changes to be real too. David Armstrong's (1997) opposite doctrine, that the basic ingredients of the world are states of affair, is wrong for a different reason: that every state is a state that a thing can be in. The correct logical sequence is: Thing-Property-State-Process.

We can finally define the concept of a material object:

Definition 9 An individual is a material object if it has a non-empty state space:

For all x in $\mathbb{M}$: (x is material $= x$ belongs to a K such that $S_K \neq \emptyset$).

To say that an individual is changeable amounts to saying that it can be in at least two states, such as the initial and the final. If materiality is coextensive with changeability, one may say that to be material is to have a state space with at least two elements:

$$|S_K| > 2.$$

Note that in characterizing material objects we have not employed any concepts of space or time. In particular, we have not stated that things are located in space, or that they change in the course of time. This makes it possible to construct relational or adjectival (as opposed to absolute or substantive) theories of space, time, and space-time – that is, theories that define space and time in terms of the concept of a changing (that is, material) thing.

Finally, a warning concerning the careless use of the concepts of state and process in the philosophy of mind. In a standard textbook (Kim 2006, 220) we read that, according to psychoneural monism, pain "arises out" of a neural *state N*, "emerges" from *N*, or "supervenes" upon *N*. This is incorrect, for mental occurrences are (identical to) neural *processes*, not *states*. For example, one can improve mood and increase sociability by inhaling oxytocin; conversely, we shut our eyes when we wish to inactivate our visual cortex.

Cognitive and emotional *processes* are *identical* to certain brain *processes* instead of "arising out" of them or "emerging" from them. Likewise boiling and freezing are identical with certain processes in liquids, namely molecular motions with average kinetic energies above or below certain critical values, instead of "arising out" of, "emerging from", or "supervening upon" certain thermal states of the liquid. This is a crucial ontological point, not trivial verbal nitpicking. Replacing processes with states amounts to substituting Zeno's static worldview for Heraclitus' dynamic one.

14.5 Ideal Objects

According to cognitive neuroscience ideation is a brain process. But ideas considered in themselves, regardless of thinkers and their circumstances, are unchangeable. For example, it makes no sense to say of an object that by construction is unchangeable, such as a number system, a Boolean algebra, or a manifold, that it is in a given state, much less that it can change from one state to another. This consideration motivates

Definition 10 An object is an ideal object if it is in no state.

Equivalently, the ideal objects constitute the kind I of individuals for which the state space is empty:

$$S_I = \emptyset.$$

In ordinary language: Whereas to be material is to become, to be ideal is to be immutable. Presumably, Plato would have concurred.

Finally, we postulate that no object is at once material and ideal:

Axiom 8 $\mathbb{M} \cap \mathbb{I} = \emptyset$.

Aristotle and his followers would disagree, for they were hylomorphists. But neither science nor technology knows of any hybrids of matter and "form" (idea) – except of course in the metaphorical sense that artifacts embody or materialize ideas.

Closing Remarks

Let us note two points. Firstly, strict nominalists have no use for any of the above, since they deny the existence of properties, or else assert that these can be defined as sets of individuals. But all knowledge consists of attributing properties or changes thereof to individuals. In particular, law statements relate properties. And identifying properties with sets of individuals amounts to confusing predicates with their extensions.

Secondly, our postulate that ideal objects are immaterial, and conversely, precludes all talk about the ontology of mathematics. One should talk instead of the reference class(es) of mathematical predicates and structures. For example, the domain D of a function *f* of a single variable is the reference class of *f*, not its ontology. The reason is that ontologies are theories about the world, not sets. Hence we do not make any ontological commitment when positing that a certain domain is non-empty. Ontology starts when specifying the nature of the members of the domain in question – e.g., material, spiritual, or hybrid.

Chapter 15
Appendix B: Truths

There are at least two quite different concepts of truth: formal and factual. Leibniz called them *vérités de raison* and *vérités de fait* respectively. Whereas the formal truths are those of logic and mathematics, the factual truths are characteristic of ordinary knowledge, science, and technology. For example, while "There are infinitely many prime numbers" is a formal truth, "There are about six billion people at this time" is a factual truth. The confusion between the two kinds of truth is even worse than the confusion between value and price.

It is generally agreed that the concept of formal truth is elucidated by model theory in the case of abstract (uninterpreted) theories, such as set theory and group theory; and that it coincides with that of theoremhood in the case of interpreted theories such as the calculus – modulo the undecidable formulas.

No comparable consensus exists with regard to factual truths. Indeed, although all clear thinkers value objective (factual) truth, no one seems to have hit on a true theory (hypothetico-deductive system) of it. So, after two and a half millennia, the correspondence theory is still a research project.

Yet, the intuitive idea is clear if fuzzy: A proposition is factually true if it fits (or matches, or corresponds to, or is adequate to) the facts it refers to. But what do the metaphorical terms "fit" (or "match", or "correspond to") mean? This is the outstanding question.

The great mathematician Alfred Tarski, in his famous 1944 article, thought he had a single formula for both kinds of truth: "The sentence '*s*' is true if and only if *s*". But this won't do, for three reasons. First, because it conflates the two kinds of truth. Second, because it confuses facts with sentences – something that only glossocentrists like Heidegger and the second Wittgenstein would accept. Third, because it does not contrast language, or rather its epistemic designatum, with extralinguistic reality – which is what "correspondence" is supposed to mean. Indeed, Tarski's formula just bridges a bit of language (the sentence *s*) to a bit of metalanguage, the name "*s*" of *s* (see Bunge 1974b). It misses the essence of factual truth, namely that it is a piece of human knowledge of the external world – something that subjectivists and phenomenalists deny themselves.

Alternative endeavors to elucidate the notion of factual truth have not fared better. In particular, the attempts by Reichenbach, Popper, and their followers, to reduce

M. Bunge, *Matter and Mind*, Boston Studies in the Philosophy of Science 287,
DOI 10.1007/978-90-481-9225-0_15,

truth to probability (or else improbability) are wrong for at least three reasons. First, truth logically precedes probability, since we must be able to evaluate the truth-value of probability formulas among others. Second, there is no justification, hence no objective criterion, for assigning probabilities to propositions, anymore than to assign them areas, temperatures, or colors. And third, whereas the truth-value of the disjunction of two mutually independent propositions equals the greatest of their truth-values, the probability that either of two mutually independent events occurs equals the sum of their separate probabilities.

Nor should factual truth, in particular partial (or approximate) truth, be mistaken for plausibility or verisimilitude. Factual truth is a feature of the knowledge-reality relation, whereas plausibility is relative to some body of knowledge. And, whereas factual truths are ascertained after the fact, plausibility judgments can be made a priori. For example, only observation can check whether the proverbial cat is on the mat. By contrast, the psychokinesis hypothesis is implausible because it contradicts energy conservation.

It is embarrassing to note that the above-mentioned attempts to elucidate the correspondence concept of truth have derived from confusion. Moreover, two of them may have originated in the everyday usage of "probable" and "Wahrscheinlich" in English and German respectively, where they mean "likely" and "plausible" as well as "chance-like".

These failures are sad indicators of the state of contemporary philosophy (Bunge 2001). By the same token, they suggest the need to make a fresh start. In this paper we shall attempt just this. We shall also examine the ontology, semantics, and methodology of the concept of partial (or approximate) truth of fact. Parts of this paper have been lifted from a previous publication (Bunge 2003a).

15.1 Ontological Concept of Factual Truth

The intuitive notion of factual truth as the fitting (or adequacy, or matching) of idea to fact is flawed if ideas are conceived in the abstract, e.g., as propositions. This is because in such case the fitting, adequacy, or matching in question is only metaphorical. Indeed, abstract ideas can be contrasted with one another, as when comparing two numbers; likewise, factual items can be contrasted with another, as when comparing two fingers. But it seems impossible to compare or confront an abstract idea with the fact it refers to. In particular, there is no similarity between a formula of optics and a light wave, any more than there is an analogy between writing and speech. However, this is not the case with concrete ideation, as a thought processes, vis-à-vis its referent in the external world – or, in the jargon of the day, the truthmaker of the truthbearer in question. Let us then glimpse at this alternative.

Consider briefly ideas the way cognitive neuroscience does, namely as brain processes. In this case we can attempt to regard truth as a property of a brain process of the mental kind. Hence in this case we can compare ideas with their external referents if any (Bunge 1980a, 1983a).

For instance, we can say that a subject has a true perception of a circular figure if, in fact, she perceives a circle when presented such a figure. Moreover, some sensational experiments with primates have shown that such mental images consist in similar configurations of activated neurons in the visual cortex (Tootell et al. 1982, 1998).

A person's perception will be false if she perceives a circle as a different figure, or if she perceives no figure at all. Obviously, in this simple case there is a single true perception and an uncounted number of false perceptions. Moreover, by observing the activity (firing rate) of certain neurons, the cognitive neuroscientist can predict quite accurately what the subject was viewing (Kreiman et al. 2000).

This approach will be judged preposterous by any believer in psychoneural dualism. But dualism has long been superseded by cognitive neuroscience, according to which "mental events are a feature of neurophysiological systems with certain properties" (Zeki 1993, 345). For example, a visual image in the brain results from the simultaneous activity of several visual areas and pathways. Another example: Damage to the lower and middle regions of the prefrontal cortex result in serious emotional impairment: the patient fails to be moved by events that elicit strong feelings in normal subjects (Damasio 1994).

That said, I hasten to admit that the realist and materialist conception of factual truth sketched above is only a research project on which too few scholars seem to be working full-time.

Let us now turn from concrete ideation, as a brain process, to ideas in themselves, that is, considered in abstraction from thought processes, the way mathematicians and idealist philosophers proceed.

15.2 The Correspondence Functions

Let $\mathbb{F}$ denote a set of really possible facts, and $\otimes$ their binary conjunction (or concatenation, or conjunction), as in "She is eating while listening". The elements of $\mathbb{F}$ may be events in either a knower's brain or in her external world. We assume that the concatenation $f \otimes g$ of any two facts f and g in $\mathbb{F}$ is a third fact rather than, say, a fiction. We also assume that fact concatenation is associative: $f \otimes (g \otimes h) = (f \otimes g) \otimes h$, for any f, g and h in $\mathbb{F}$. And we define the null fact O as that which, when concatenated with an arbitrary fact f, leaves it unaltered: $O \otimes f = f \otimes O = f$. That is, O plays the role of the unit element. Clearly, $<\mathbb{F}, \otimes, O>$ is a monoid (or semigroup with identity).

Note that we do not assume that facts can disjoin: Real things and changes in them (events or processes) conjoin but do not disjoin. Reality does not practice negation either: Negation, like disjunction, is a purely conceptual process. In other words, the operator $\neg$ applies to propositions and predicates, not to facts. (See Bunge 1977a.)

Next, let $\mathbb{P}$ stand for the set of all propositions, and $\wedge \vee, \neg$ for the standard propositional connectives. As is well-known, $<\mathbb{P}, \wedge, \vee, \neg>$ is a complemented lattice.

We also assume that this lattice is distributive. (The rumor that quantum mechanics has destroyed distributivity stems from the confusion of propositions with operators.)

So much for definitions. Let us now put them to work. We stipulate that a map from $<\mathbb{F}, \otimes, O>$ into $< \mathbb{P}, \wedge, \vee, \neg >$ formalizes the concept of representation of facts by propositions. In turn, a partial map from $<\mathbb{P}, \neg, \wedge>$ into the unit interval [0,1] of the real line is a truth-valuation function *V*. This second map is partial because not every proposition in $\mathbb{P}$ is assigned a truth-value: Think of the undecidable propositions. Furthermore, we assume that the truth-values range between 0 and 1, because we admit approximate (partial) truths and falsities, such as "There are now six billion people".

The two maps compose as follows:

$$<\mathbb{F}, \otimes, O, > \xrightarrow{\textit{Representation } \mathcal{R}^{-1}} <\mathbb{P}, \wedge, \vee, \neg > \xrightarrow{\textit{Truth valuation } \mathcal{V}} [0,1]$$

The theory should specify the maps R^{-1} and *V* in such a way that

(1) all the propositions representing the null fact O are false:
If $R^{-1}(p) = O$, then $V(p) = 0$ for all p in $\mathbb{P}$;
(2) for all $f \in \mathbb{F}, R^{-1}(f) = p \in \mathbb{P}$; and, for some p, $V(p) = u \in [0,1]$;
(3) for all $f, g \in \mathbb{F}, R^{-1}(f \otimes g) = p \wedge q \in \mathbb{P}$; and, for some $p \wedge q$, $V(p \wedge q) \in [0,1]$.

Note the occurrence of "some" instead of "all". This is due to the existence of utterly unknown facts. On the other hand we need not assume gaps in $\mathbb{P}$, since a factual proposition may be objectively true or false even if it has not yet been put to the test. For example, Le Verrier's prediction of the existence of Neptune was true before Galle saw the planet through the telescope.

The map R^{-1} is the inverse of the reference function defined elsewhere (Bunge 1974a). It can be analyzed as the composition of two maps: the facts-thoughts, and the thoughts-propositions maps. (A particular thought is construed here as a particular brain process, whereas a proposition is conceived of as an equivalence class of thoughts: see Bunge 1980a, 1983a. No two thoughts are strictly identical even if they consist in thinking of the same proposition.)

The analysis in question is the composition of two maps: imaging, or *Im*, from facts $\mathbb{F}$ into thoughts Θ, and conceptualization, or *Con*, from thoughts to propositions $\mathbb{P}$:

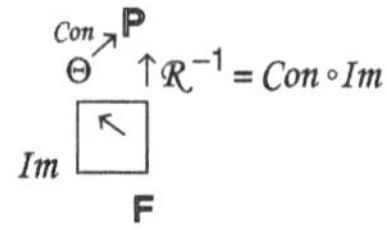

where R^{-1} is the inverse of the reference map. However, as long as the representation map R^{-1} remains undefined, we have no right to talk about a correspondence *theory* of truth. We can only talk about the *project* of crafting such a theory.

In Section 15.4 we shall have something to say about *V*. Still, we must first recall the dual of truth, namely error, because it is a mark of factual knowledge.

15.3 Methodological Concept of Truth

Scientists distinguish between theoretical and empirical propositions. For instance, they contrast the calculated to the observed orbit of a planet. And in practice they use the following truth criteria concerning a quantitative property (magnitude) *M*, such as mass, transition probability, metabolic rate, GDP, or what have you.

Criterion 1 The true empirical (measured) value of *M* measured with technique *t* is

$$\mu_t M = e \pm \varepsilon,$$

where *e* is the average of a large sequence of high-precision measurement results, and ε is the experimental error (mean-standard deviation) of that sequence.

Criterion 2 The theoretical (calculated) value θ of *M* is true provided the absolute value of the discrepancy between the two is smaller than the experimental error:

$$|\theta - e| < \varepsilon$$

Criterion 1 tells us that basic quantitative truths result from measurements, in particular pointer readings. More precisely, it tells us that, to extract the true value from a mass of such readings, we must take their arithmetic average. It starts then with a set of numbers and ends up in a single number. (The underlying assumption is that the experimental errors have a bell-shaped distribution.) And Criterion 2 contrasts two numbers, the said average and the calculated number.

We emphasize that the above are truth criteria, not definitions. (This remark is necessitated by the positivist confusion of the concept of truth with empirical truth criteria, a conflation parallel to that of reference with evidence.) The truth criteria suggest how to find truth-values, but do not tell us what a truth value is. (Parallel: A cooking recipe instructs how to make a cake of a particular kind, but it does not define the general concept of a cake.) However, the above criteria suggest that no theory of partial truth can be satisfactory if it ignores the concept of error. Let us proceed to sketching such a theory.

15.4 Partial Truth

Asking how accurate a proposition is, presupposes that there are truth-values other than 0 and 1. This is a standard assumption in applied mathematics, factual science, and technology. Indeed, in all these fields it is taken for granted that the best one may ordinarily come up with is a good approximation to the truth – one that may eventually be perfected, though.

This is what approximation theory (pioneered by Archimedes) and the calculus of errors (fathered by Gauss) are all about. A methodological consequence of the thesis that truth is graded, is that so is falsification (or refutation). For instance, the popular opinion that classical mechanics has been falsified is utterly false. In fact, that theory is an excellent approximation for medium-size bodies in slow motion. This is why physicists, astronomers and mechanical engineers continue to use that theory when appropriate. And this is why Thomas Kuhn's opinion, that all past beliefs about nature have turned out to be false, is itself utterly false. Furthermore, it is an opinion that discourages all attempts to increase the accuracy, extension, and depth of knowledge.

In other words, one assumes, usually tacitly, that there is a truth valuation function V from some set $\mathbb{P}$ of propositions to a numerical interval, which can be taken to be the unit real interval [0, 1]. That is, we may set $V: \mathbb{P} \to [0, 1]$. Our problem is to come up with a plausible system of conditions (postulates) defining V.

We want such postulates to make room for such half-truths as "Aristotle was a Spartan philosopher". Regrettably, the standard valuation function defined by

$$V(p \wedge q) = \min \{V(p), V(q)\},$$
$$V(p \vee q) = \max \{V(p), V(q)\}$$

has the counterintuitive consequence that the above proposition about Aristotle is totally false rather than half-true. Hence the standard valuation function cannot be regarded as exactifying the intuitive notion of partial truth. However, this criticism only affects the conjunction.

The following is a somewhat more plausible, yet tentative, set of desiderata for V.

D1 If p is a quantitative proposition that has been found to be true to within the relative error ε, then $V(p) = 1-\varepsilon$.

Example $p =$ "There are 9 people in this room", while the actual count shows that there are 10 people. Relative error $= \varepsilon = 1/10$. $V(p) = 1-1/10 = 9/10$, a pretty good approximation.

D2 If p is not the negate of another proposition, then

$$V(\neg p) = \begin{cases} 0 \text{ iff } V(p) = 1 \\ 1 \text{ iff } V(p) < 1 \end{cases}$$

Otherwise, that is, if q is the negate of a proposition p, which in turn is not the negate of another proposition, then

$$V(\neg p) = V(q).$$

Example if p is the example of D1 above, then $V(\neg p) = 1$. That is, the statement that there are not nine people in the room is completely true – a cheap truth, though.

The second part of the previous axiom says that the negation of a partial truth is not another partial truth but a plain falsehood. This result contradicts the opinion of the great physicist Niels Bohr that, whereas the opposite of an ordinary truth is a falsehood, that of a deep truth is another deep truth. Obviously, anyone who wished to formalize this witty but obscure opinion would have to construct a completely different theory of truth. However, let us move on.

D3 For any two propositions p and q:

$$\text{if } p \Leftrightarrow q, \text{ then } V(p) = V(q).$$

This is not offered as a deep insight but as an obvious control.

D4 If p is not the negate of q, then

$$V(p \wedge q) = 2^{-1}[V(p) + V(q)].$$

Otherwise, $V(p \wedge \neg p) = 0$.

Example $p =$ "Aristotle was a Spartan philosopher". This is the conjunction of two propositions, one true and the other false. Hence, $V(p) = 1/2$.

D5 For any two propositions p and q: $V(p \vee q) = \max\{V(p), V(q)\}$.

Example $p \vee q =$ "Heidegger was a philosopher or a scribbler." $V(p \vee q) = 1$.

A corollary of D5 is that

$$V(p \Rightarrow q) = \max\{V(\neg p), V(q)\}.$$

In particular,

$$\text{if } V(p) = 1, \text{ then } V(\neg p) = 0, \text{ and } V(p \Rightarrow q) = V(q); \text{ and}$$
$$\text{if } V(p) < 1, \text{ then } V(\neg p) = 1, \text{ and } V(p \Rightarrow q) = 1.$$

In this system negation, like death, does not come in degrees: it is abrupt, equalizing, and cheap. This is why critics can be right more often than their targets. Still, contradictions, though false, are not totally worthless, for they may act as alarm bells. Without them we could not use the reductio ad absurdum principle. They have also some heuristic value, since one of the two mutually exclusive options may be true.

Despite these redeeming qualities, contradiction is of course a bane, since ordinarily it stops reasoning in its track and it paralyzes action. Yet, it is not as bad as confusion, let alone nonsense. Indeed, a contradiction can be "solved" (eliminated) by just dropping one of its constituents; and confusion can be clarified by analysis, whereas nonsense is intractable. The correct semantic ranking is this:

Nonsense< Confusion< Contradiction< Partial truth< Total truth.

And the correct methodological ranking is:

Meaningful statement< Plausibility judgment< Test< Truth-value assignment.

(Contrary to the verifiability theory, meaning precedes test: see Bunge 1974b.)

Example The conditional "If *A*, then *B*" is scrutinized, and it is found that *B* is indeed true in many cases, and false in none – so far. This result renders *A* plausible (not probable though), and as such worthy of further investigation. How plausible is *A*? There is no ground to assign a numerical value to this plausibility. We could at most say that, in the light of the available evidence, *A* is more plausible than not-*A*. Only future research may be able to substantiate ("prove") *A*, or at least show that its truth-value is high.

15.5 The Problem is Still Open

The next task is to craft a consistent postulate system incorporating some or all of the above desiderata. Three caveats are in place. The first is that, if truth and falsity are regarded as mutually complementary, one may be tempted to postulate that $V(\neg p) = 1 - V(p)$. However, this assumption entails that the negation of a half-truth (= half-falsity), which should be completely true, is worth the same as its assertion. Besides, jointly with D4, it leads to the unacceptable result that conjunctions and disjunctions have the same truth-value. Allow me to repeat: Denying is far cheaper than asserting.

The second caveat is that one should resist the temptation to define partial truth in terms of probability. One reason is that truth and probability are not interdefinable, if only because truth is predicated of propositions, whereas probabilities can only be predicated of facts of a certain type. A second reason is that the concept of truth is logically prior to that of probability because, when checking probabilistic statements, whether theoretically or experimentally, we take it for granted that they can be true to some extent. Finally, truth-values do not combine like probabilities. For instance, the truth-value of the conjunction of two independent propositions with the same truth-value equals the latter, whereas the probability of the conjunction of two independent equiprobable events equals the product of their probability.

A final suggestion is that the theory should include the concept of reliability of the truth source, such as the testing technique. Indeed, it often happens that one assigns a high truth-value when employing a coarse method, only to discover error when a more exacting procedure is employed. This suggests adding the following desideratum:

D6 If a proposition p can be assigned different truth-values on the strength of tests with different reliabilities $r(p)$, choose the assignment that maximizes the product of the two values:

$$r(p).V(p) = \max.$$

where the reliability coefficient ranges between 0 and 1.

So much for the desiderata for the valuation function. The enumeration of these conditions is proposed as a research project. This project consists in figuring out a consistent set of postulates satisfying the above desiderata, some of which should occur as axioms, others as theorems, and perhaps still others in altered form or even not at all.

Closing Remarks

Everyone uses the correspondence concept of truth, but nobody seems to know exactly what it is. Therefore, it behooves philosophers to elucidate it through a theory proper, that is, a hypothetico-deductive system. However, an adequate theory of factual truth as adequacy of thought to fact should not be a priori. Instead, it should jibe with the way scientists and technologists go about finding truths and refining them. In particular, it should include the notions of adequacy of the fact-thought and the thought-proposition relations, as well as that of closeness of fit (the dual of error). Surely this project should be more interesting, exciting, and rewarding than either a many-worlds fantasy or one more constructivist-relativist manifesto against the very possibility of finding objective truths.

References

Adolphs, Ralph. 2009. The social brain: Neural basis of social knowledge. *Annual Reviews of Psychology* 60: 693–716.

Alexander, Samuel. 1920. *Space, time, and deity*, 2 vols. New York: Humanities Press.

Alkire, Michael T., Anthony G. Hudetz, and Giulio Tononi. 2008. Consciousness and anesthesia. *Science* 322: 876–80.

Allman, John Morgan. 1999. *Evolving brains*. New York: Scientific American Library.

Appiah, Kwame Anthony. 2008. *Experiments in ethics*. Cambridge, MA: Harvard University Press.

Aristotle. 1941. In *Basic works*, ed. Richard McKeon. New York: Random House.

Armstrong, D[avid]. M. 1968. *A materialist theory of the mind*. London: Routledge & Kegan Paul.

———. 1997. *A world of states of affair*. Cambridge: Cambridge University Press.

Ball, Philip. 2001. *The self-made tapestry: Pattern formation in nature*. Oxford: Oxford University Press.

Barber, Benjamin. 2003 [1984]. *Strong democracy*. Berkeley, CA: University of California Press.

Barkow, Jerome H. 2006. Introduction: Sometimes the bus does wait. In *Missing the revolution: Darwinism for social scientists*, 3–60. Oxford: Oxford University Press.

Barkow, Jerome H., Leda Cosmides, and John Tooby, eds. 1992. *The adapted mind: Evolutionary psychology and the generation of culture*. New York: Oxford University Press.

Barraclough, Geoffrey. 1979. *Main trends in history*. New York and London: Holmes & Meier.

Barrow, John D., Paul C. W. Davies, and Charles L. Harper, Jr., eds. 2004. *Science and ultimate reality: Quantum theory, cosmology, and complexity*. Cambridge: Cambridge University Press.

Bartlett, Frederic. 1932. *Remembering: A study in experimental and social psychology*. Cambridge: Cambridge University Press.

Bass, Andrew H., Edwin H. Gilland, and Robert Baker. 2008. Evolutionary origin for social vocalization in a vertebrate hindbrain-spinal compartment. *Science* 321: 417–21.

Baynes, Tim. 2008. The unity of consciousness and the split-brain syndrome. *Journal of Philosophy* 105: 277–300.

Bechtel, William. 2008. *Mental mechanisms: Philosophical perspectives on cognitive neuroscience*. London: Routledge.

Bedau, Mark A., and Paul Humphreys, eds. 2007. *Emergence: Contemporary readings in philosophy and science*. Cambridge, MA: MIT Press.

Beer, Jennifer S. 2006. Orbitofrontal cortex and social regulation. In ed. Cacioppo et al., *Social Neuroscience* 153–65.

Békésy, Georgy von. 1967. *Sensory inhibition*. Princeton, NJ: Princeton University Press.

Bekinschtein, Tristán A., Diego E. Shalom, Cecilia Forcato, María Herrera, Martin R. Coleman, Facundo F. Manes, and Mariano Sigman. 2009. Classical conditioning in the vegetative and minimally conscious state. *Nature Neuroscience* 12: 1343–9.

Bell, Graham. 2008. *Selection: The Mechanism of Evolution*, 2nd ed. Oxford: Oxford University Press.

M. Bunge, *Matter and Mind*, Boston Studies in the Philosophy of Science 287,
DOI 10.1007/978-90-481-9225-0, © Springer Science+Business Media B.V. 2010

Benedetti, Fabrizio. 2009. *Placebo effects: Understanding the mechanisms in health and disease*. Oxford: Oxford University Press.

Benedetti, Fabrizio, Helen S. Mayberg, Tor D. Wager, Christian S. Stohler, and Jon-Kar Zubieta. 2005. Neurobiological mechanisms of the placebo effect. *Journal of Neuroscience* 25: 10390–402.

Bennett, Max, and Peter Hacker. 2003. *Philosophical foundations of neuroscience*. Oxford: Blackwell.

Berkeley, George. 1901 [1710]. Principles of human knowledge. In *Works*, ed. A. Campbell Fraser, vol. 1. Oxford: Clarendon Press.

Berman, Sheri. 2006. *The primacy of politics*. Cambridge: Cambridge University Press.

Bestmann, Sven, Jürgen Budewig, Hartwig R. Siebner, John C. Rothwell, and Jens Frahm. 2004. Functional MRI of the immediate impact of transcranial magnetic stimulation on cortical and subcortical motor circuits. *European Journal of Neuroscience* 19: 1950–62.

Bindra, Dalbir, ed. 1980. *The brain's mind*. New York: Gardner Press.

Bjorklund, David F., and Anthony D. Pellegrini. 2002. *The origins of human nature: Evolutionary developmental psychology*. Washington, D.C.: American Psychological Association.

Blatt, John. 1983. How economists misuse mathematics. In *Why economics is not yet a science* ed. Alfred S. Eichner, 166–86.

Blitz, David. 1992. *Emergent evolution: Qualitative novelty and the levels of reality*. Dordrecht, NL: Kluwer Academic Publishers.

Blitz, David, and Mario Bunge. 1989. Gradualism vs. saltationism in evolutionary biology: From Darwin to Gould. *Proceedings of the 13th International Wittgenstein Symposium* pp. 297–301. Wien: Hölder-Pichler-Tempsky.

Bloch, Marc. 1949. *Apologie pour l'histoire, ou Métier d'historien*. Paris: Armand Colin.

Blok, Ned, Owen Flanagan, and Güven Güzeldere, eds. 1997. *The nature of consciousness*. Cambridge, MA: MIT Press.

Botvinick, Matthew, Leigh E. Nystrom, Kate Fissell, Cameron S. Carter, and Jonathan D. Cohen. (1999). Conflict monitoring versus selection-for-action in anterior cingulate cortex. *Nature* 402: 179–81.

Boudon, Raymond. 2001. *The origin of values*. New Brunswick, NJ: Transaction Publishers.

Braudel, Fernand. 1969. *Ecrits sur l'histoire*. Paris: Flammarion.

Brentano, Franz. 1960 [1874]. The distinction between mental and physical phenomena. In *Realism and the background of phenomenology*, ed. Roderick M. Chisholm, 39–61. Glencoe, IL: The Free Press.

Bressler, Steven L., Wei Tang, Chad M. Sylvester, Gordon L. Shulman, and Maurizio Corbetta. 2008. Top-down control of human visual cortex by frontal and parietal cortex in anticipatory visual spatial attention. *Journal of Neuroscience* 28: 10056–61.

Bridgham, Jamie T., Eric A. Ortlund, and Joseph W. Thornton. 2009. An epistatic ratchet constrains the direction of glucocorticoid receptor evolution. *Nature* 461: 515–9.

Brune, M., E. Hagley, J. Dreyer, X. Maître, A. Maali, C. Wunderlich, J. M. Raimond, and S. Haroche, 1996. Observing the progressive decoherence of the "meter" in a quantum measurement. *Physical Review Letters* 77: 4887–90.

Brush, Stephen G. 1983. *Statistical physics and the atomic theory of matter*. Princeton, NJ: Princeton University Press.

Buller, David J. 2005. *Adapting minds: Evolutionary psychology and the persistent quest for human nature*. Cambridge, MA: MIT Press.

Bunge, Mario. 1951. What is chance? *Science & Society* 15: 209–31.

——. 1955. Strife about complementarity. *British Journal for the Philosophy of Science* 6: 1–12; 6: 141–54.

——. 1956a. Do computers think? *British Journal for the Philosophy of Science* 7: 139–48; 7: 212–9. Reprinted in Bunge 1959b.

——. 1956b. A survey of the interpretations of quantum mechanics. *American Journal of Physics* 24: 272–86.

———. 1959a. *Causality: The place of the causal principle in modern science*, 4th rev. ed. New Brunswick, NJ: Transaction Publishers, 2008.
———. 1959b. *Metascientific queries*. Springfield, IL: Charles C. Thomas.
———. 1960a. Levels: A semantic preliminary. *Review of Metaphysics* 13: 396–406.
———. 1960b. *Cinemática del electrón relativista*. Tucumán: Universidad Nacional de Tucumán.
———. 1961. Laws of physical laws. *American Journal of Physics* 29: 518–29.
———. 1962. Cosmology and magic. *The Monist* 44: 116–41.
———. 1966. Mach's critique of Newtonian mechanics. *American Journal of Physics* 34: 585–96.
———. 1967a. *Scientific research*, 2 vols. New York: Springer-Verlag Rev. ed.: *Philosophy of science*, 2 vols. New Brunswick, NJ: Transaction Publishers.
———. 1967b. *Foundations of physics*. New York: Springer-Verlag.
———. 1967c. Analogy in quantum mechanics: From insight to nonsense. *British Journal for the Philosophy of Science* 18: 265–86.
———. 1967d. The structure and content of a physical theory. In *Delaware seminar in the foundations of physics*, ed. M. Bunge, 15–27. New York: Springer-Verlag.
———. 1969. The metaphysics, epistemology and methodology of levels. In *Hierarchical levels*, ed. L. L. Whyte, A. G. Wilson, and D. Wilson, 17–28. New York: American Elsevier.
———. 1970. The so-called fourth indeterminancy relation. *Canadian Journal of Physics* 48: 1410–1.
———. 1973a. *Philosophy of physics*. Dordrecht, NL: D. Reidel.
———. 1973b. *Method, model, and matter*. Dordrecht, NL: D. Reidel.
———. 1974a. *Treatise on basic philosophy*, vol. 1: *Sense and reference*. Dordrecht, NL: Reidel.
———. 1974b. The relations of logic and semantics to ontology. *Journal of Philosophical Logic* 3: 195–210.
———. 1974c. On confusing 'measure' with measurement' in the methodology of the behavioral sciences. In *The methodological unity of science*, ed. M. Bunge, 105–22. Dordrecht, NL: Reidel.
———. 1977a. *Treatise on basic philosophy, vol. 3: The furniture of the world*. Dordrecht, NL: Reidel.
———. 1977b. Emergence and the mind. *Neuroscience* 2: 501–9, 1977.
———. 1977c. Levels and reduction. *American Journal of Physiology: Regulatory, Integrative and Comparative Physiology* 2: 75–82.
———. 1979a. *Treatise on basic philosophy, vol. 4: A world of systems*. Dordrecht, NL: Reidel.
———. 1979b. The Einstein-Bohr debate over quantum mechanics: Who was right about what? *Lecture Notes in Physics* 100: 204–19.
———. 1979c. The bankruptcy of psychophysical dualism (in Russian). *Filosofskie Nauki* 2: 77–87, 1979. Comment by D. I. Dubrosvskii, ibid., 88–97.
———. 1979d. The mind-body problem in an evolutionary perspective. In *Brain and mind*, Ciba Foundation Series 69, 53–63. Amsterdam: Excerpta Medica.
———. 1980a. *The mind-body problem*. Oxford: Pergamon Press.
———. 1980b. From neuron to behavior and mentation: an exercise in levelmanship. In *Information processing in the nervous system*, ed. H. M. Pinsker and W. D. Williams, 1–16. New York: Raven Press.
———. 1981. *Scientific materialism*. Dordrecht, NL: Reidel.
———. 1982a. Is chemistry a branch of physics? *Zeitschrift für allgemeine Wissenschaftstheorie* 13: 209–33.
———. 1982b. A pszichoneuralist azonossag elmélete. Transl. of the 1979 paper. *Magyar filozofiai szemle* 1982: 540–53. Comment by Szentagothai Janos, ibid., 554–7.
———. 1983a. *Treatise on basic philosophy, vol. 5: Exploring the world*. Dordrecht, NL: Reidel.
———. 1983b. *Treatise on basic philosophy, vol. 6: Understanding the world*. Dordrecht, NL: Reidel.
———. 1985. *Treatise on basic philosophy*, vol. 7, part II. Dordrecht, NL: Reidel.
———. 1989. *Treatise on basic philosophy*, vol. 8: *Ethics*. Dordrecht, NL: Reidel.

———. 1996. *Finding philosophy in social science*. New Haven, CT: Yale University Press.

———. 1998. *Social science under debate*. Toronto: University of Toronto Press.

———. 1999. *The sociology-philosophy connection*. New Brunswick, NJ: Transaction Publishers.

———. 2000. Energy: Between physics and metaphysics. *Science and Education* 9: 457–61.

———. 2001. *Philosophy in crisis: The need for reconstruction*. Amherst, NY: Prometheus Books.

———. 2003a. *Emergence and convergence*. Toronto: University of Toronto Press.

———. 2003b. Twenty-five centuries of quantum physics: From Pythagoras to us, and from subjectivism to realism. *Science & Education* 12: 445–66.

———. 2003c. Velocity operators and time-energy relations in relativistic quantum mechanics. *International Journal of Theoretical Physics* 42: 135–42.

———. 2003d. *Philosophical dictionary*, enlarged ed. Amherst, NY: Prometheus Books.

———. 2004. The pseudoscience concept, dispensable in professional practice, is required to evaluate research projects. *Scientific Review of Mental Health Practice* 2: 111–4.

———. 2006a. *Chasing reality*. Toronto: University of Toronto Press.

———. 2006b. Naturalism and mathematics: A comment om Kanitscheider's paper. *Erwägen, Wissen, Ethik* 17(3): 341–2.

———. 2006c. The philosophy behind pseudoscience. *The Skeptical Inquirer* 30(4): 29–37.

———. 2007a. Max Weber did not practise the philosophy he preached. In *Max Weber's "Objectivity" revisited*, ed. Lawrence McFalls, 119–34. Toronto: University of Toronto Press.

———. 2007b. Blushing and the philosophy of mind. *Journal of Physiology (Paris)* 101: 247–56.

———. 2008a. *Political philosophy: Fact, fiction, and vision*. New Brunswick, NJ: Transaction Publishers.

———. 2008b. Bayesianism: Science or pseudoscience? *International Review of Victimology* 15: 169–82.

———. 2009. *Political philosophy*. New Brunswick, NJ: Transaction Publishers.

Bunge, Mario, and Adalberto García-Máynez. 1977. A relational theory of physical space. *International Journal of Theoretical Physics* 15: 961–72.

Bunge, Mario, and Andrés J.Kálnay. 1983a. Solution to two paradoxes in the quantum theory of unstable systems. *Nuovo Cimento B* 77: 1–9, 1983.

———. 1983b. Real successive measurements on unstable quantum systems take nonvanishing time intervals and do not prevent them from decaying. *Nuovo Cimento B* 77: 10–8, 1983.

Bunge, Mario, and Martin Mahner. 2004. *Ueber die Natur der Dinge*. Düsseldorf: Hirzel.

Bunge, Mario, and Rubén Ardila. 1987. *Philosophy of psychology*. New York, NY: Springer-Verlag.

Bunge, Silvia A., and I. Kahn. 2009. Cognition: An overview of neuroimaging techniques. In *Encyclopedia of neuroscience*, ed. L. R. Squire, vol. 2, 1063–7. Oxford: Academic Press.

Buss, David M. (2004). *Evolutionary psychology: The new science of the mind*, 2nd ed. Boston: Pearson.

Cabanac, Michel. 1999. Emotion and philogeny. *Japanese Journal of Physiology* 49: 1–10.

Cacioppo, John T., Penny S. Visser, and Cynthia I. Pickett, eds. 2006. *Social neuroscience: People thinking about thinking people*. Cambridge, MA: MIT Press.

Calvin, William H., and Derek Bickerton. 2000. *Lingua ex Machina*. Cambridge, MA: MIT Press.

Camerer, Colin F. 2003. Strategizing in the brain. *Science* 300: 1673–5.

Campbell, Donald T. 1974a. Evolutionary epistemology. In *The philosophy of Karl Popper*, ed. P. A. Schilpp, vol. I, 413–63. La Salle, IL: Open Court.

———. 1974b. Downward causation in hierarchically organised biological systems. In *Studies in the philosophy of biology: Reduction and related problems*, ed. F. J. Ayala and T. Dobzhansky, 179–86. Berkeley: University of California Press.

Campbell, Norman Robert. 1957 [1920]. *Foundations of science* [formerly titled *Physics, The Elements*]. New York: Dover Publications.

Carnap, Rudolf. 1967 [1928]. *The logical structure of the world*. Berkeley and Los Angeles: University of California Press.

Carreiras, Manuel, Mohamed L. Seghier, Silvia Baquero, Adelina Estévez, Alfonso Lozano, Joseph T. Devlin, and Cathy J. Price. 2009. An anatomical signature for literacy. *Nature* 461: 983–6.

Cattell, Raymond B. 1987. *Intelligence: Its structure, growth, and action*. Amsterdam: North Holland.

Cavalli-Sforza, Luigi L., and Marcus W. Feldman. 1981. *Cultural transmission and evolution: A quantitative approach*. Princeton, NJ: Princeton University Press.

Cavalli-Sforza, Luigi L., Paolo Menozzi, and Alberto Piazza. 1994. *The history and geography of human genes*. Princeton, NJ: Princeton University Press.

Chaitin, Gregory. 2006. *Meta Math!* New York: Viking.

Chalmers, David. (1995) Facing up to the problem of consciousness. *Journal of Consciousness Studies* 2: 200–19.

———. 1996. *The conscious mind: In search of a fundamental theory*. Oxford: Oxford University Press.

Changeux, Jean-Pierre. 2004. *The physiology of truth: Neuroscience and human knowledge*. Cambridge, MA: Belknap Press/Harvard University Press.

Chapman, H. A., D. A. Kim, J. M. Susskind, and A. K. Anderson. 2009. In bad taste: Evidence for the oral origins of moral disgust. *Science* 323: 1222–6.

Charbonnat, Pascal. 2007. *Histoire des philosophies matérialistes*. Paris: Syllepse.

Chomsky, Noam. 1995. Language and nature. *Mind* 104: 1–61.

———. 2009. The mysteries of nature: How deeply hidden? *Journal of Philosophy* 106: 167–200.

Churchland, Patricia Smith, and Terrence J. Sejnowski. 1993. *The computational brain*. Cambridge, MA: MIT Press.

Churchland, Paul M. 1984. *Matter and consciousness: A contemporary introduction to the philosophy of mind*. Cambridge, MA: MIT Press.

Clark, Andy. 2008. *Supersizing mind: Embodiment, action, and the cognitive extension*. Oxford: Oxford University Press.

Cochran, Gregory, and Thomas Harpending. 2009. *The 10,000 year explosion: How civilization accelerated human evolution*. New York: Basic Books.

Cole, Michael. 1996. *Cultural psychology: The once and future discipline*. Cambridge, MA: Belknap Press/Harvard University Press.

Coleman, James S. 1990. *Foundations of social theory*. Cambridge, MA: Belknap Press/Harvard University Press.

Condorcet [Marie–Jean–Antonie–Nicolas, Caritat, Marquis de]. 1976. *Condorcet: Selected writings*, ed. K. M. Baker. Indianapolis: Bobbs–Merrill.

Conway, John H., and Simon Kochen. 2009. The strong free will theorem. *Notices of the American Mathematical Society* 56: 226–32.

Cooper, Steven J. 2005. Donald O, Hebb's synapse and learning rule: A history and commentary. *Neuroscience and Behavioral Reviews* 28: 851–74.

Corballis, Michael C., and Stephen E.G. Lea, eds. 1999. *The descent of mind: Psychological perspectives on homimid evolution*. New York, NY: Oxford University Press.

Cornforth, Maurice. 1954. *Dialectical materialism*, 3 vols. London: Lawrence & Wishart.

Cornwell, John. 2003. *Hitler's scientists: Science, war, and the devil's pact*. New York: Viking.

Cosmides, Leda, and John Toby. 1992. Cognitive adaptations for social exchange. In *The adapted mind*, ed. Barkow, Cosmides, and Tooby, 163–228.

Covarrubias, G. M. 1993. An axiomatization of general relativity. *International Journal of Theoretical Physics* 32: 1235–54.

Craver, Carl F. 2009. *Explaining the brain*. Oxford: Clarendon Press.

Crick, Francis, and Christof Koch. (2002). A framework for consciousness. *Nature Neuroscience* 6: 119–26.

Crockett, Molly J., Luke Clark, Golnaz Tabibniam Matthew D. Lieberman, and Trevor W. Robbins. 2008. Serotonin modulates behavioral reactions to unfairness. *Science* 320: 1739.

D'Abro, A[braham].1939. *The decline of mechanism (in modern physics)*. New York: Van Nostrand.

Damasio, Antonio R. 1994. *Descartes' error: Emotion, reason, and the human brain*. New York: G. P. Putnam.
Damasio, Antonio R., and Kaspar Meyer. 2009. Consciousness: An overview of the phenomenon and its possible neural basis. In *The neurology of consciousness*, ed. Laureys and Tonini, 3–14.
Darwin, Charles. 1911 [1871]. *The descent of man*. New York, NY: P. F. Collier & Son.
Daston, Lorraine, and Gregg Mitman, eds. (2005). *Thinking with animals: New perspectives on anthropomorphism*. New York: Columbia University Press.
Davies, Paul C. W. 2004. John Archibald Wheeler and the clash of ideas. In *Science and ultimate reality*, ed. Barrow, Davies, and Harper, 3–23.
Dawkins, Richard. 1976. *The selfish gene*. Oxford: Oxford University Press.
De Caro, Mario, and David Macarthur, eds. 2004. *Naturalism in question*. Cambridge, MA: Harvard University Press.
Dehaene, Stanislas, Véronique Izard, Elizabeth Spelke, and Pierre Pica. 2008. Log or linear? *Science* 320: 1217–20.
Dehaene, Stanislas, Michael Kerszberg, and Jean-Pierre Changeux. (1998). A neuronal model of a global workspace in effortless cognitive tasks. *Proceedings of the National Academy of Sciences of the USA* 95: 14529–34.
Dehaene, Stanislas, and Lionel Naccache. 2001. Towards a cognitive neuroscience of consciousness: basic evidence and a workspace framework. *Cognition* 79: 1–37.
Dennett, Daniel. 1991. *Consciousness explained*. Boston: Little, Brown.
———. 1995. *Darwin's dangerous idea*. New York: Simon & Shuster.
Descartes, René. 2008 [1641]. *Meditations on first philosophy*. Oxford: Oxford University Press.
———. 1974 [1664]. *Le monde*, in *Oeuvres de Descartes*, ed. C. Adam and P. Tannery, vol. XI. Paris: Vrin.
Desmurget, Michel, Karen T. Reilly, Nathalie Richard, Alexandru Szathmari, Carmine Mottolese, and Angela Sirigu. 2009. Movement intention after parietal cortex stimulation in humans. *Science* 324: 811–3.
Deutch, David. 2004. It from qubit. In *Science and ultimate reality*, ed. Barrow, Davies, and Harper, 90–102.
De Waal, Frans. 1996. *Good natured: The origin of right and wrong*. Cambridge, MA: Harvard University Press.
———. 1998. *Chimpanzee politics: Power and sex among apes*. Baltimore, MD: Johns Hopkins University Press.
———. 2008. Putting the altruism back into altruism: The evolution of empathy. *Annual Review of Psychology* 59: 279–300.
Dewey, John. 1938. *Logic: The theory of inquiry*. New York: H. Holt.
———. 1958 [1925]. *Experience and nature*. La Salle, IL: Open Court.
Dewey, John, Sydney Hook, and Ernest Nagel. 1945. Are naturalists materialists? *Journal of Philosophy* 42: 515–30.
Dijksterhuis, E[duard]. J[an]. 1986 [1959]. *The mechanization of the world picture from Pythagoras to Newton*. Princeton, NJ: Princeton University Press.
DiLorenzo, Daniel J., and Joseph D. Bronzino, eds. 2008. *Neuroengineering*. Boca Raton, FL: CRC Press.
Dilthey, Wilhelm. 1959 [1883]. *Einleitung in die Geiteswissenschaften*. In *Gesammelte Schriften*, vol. 1. Stuttgart: Teubner; Göttingen: Vanderhoeck und Ruprecht.
Donald, Merlin. 2001. *Origins of the modern mind*. Cambridge, MA: Harvard University Press.
Dunbar, R[obin]. I. M. 2003. The social brain: Mind, language, and society in evolutionary perspective. *Annual Review of Anthropology* 32: 163–81.
Dworkin, Ronald. 1986. *Law's empire*. Cambridge, MA: Belknap/Harvard University Press.
Eccles, John C. 1951. Hypotheses relating to the brain-mind problem. *Nature* 168: 53–64.
Edel, Abraham. 1944. Naturalism and ethical theory. In ed. Krikorian, 65–95.
Eichner, Alfred S. 1983. *Why economics is not yet a science*. Armonk, NY: M. E. Sharpe.
Einstein, Albert. 1934. *Mein Weltbild*. Amsterdam: Querido Verlag.

Einstein, Albert, Boris Podolsky, and Nathan Rosen. 1935. Can quantum-mechanical description of physical reality be considered complete? *Physical Review* 47: 777–89.

Elman, Jeffrey K., Elizabeth A. Bates, Mark H. Johnson, Annette Karmiloff-Smith, Domenico Parisi, and Kim Plunkett. 1998. *Rethinking innateness: A connectionist perspective on development*. Cambridge, MA: MIT Press.

Enfield, N. J., and Stephen C. Levinson. 2006. Introduction: Human sociality as a new interdisciplinary field. In *Roots of human sociality*, ed. Enfield and Levinson, 1–38.

Enfield, N. J., and Stephen C. Levinson, eds. 2006. *Roots of human sociality: Culture, cognition and interaction*. Oxford and New York: Berg.

Engels, Frederick. 1876. The part played by labour in the transition from ape to man. In K. Marx and F. Engels, *Selected works* 1962, vol. II, 80–92.

———. 1954 [1894]. *Anti-Dühring*. Moscow: Foreign Languages Publishing House.

———. 1940 [1883]. *Dialectics of nature*. New York: International Publishers.

Evarts, Edward V., Yoshikazu Shinoda, and Steven P. Wise. 1984. *Neurophysiological approaches to higher brain function*. New York: Wiley (Interscience).

Everett, Hugh, III. 1957. "Relative state" formulation of quantum mechanics. *Reviews of Modern Physics* 29: 454–62.

Feferman, Solomon. 2009. Gödel, Nagel, minds, and machines. *Journal of Philosophy* 106: 201–19.

Fehr, Ernst, and Urs Fischbacher. 2003. The nature of human altruism. *Nature* 425: 785–91.

Fehr, Ernst, Helen Bernhard, and Bettina Rockenbach. 2008. Egalitarianism in young children. *Nature* 454: 1079–81.

Feigl, Herbert. 1967 [1958]. *The "mental" and the "physical"*. Minneapolis, MI: University of Minnesota Press.

Feldman, Daniel E. 2009. Synaptic mechanisms for plasticity in neocortex. *Annual Review of Neuroscience* 32: 33–55.

Ferrer, Emilio, Elizabeth D. O'Hare, and Silvia A. Bunge. 2009. Fluid reasoning and the developing brain. *Frontiers in Neuroscience* 3: 46–51.

Flew, Antony, ed. 1987. *Readings in the philosophical problems of parapsychology*. Buffalo, NY: Prometheus Books.

Frazier, Kendrick, ed. 2009. Science under siege: *Defending science, exposing pseudoscience*. Amherst, NY: Prometheus Books.

Friedman, Milton. 1991. Old wine in new bottles. *Economic Journal* 101: 33–40.

Fodor, Jerry A. 1983. *The modularity of mind*. Cambridge, MA: MIT Press.

Freud, Sigmund. 2006 [1900]. *The interpretation of dreams*. London: Penguin.

Galbraith, John Kenneth. 1983. *The anatomy of power*. Boston: Houghton Mifflin.

Galilei, Galileo. 1953 [1623]. *Il saggiatore*. In *Opere*, ed. F. Flora. Milano-Napoli: Riccardo Ricciardi.

Gallistel, C[harles]. R., and Adam Philip King. 2009. *Memory and the computational brain: Why cognitive science will transform neuroscience*. New York: Wiley and Blackwell.

Gardner, Martin. 1983. *Science: Good, bad, and bogus*. Oxford: Oxford University Press.

Gazzaniga, Michael S. 2008. *Human: The science behind what makes us unique*. New York: HarperCollins.

Genty, Emilie, and Jean-Jacques Roeder. 2006. Self-control: why should sea lions, *Zalophus californianus*, perform better than primates? *Animal Behavio*r 72: 1241–7.

Geschwind, Norman. 1974 (1965). Disconnexion syndromes in animals and man. In *Selected papers on language and the brain*, 105–236. Dordrecht, The Netherlands: Reidel.

Gintis, Herbert, Samuel Bowles, Robert Boyd, and Ernst Fehr, eds. 2005. *Moral sentiments and material interests: The foundations of cooperation in economic life*. Cambridge MA: MIT Press.

Glass, Leon, and Michael C. Mackey. 1988. *From clocks to chaos: The rhythms of life*. Princeton, NJ: Princeton University Press.

Gonseth, Ferdinand. 1937. *Qu'est-ce que la logique?* Paris: Hermann.

Goodale, Melvyn, and David Milner. (2005). *Sight unseen*. Oxford: Oxford University Press.
Gottfried, Kurt, and Tung-Mow Yan. 2003. *Quantum mechanics: Fundamentals*. New York: Springer-Verlag.
Gottlieb, Gilbert. 1992. *Individual development and evolution: The genesis of novel behavior*. New York: Oxford University Press.
Gould, Stephen J. (1997a). Evolution: The pleasures of pluralism. *New York Review of Books* 44(11): 47–52.
———. 1997b. Nonoverlapping magisteria. *Natural History* 106: 16–22.
———. 2002. *The structure of evolutionary theory*. Cambridge, MA: Belknap Press/Harvard University Press.
Graham, Loren R. 1981. *Between science and values*. New York: Columbia University Press.
Gray, Jeremy R., and Paul M. Thompson. 2004. *Nature Reviews Neuroscience* 5: 471–82.
Gregory, Frederick. 1977. *Scientific materialism in nineteenth century Germany*. Dordrecht, NL: D. Reidel.
Gross, James J., ed. 2007. *Handbook of emotion regulation*. New York: Guilford Press.
Haack, Susan. 1996. *Deviant logic, fuzzy logic*. Chicago, IL: University of Chicago Press.
Haggard, Patrick. 2005. Conscious intention and motor cognition. *Trends in Cognitive Sciences* 9: 290–95.
Hare, Todd, John O'Doherty, Colin F. Camerer, Wolfram Schutz, and Antonio Rangel. 2008. Dissociating the role of the orbitofrontal cortex and the striatum in the computation of goal values and prediction errors. *Journal of Neuroscience* 28: 5623–30.
Harris, Marvin. 1979. *Cultural materialism: The struggle for a science of culture*. New York: Random House.
Hartmann, Nicolai. 1949. *Neue Wege der Ontologie*, 3rd ed. Stuttgart: W. Kohlhammer Verlag.
He, Biyu J., and Marcus E. Raichle. 2009. The fMRI signal, slow cortical potential and consciousness. *Trends in Cognitive Science* 13: 392–09.
Hebb, Donald O. 1949. *The organization of behavior*. New York: Wiley.
———. 1980. *Essay on mind*. Hillsdale, NJ: Erlbaum.
Hegel, Friedrich. 1929 [1812]. *Science of logic*, 2 vols. London: George Allen & Unwin.
Heidegger, Martin. *Sein und Zeit*, 17th ed. Tübingen: Max Niemeyer Verlag.
———. 1954 [1947]. Brief über den "Humanismus." In *Platos Lehre von der Wahrheit*, 2nd ed., 53–119. Bern: Francke Verlag.
Heisenberg, Werner. 1930. *The physical principles of the quantum theory*. Chicago, IL: University of Chicago Press.
———. 1969. *Der Teil und das Ganze*: *Gespräche im Umkreis der Atomphysik*. München: R. Piper.
Helvétius, Claude A. 1988 [1758]. *De l'esprit*. Repr. Paris: Fayard.
Henderson, Lawrence J. 1970 [1913]. *The fitness of the environment: An inquiry into the biological significance of the properties of matter*. Boston: Beacon Press.
Henrich, J., R. Boyd, S. Bowles, C. Camerer, E. Fehr, and H. Gintis, eds. 2004. *Foundations of human sociality: Economic experiments and ethnographic evidence from fifteen small-scale societies*. Oxford: Oxford University Press.
Henry, Richard Conn. 2005. The mental universe. *Nature* 436: 29.
Herodotus. 1972 [415 B.C.E]. *The histories*. London: Penguin Books.
Hickok, Gregory. 2009. Eight problems for the mirror theory of action understanding in monkeys and humans. *Journal of Cognitive Neuroscience* 21: 1229–43.
Hilbert, David, and Paul Bernays. 1968, *Grundlagen der Mathematik*, 2 vols., 2nd ed. Berlin, Heidelberg, and New York: Springer-Verlag.
Hobson, J. Allan. 1999. *Consciousness*. New York: Scientific American Library.
Hobsbawm, Eric. 1997. *On history*. New York: The New Press.
Holbach, Paul-Henry Thiry, Baron d'. 1770. *Système de la nature*. 2 vols. Repr. Hildesheim and New York: Georg Olms, 1966.
———. 1773. *Système social*. 3 vols. Repr. Hildesheim and NewYork: Georg Olms, 1969.

Hsu Ming, Cédric Anen, and Steven R. Quartz. 2008. The right and the good: Distributive justice and neural encoding of equity and efficiency. *Science* 320: 1092–95.
Humphrey, Nicholas. 1983. *Consciousness regained.* Oxford: Oxford University Press.
Husserl, Edmund. 1960 [1931]. *Cartesian meditations: An introduction to phenomenology.* The Hague: M. Nijhoff.
———. 1970. [1936]. *The crisis of European sciences and transcendental phenomenology: An introduction to phenomenological philosophy.* Evanston, IL: Northwestern University Press.
Huxley, Thomas Henry. 1893. Evolution and ethics. In *Evolution and ethics*, ed. T. H. Huxley and J. Huxley, 60–84. London: Pilot Press.
Ingenieros, José. 1946 [1919]. *Principios de psicología*, 6th ed. Buenos Aires: Losada.
Iversen, Leslie L., Susan D. Iversen, Floyd D. Bloom, and Robert H. Roth. 2008. *Introduction to neuropsychopharmacology.* Oxford: Oxford University Press.
Jacob, François. 1977. Evolution and tinkering. *Science* 196: 1161–6.
James, William. 1890. *The principles of psychology*, 2 vols. Repr. New York: Dover, 1950.
Jarochewski, Michail. 1975. *Psychologie im 20. Jahrhundert.* Berlin: Volk und Wissen.
Jaynes, Edwin T. 1967. Foundations of probability theory and statistical mechanics. In *Delaware seminar in the foundations of physics*, ed. Mario Bunge, 77–101. New York: Springer-Verlag.
Jensen, Keith, Joseph Call, and Michael Tomasello. 2007. Chimpanzees are rational maximizers. *Science* 318: 107–9.
Johnson-Laird, P. N., and P. C. Wason, eds. (1977) *Thinking: Readings in cognitive science.* Cambridge: Cambridge University Press.
Jones, Ernest. 1961. *The life and work of Sigmund Freud*, ed. and Abridged by L. Trillling and S. Marcus. New York: Basic Books.
Kanitscheider, Bernulf. 1996. *Im Innern der Natur: Philosophie und moderne Physik.* Darmstadt: Wissenschaftliche Buchgesellschaft.
Kandel, Eric, 2006. *In search of memory: The emergence of a new science of mind.* New York: W. W. Norton.
Kanitscheider, Benulf. 2006. Naturalismus und logtisch-mathematische Grundlagenprobleme. *Erwägen, Wissen, Ethik* 17: 325–38.
———. 2007. *Die Materie und ihre Schatten: Naturalistische Wissenschaftstheorie.* Aschaffenburg: Alibri Verlag.
Kant, Immanuel. 1780. *Metaphysik II, Reflexionen auf losen Blättern.* In *Gesammelte Schriften, AAXVIII: Handschriftlicher Nachlass.* Akademie Ausgabe. Online: ikp.uni-bonn.de/kant/aa
———. 1787. *Kritik der reinen Vernunft*, 2nd ed. Hamburg: Felix Meiner.
Kaplan, Hillard, Kim Hill, Jane Lancater, and A. Magdalena Hurtado. 2000. A theory of human life history evolution: Diet, intelligence, and longevity. *Evolutionary Anthropology* 8: 156–85.
Karmiloff-Smith, Annette. 2006. The tortuous route from genes to behavior: A neuroconstructivist approach. *Cognitive, Affective, & Behavioral Neuroscience* 6: 9–17.
Kemeny, Margaret E. 2009. Psychobiological responses to social threat: Evolution of a psychological model in psychneuroimmunology. *Brain, Behavior, and Immunity* 23: 1–9.
Keyfitz, Nathan, ed. 1984. *Population and biology: Bridge between disciplines.* Liège: Ordina Editions.
Kim, Jaegwon. 1993. *Supervenience and mind.* Cambridge: Cambridge University Press.
———. 2006. *Philosophy of mind*, 2nd ed. Cambridge, MA: Westview.
Kishiyama, Mark M., W. Thomas Boyce, Amy M. Jimenez, Lee M. Perry, and Robert T. Knight. 2009. Socioeconomic disparities affect prefrontal function in children. *Journal of Cognitive Neuroscience* 21: 1106–15.
Koppelberg, Dirk. 1999. Naturalismus/Naturalisierung. In *Enzyklopädie Philosophie*, ed. H. J. Sandkühler, vol. I, 904–14. Hamburg: Felix Meiner.
Kosfeld, M., M. Heinrichs, P. Zak, U. Fischbacher, and E. Fehr. 2005. Oxytocin increases trust in humans. *Nature* 435: 673–6.
Koslicki, Kathrin. 2008. *The structure of objects.* New York: Oxford University Press.

Kreiman, Gabriel, Christof Koch, and Itzhak Fried (2000). Imagery neurons in the human brain. *Nature* 408: 357–61.
Krikorian, Yervant V., ed. 1944. *Naturalism and the human spirit*. New York: Columbia University Press.
Kripke, Saul. 1971. "Identity and Necessity". In *Identity and individuation*, ed. Milton K. Munitz, 135–64.
Kurtz, Paul. 2001. *Skeptical odysseys*. Amherst, NY: Prometheus Books.
Kurtz, Paul, ed. 1985. *A skeptic's handbook of parapsychology*. Buffalo, NY: Prometheus Books.
Lachs, John. 1963. Epiphenomenalism and the notion of cause. *Journal of Philosophy* 60: 141–6.
Lalande, André. 1938. *Vocabulaire technique et critique de la philosophie*, 3 vols. Paris: Félix Alcan.
Lanczos, Cornelius. 1949. *The variational principles of mechanics*. Toronto: University of Toronto Press.
Lang, Serge. 1981. *The file*. New York: Springer-Verlag.
Lange, Friedrich Albert. 1905 [1873]. *Geschichte des Materialismus und Kritik Seiner Bedeutung in der Gegenwart*. Leipzig: Philipp Reclam.
Laureys, Steven, and Giulio Tononi, eds. 2009. *The neurology of consciousness: cognitive neuroscience and pathology*. Amsterdam: Academic Press.
Lazcano, Antonio. 2007a. What is life? A brief historical overview. *Chemistry and Biodiversity* 4: 1–15.
———. 2007b. Prebiotic evolution and the origin of life: Is a system-level understanding feasible? In ed. I. Rigoutsos and G. Stephanopoulos, 57–78.
LeDoux, Joseph. 2002. *Synaptic self: How our brains become who we are*. New York: Penguin Books.
Leibniz, Gottfried Wilhelm. 1956 [1714]. *The principles of nature and of grace, based on reason*. In *Philosophical papers and letters*, ed. L. E. Loemke, vol. II, 1033–43. Chicago, IL: University of Chicago Press.
Lenin, V[ladimir]. I[lich]. 1947 [1908]. *Materialism and empiriocriticism*. Moscow: Foreign Languages Publishing House.
———. 1981. *Collected works, vol. 38: Philosophical Notebooks* [1914–1915]. Moscow: Foreign Languages Publishing House.
Levitis, Daniel A. William Z. Lickider, and Glenn Freund. 2009. Biologists do not agree on what constitutes behaviour. *Animal Behaviour* 78: 103–10.
Lévy-Leblond, Jean-Marc. 1990. Did the big bang begin? *American Journal of Physics* 58: 156–9.
———. 2006. *De la matière–quantique, relativiste, interactive*. Paris: Seuil.
Lévy-Leblond, Jean-Marc, and Françoise Balibar. 1990. *Quantics*. Amsterdam: North-Holland.
Lewes, George Henry. 1874. *Problems of life and mind*. London: Trueubner.
Lewis, David. 1986. *On the plurality of worlds*. Oxford: Blackwell.
Lewontin, Richard, and Richard Levins. 2007. *Biology under the influence: Dialectical essays on ecology, agriculture, and health*. New York: Monthly Review Press.
Lickleiter, Robert, and Hunter Honeycutt. 2003. Developmental dynamics: Toward a biologically plausible evolutionary psychology. *Psychological Bulletin* 129: 819–35.
Lieberman, Matthew D., and Naomi I. Eisenberger. 2006. A pain by any other name (rejection, exclusion, ostracism) still hurts the same. In *Social neuroscience*, ed. Cacioppo, Visser, and Pickett, 167–88.
Llinás, Rodolfo R. 2001. *i of the vortex: From neurons to self*. Cambridge, MA: MIT Press.
Lloyd, Elisabeth. 1999. Evolutionary psychology: The burdens of proof. *Biology and Philosophy* 14: 211–34.
Lloyd Morgan, C[onwy]. *Emergent evolution*. London: Williams and Norgate.
Loeb, Jacques. 1912. *The mechanistic conception of life: Biological* essays. Chicago, IL: University of Chicago Press.
Locke, John. 1690. *An essay concerning human understanding*. London: George Routledge & Sons, s.d.

Loeber, Rolf, and Dustin Pardini. 2008. Neurobiology and the development of violence: common assumptions and controversies. *Philosophical Transactions of the Royal Society* 363: 2491–503.

Logothetis, Nikos K. 2008. What we can do and what we cannot do with fMRI. *Nature* 453: 869–78.

London, F., and E. Bauer. 1939. *La théorie de l'observation en mécanique quantique*. Paris: Hermann.

Lorenz, Konrad. 1971 [1954]. Psychology and phylogeny. In *Studies in Animal and Human Behavior*, vol II, 196–245. Cambridge, MA: Harvard University Press.

Lormand, Eric. 2004. The explanatory stopgap. *Philosophical Review* 113: 303–57.

Lovejoy, Arthur O. 1953 [1936]. *The great chain of being*. Cambridge, MA: Harvard University Press.

———. 1955. *The revolt against dualism: An inquiry concerning the existence of ideas*. La Salle, IL.: Open Court Pub. Co.

Luisi, Pier Luigi. 2006. *The emergence of life: From chemical origins to synthetic biology*. Cambridge: Cambridge University Press.

Luria, Alexander R.1976. *Cognitive development. Its cultural and social foundations*. Cambridge, MA: Harvard University Press.

Lycan, William G., and George S. Pappas. 1976. Quine's materialism. *Philosophia* 6: 101–30.

Mach, Ernst. 1914 [1900]. *The analysis of sensations and the relation of the physical to the psychical*. Chicago, IL: Open Court.

———. 1942 [1893]. *The science of mechanics*. La Salle, IL: Open Court.

Machamer, Peter, Lindley Darden, and Carl F. Craver. 2000. Thinking about mechanisms. *Philosophy of Science* 67: 1–25.

Mahner, Martin. 2007a. Kann man als metaphysischer Naturalist zugleich erkenntistheoretischer Naturalist sein? In *Naturalismus: Positionen, Perspektiven, Probleme*, ed. T. Sukopp and G. Vollmer, 115–36. Tübingen: Mohr Siebeck.

———. 2007b. Unverezichtbarkeit und Reichweite des ontologischen Naturalismus. In *Zufall Mensch?*, ed. L. Klinnert, 77–90. Darmstadt: Wissenschaftliche Buchgesellschaft.

———. 2007c. Demarcating science from pseudoscience. In *Handbook of the philosophy of science: General philosophy of science – Focal issue*, ed. T. Kuipers, 515–75. Amsterdam: Elsevier.

Mahner, Martin, and Mario Bunge. 1996. Is religion education compatible with science education? *Science & Education* 5: 101–23.

———. 1997. *Foundations of biophilosophy*. Berlin, Heidelberg, and New York: Springer-Verlag.

———. 2001. Function and functionalism: A synthetic perspective. *Philosophy of Science* 68: 75–94.

Mahon, Bradford Z., and Adolfo Caramazza. 2009. Concepts and categories: A cognitive neuropsychological perspective. *Annual Reviews of Psychology* 60: 27–51.

Mahon, Bradford Z., Stefano Anzellotti, Jens Schwarzbach, Massimiliano Zampini, and Alfonso Caramazza. 2009. Category-specific organization in the human brain does not require visual experience. *Neuron* 63: 397–405.

Mameli, Matteo. 2008. On innateness. *Journal of Philosophy* 105: 719–36.

Marks, Jonathan. 2002. *What it means to be 98% Chimpanzee*. Berkeley and Los Angeles: University of California Press.

Marx, Karl, and Frederick Engels. *Selected works*, 2 vols. Moscow: Foreign Languages Publishing House.

Maudlin, Tim. 2003. Distilling metaphysics from quantum physics. In *The Oxford handbook of metaphysics*, ed. M. J. Loux and D. W. Zimmerman, 461–90. Oxford: Oxford University Press.

McCulloch, Warren S. 1965. *Embodiments of mind*. Cambridge, MA: MIT Press.

McGinn, Colin. 1993. *Problems in philosophy*. Oxford: Blackwell.

———. 2004. *Consciousness and its objects*. Oxford: Clarendon Press.

McGowan, Patrick O., Aya Sasaki, Ana C. D. D'Alessio, Sergiy Dymov, Benoit Labonté, Moshe Szyf, Gustavo Turecki, and Michael J. Meaney. (2009) Epigenetic regulation of the glucocorticoid receptor in human brain associates with childhood abuse. *Nature Neuroscience* 12: 342–8.

McLane, Saunders. 1998. *Categories for the working mathematician*, 2nd ed. New York: Springer-Verlag.

MacLean, R. Craig, and Graham Bell. 2003. Divergent evolution during an experimental adaptive radiation. *Proceedings of the Royal Society of London B* 270: 1645–50.

Medawar, Peter. 1957. *The uniqueness of the individual*. New York: Basic Books.

Mellor, D. H. 1991. *Matters of metaphysics*. Cambridge: Cambridge University Press.

Melzack, Ronald. 1989. Phantom limbs, the self and the brain. *Canadian Journal of Psychology* 30: 1–16.

Mermin, N. David. 1981. Quantum mysteries for anyone. *Journal of Philosophy* 78: 397–408.

———. 2009. What's bad this habit. *Physics Today* 62(5): 8–9.

Merton, Robert K. 1968. *Social theory and social structure*, enlarged ed. New York: The Free Press.

———. 1973. *The sociology of science*. Chicago, IL: University of Chicago Press.

———. 1976 [1936]. The unanticipated consequences of social action. In *Sociological ambivalence and other essays*, 145–55. New York: The Free Press.

———. 2001 [1938]. *Science, technology & society in seventeenth-century England*. New York: Howard Fertig.

Merton, Robert K., and Elinor Barber. 2004. *The travels and adventures of serendipity*. Princeton, NJ: Princeton University Press.

Miller, David, ed. 1985. *Popper selections*. Princeton, NJ: Princeton University Press.

Misner, Charles W., Kip S. Thorne, and John Archibald Wheeler. 1973. *Gravitation*. San Francisco, CA: W. H. Freeman.

Mithen, Steven. 1994. From domain specific to generalized intelligence: A cognitive interpretation of the middle/upper palaeolithic transition. In *The ancient mind*, ed. Renfrew and Zubrov, 29–39.

———. 1996. *The prehistory of the mind*. London: Thames & Hudson.

———. 1999. The evolution of deep social mind in humans. In *The descent of mind*, ed. Corballis and Lea, 173–93.

Moessinger, Pierre. 2008. *Voir la société: Le micro et le macro*. Paris: Hermann.

Morgan, C. Lloyd. 1933. *The emergence of novelty*. London: Williams & Norgate.

Morris, J. S., A. öhman, and R. J. Dolan. 1998. Conscious and unconscious emotional learning in the human amygdala. *Nature* 393: 467–70.

Mountcastle, Vernon. 1998. *Perceptual neuroscience: The cerebral cortex*. Cambridge, MA: Harvard University Press.

Nader, Karim, G. E Schafe, and Joseph E. Ledoux. 2000. Fear memories require protein synthesis in the amygdala for reconsolidation after retrieval. *Nature* 406: 722–6.

Nagel, Ernest. 1956. *Logic without metaphysics*. Glencoe, IL: The Free Press.

Nagel, Thomas. 1974. What is it like to be a bat? *Philosophical Review* 83: 435–50.

Nauenberg, Michael. 2007. Critique of "Quantum Enigma: Physics Encounters Consciousness". *Foundations of Physics* 37: 1612–7.

Needham, Joseph. 1943. *Time: The refreshing river*. London: George Allen & Unwin.

Nesse, Randolph M., and George C. Willliams. 1994. *Why we get sick*. New York: Times Books Random House.

Neurath, Otto. 1981. In *Gesammelte philosophische und methodologische Schriften*, 2 vols. Ed. R. Haller and H. Rutte. Wien: Hölder-Pichler-Tempsky.

Norsen, Travis. 2007. Against 'realism'. *Foundations of physics* 37: 311–40.

Novikoff, Alex. 1945. The concept of integrative levels in biology. *Science* 101: 209–15.

O'Connor, John, ed. 1969. *Modern materialism: Readings on mind-body identity*. New York: Harcourt, Brace & World.

Odling-Smee, F. John, Kevin N. Laland, and Marcus W. Feldman. 2003. *Niche Construction*. Princeton, NJ: Princeton University Press.

OECD. 2008. *Growing unequal? Income distribution and poverty in OECD countries*. Paris: OECD.

Oparin, A[lexandr]. I. 1953. *The origin of life*. New York: Dover.

Ostwald, Wilhelm. 1902. *Vorlesungen über Naturphilosophie*. Leipzig: Veit & Comp.

Panasiuk, Ryszard, and Leszek Nowak, eds. 1998. *Marx's theories today*. Amsterdam-Atlanta: Rodopi.

Papineau, David. 1993. *Philosophical naturalism*. Oxford: Basil Blackwell.

———. 2003. Theories of consciousness. In *Consciousness*, ed. Smith and Jokic, 353–83.

Paradis, Michel. (2004). *A neurolinguistic theory of bilingualism*. Amsterdam: John Benjamin.

Park, Robert L. 2008. *Superstition*. Princeton, NJ: Princeton University Press.

Passingham, Richard. 2008. *What is special about the human brain?* Oxford: Oxford University Press.

Pastalkova, Eva, Vladimir Itskov, Asohan Amarasingham, and György Buzsaki. 2008. Internally generated cell assembly sequences in the rat hippocampus. *Nature* 321: 1322–7.

Pauli, Wolfgang. 1958. Die allgemeinen Principien der Wellenmechanik. In *Handbuch der Physik*, ed. S. Flügge, vol. V, Teil 1, 1–168. Berlin: Springer-Verlag.

Pavlov, I. P. 1955. *Selected works*. Moscow: Foreign Languages Publishing House.

Peirce, Charles S. 1935 [1898]. Scientific metaphysics. In *Collected papers*, ed. C. Hartshorne and P. Weiss, vol. 6. Cambridge, MA: Harvard University Press.

Penfield, Wilder, and Waldo Rasmussen. 1968. *The cerebral cortex of man: A clinical study of localization of function*. New York: Hafner.

Penn, Derek C., Keith J. Holyoak, and Daniel J. Povinelli. 2008. Darwin's mistake: Explaining the discontinuity between human and nonhuman minds. *Behavioral and Brain Sciences* 31: 109–78.

Pérez-Bergliaffa, Santiago E., Gustavo Romero, and Héctor Vucetich. 1993. Axiomatic foundations of non-relativistic quantum mechanics: A realistic approach. *Journal of Theoretical Physics* 32: 1507–25.

———. 1996. Axiomatic foundations of quantum mechanics revisited: A case for systems. *International Journal of Theoretical Physics* 35: 1805–19.

———. 1998. Toward an axiomatic pregeometry of space-time. *International Journal of Theoretical Physics* 37: 2281–99.

Pesaran, Bijan, Matthew J. Nelson, and Richard A. Andersen. 2008. Free choice activates a decision circuit between frontal and parietal cortex. *Nature* 453: 406–9.

Pessoa, Luiz. 2008. On the relationship between emotion and cognition. *Nature Reviews/ Neuroscience* 9: 148–58.

Phelps, Elizabeth A. 2006. Emotion and cognition. *Annual Reviews of Psychology* 57: 27–53.

Phillips, Melba. 1949. Quantum mechanics. In *Philosophy for the future*, ed. R. Wood Sellars, V. J. McGill, and M. Farber, 188–201.

Piaget, Jean. 1965. *Etudes sociologiques*. Genève: Librairie Droz.

Pinker, Steven. 1994. *The language instinct: How the mind creates language*. New York: Morrow.

———. 1997. *How the mind works*. New York: Norton.

———. 2003. *The blank slate: The modern denial of human nature*. New York: Penguin Books.

Place, U[llian]. T. 1956. Is consciousness a brain process? *British Journal of Psychology* 47: 44–50.

Plato. *Laws*. 1937. In *The dialogues of plato*, 2 vols., transl. B. Jowett. New York: Random House.

Platt, Michael L., and Elizabeth S. Spelke. 2009. What can developmental and comparative cognitive neuroscience tell us abut the adult human brain? *Current Opinion in Neurobiology* 19: 1–5.

Plekhanov, Georgii Valentinovich. 1967 [1896]. *Essays in the history of materialism*. New York: H. Fertig.

Popper, Karl R. 1961. Personal Communication.

———. 1967. Knowledge: Subjective versus objective. In *Popper selections*, ed. Miller, 1985, 58–77.
———. 1972. *Objective knowledge: An evolutionary approach*. Oxford: Clarendon Press.
———. 1978. Natural selection and the evolution of mind. *Dialectica* 32: 339–55.
Popper, Karl R., and John C. Eccles. 1977. *The mind and its brain*. New York: Springer International.
Pound, Roscoe. 1954 [1924]. *An introduction to the philosophy of law*, Rev. ed. New Haven, CT: Yale University Press.
Prather, J. F., S. Peters, S. Nowicki, and R. Mooney. 2008. Precise auditory-vocal mirroring in neurons for learned vocal communication. *Nature* 451: 305–10.
Preston, Stephanie D., and Frans B. M. de Waal. 2002. Empathy: its ultimate and proximate bases. *Behavioral and Brain Sciences* 25: 1–71.
Preuss, Todd M. 2007. Primate brain evolution in phylogenetic context. In *Evolution of nervous systems, vol. 4: Primates*, ed. J. H, Kaas and T. M. Preuss, 1–34. Oxford: Elsevier.
Pribram, Karl H. 1971. *Languages of the Brain: Experimental paradoxes and principles in neuropsychology*. Englewood Cliffs, NJ: Prentice-Hall.
Purves, Dale, Elizabeth M. Brannon, Roberto Cabeza, Scott A. Huettel, Kevin S. LaBar, Michael L. Platt, and Marty G. Woldorff. 2008. *Principles of cognitive neuroscience*. Sunderland, MA: Sinauer Associates.
Putnam, Hilary. 1960. Minds and machines. In *Dimensions of mind*, ed. S. Hook, 148–79. New York: New York University Press.
———. 1975. *Philosophical papers, vol. 2: Mind, language and reality*. Cambridge: Cambridge University Press.
———. 1978. *Meaning and the moral sciences*. London: Routledge.
———. 2004. *Ethics without ontology*. Cambridge, MA: Harvard University Press.
Pylyshyn, Zenon W. 1984. *Computation and cognition*. Cambridge, MA: MIT Press.
Quine, W. V. 1953. *From a logical point of view*. Cambridge, MA: Harvard University Press.
———. 1968. Ontological relativity. *Journal of Philosophy* LXV: 185–212.
———. 1973. *The roots of reference*. La Salle, IL: Open Court.
Raichle, Marcus. 2006. The brain's dark energy. *Science* 314: 1249–50.
Railton, Peter. 2003. *Fact, values, and norms: Essays towards a morality of consequence*. Cambridge: Cambridge University Press.
Randall, John Herman, Jr. 1958. *Nature and historical experience*. New York: Columbia University Press.
Randi, James. 1982. *Flim-Flam!* Buffalo, NY: Prometheus Books.
Renan, Ernest. 1949 [1852]. *Averroès et l'averroisme. Oeuvres complètes*, vol. III. Paris: Calmann-Lévy.
Renfrew, Colin, and Ezra B. Zubrov, eds. 1994. *The ancient mind: Elements of cognitive archaeology*. Cambridge: Cambridge University Press.
Rescher, Nicholas. 2009. *Free will: A philosophical reappraisal*. New Brunswick, NJ: Transaction Publishers.
Richardson, Robert C. 2007. The adaptive programme of evolutionary psychology. In *Philosophy of psychology and cognitive science*, ed. Thagard, 369–411.
———. 2007. *Evolutionary psychology as maladapted psychology*. Cambridge, MA: MIT Press.
Richerson, Peter J., and Robert Boyd. 2005. *Not by genes alone: How culture transformed human evolution*. Chicago, IL: University of Chicago Press.
Rigoutsos, Isidore, and Gregory Stephanopoulos, eds. 2007. *Systems biology*. Oxford: Oxford University Press.
Rizzolatti, Giacomo, and Laila Craighero. 2004. The mirror-neuron system. *Annual Review of Neuroscience* 27: 169–92.
Rizzolatti, Giacomo, and Corrado Sinigaglia. 2008. *Mirrors in the brain – How our minds share actions and emotions*. Oxford: Oxford University Press.

Robinson, Joan, and John Eatwell. 1974. *An introduction to modern economics*, Rev. ed. London: McGraw-Hill.
Rosenberg, Alexander. 1980. *Sociobiology and the preemption of social science*. Baltimore, MD: Johns Hopkins University Press.
Rosental, M., and P. Yudin. 1945 [1940]. *Diccionario de filosofía*. Santiago de Chile: Nueva América.
Rozental, M. M., and P. Yudin, eds. 1967. *A dictionary of philosophy*. Moscow: Progress Pub.
Rumbaugh, Duane M., and Timothy V. Gill. 1976. The mastery of language-type skills by the chimpanzee (*Pan*). *Annals of the New York Academy of Sc*iences 280: 562–78.
Russell, Bertrand. 1914. *Our knowledge of the external world*. London: Allen & Unwin.
———. 1954 [1927]. *The analysis of matter*. New York: Dover.
Ryder, John, ed. 1994. *American philosophical naturalism in the twentieth century*. Amherst, NY: Prometheus Books.
Sapolsky, Robert. 2005. The influence of social hierarchy on primate health. *Science* 308: 648–52.
Savage, J. L. 1954. *The foundations of statistics*. New York: Wiley.
Schacter, Daniel L. 2001. *The seven sins of memory*. Boston, MA: Houghton Mifflin.
Schelling, Thomas C. 1978. *Micromotives and macrobehavior*. New York: W. W. Norton.
Schlick, Moritz. 1974 [1925]. *General theory of knowledge*. Wien: Springer-Verlag.
Schlosshauer, Maximilian. 2007. *Decoherence and the Quantum-to-Classical Transition*. Berlin-Heidelberg-New York: Springer-Verlag.
Schmidt, Michael, and Hod Lipson. 2009. Distilling free-form natural laws from experimental data. *Science* 324: 81–5.
Schrödinger, Erwin. 1935. Die gegenwärtige Situation in der Quantenmechanik. *Die Naturwissenschaften* 23: 807–12; 823–8; 844–89.
Schwartz, Charlie, et al. 1972. *Science against the people: The story of Jason*. http://socrates.berkeley.edu
Searle, John. 1980. Minds, brains and programs. *Behavioral and Brain Sciences* 3: 417–57.
———. 1995. *The construction of reality*. New York: The Free Press.
———. 1997. *The mystery of consciousness*. New York: New York Review.
———. 2007. *Freedom & Neurobiology*. New York, NY: Columbia University Press.
Sejnowski, Terrence J., Christof Koch, and Patricia S. Churchland. 1988. Computational neuroscience. *Science* 241: 1299–306.
Sellars, Roy Wood. 1969 [1922]. *Evolutionary naturalism*. New York: Russell & Russell.
———. 1970. In *Principles of emergent realism*, ed. W. Preston Warren. St. Louis, MO: Warren H. Green.
Sellars, Roy Wood, V. J. McGill, and Marvin Farber, eds. 1949. *Philosophy for the future: The quest of modern materialism*. New York: Macmillan.
Seth, Anil K., Zoltán Dienes, Axel Cleeremans, Morten Overgaard, and Luiz Pessoa. 2008. Measuring consciousness: relating behavioral and neurophysiological approaches. *Trends in Cognitive Sciences* 12: 314–21.
Shear, Jonathan, ed. 1995. *Explaining consciousness: The 'Hard Problem'*. Cambridge, MA: MIT Press.
Sherrington, Charles. 1964. *Man on his nature*. New York: New American Library.
Shimony, Abner. 1993. *Search for a naturalistic world view*, 2 vols. Cambridge: Cambridge University Press.
Shirokov, M., ed. 1937. *A textbook of marxist philosophy*. London: Victor Gollancz.
Shorter, Edward. 1997. *A history of psychiatry*. New York: Wiley.
Shubin, Neil, Cliff Tabin, and Sean Carroll. 2009. Deep homology and the origins of evolutionary novelty. *Nature* 457: 818–23.
Simon, Herbert A. 1962. The architecture of complexity. *Proceedings of the American Philosophical Society* 106: 467–82.

Singer, Wolf. 2009. Consciousness and neuronal synchronization. In *The neurology of consciousness*, ed. Laureys and Tononi, 43–52.

Smail, Daniel Lord. 2008. *On deep history and the brain*. Berkeley: University of California Press.

Smart, J. C. C. 1963a. Materialism. *Journal of Philosophy* 60: 651–62.

———. 1963b. *Philosophy and scientific realism*. New York: The Humanities Press.

Smith, Adam. 1976 [1776]. *The wealth of nations*, 2 vols. Chicago, IL: University of Chicago Press.

Smith, Quentin. 2003. Why cognitive scientists cannot ignore quantum mechanics. In *Consciousness*, ed. Smith and Jokic, 409–46.

Smith, Quentin, and Aleksandar Jokic, eds. 2003. *Consciousness: New philosophical perspectives*. Oxford: Clarendon Press.

Smolin, Lee. 2006. *The trouble with physics: The rise of string theory, the fall of science, and what comes next*. Boston, MA: Houghton-Mifflin.

Sobel, N., V. Prabhakaran, E. E. Desmond, G. H. Glover, R. L. Goode, E. V. Sullivan, and J. D. Gabrieli. 1998. Sniffing and smelling: different subsystems in the human olfactory cortex. *Nature* 392: 282–8.

Sober, Elliott, and David Sloan Wilson. 1998. *Unto others: The evolution and psychology of unselfish behavior*. Cambridge, MA: Harvard University Press.

Stich, Stephen P. 1999. *Deconstructing the mind*. New York: Oxford University Press.

Stoljar, Daniel. 2006. *Ignorance and imagination*. Oxford: Oxford University Press.

Stonor Saunders, Frances. 2000. *Who paid the pipers?: The CIA and the cultural cold war*. London: Granta Books.

Stove, David. 1991. *The plato cult and other philosophical follies*. Oxford: Basil Blackwell.

Suppes, Patrick. 1957. *Introduction to logic*. Princeton, NJ: D. Van Nostrand.

———. 1970. A probabilistic theory of causality. *Acta Philosophica Fennica 24*. Amsterdam: North Holland.

———. 1974. Aristotle's concept of matter and its relation to modern concepts of matter. *Synthese* 28: 27–50.

Takahashi, Hidehiko, Motoichiro Kato, Masato Matsuura, Dean Mobbs, Tetsuya Suhara, and Yoshiro Okubo. 2009. When your gain is my pain and your pain is my gain: Neural correlates of envy and Schadenfreude. *Science* 323: 937–9.

Tarski, Alfred. 1944. The semantical concept of truth and the foundations of semantics. *Philosophy and Phenomenological Research* 4: 341–75.

Tegmark, Max. 2004. Parallel universes. In *Science and ultimate reality*, ed. Barrow, Davies, and Harper, 459–91.

———. 2008. The mathematical universe. *Foundations of Physics* 38: 101–50.

Thagard, Paul, ed. 2007. *Philosophy of psychology and cognitive science*. Oxford: Elsevier.

Thaler, Richard H. 1992. *The winner's curse: Paradoxes and anomalies of economic life*. Princeton, NJ: Princeton University Press.

Thompson, Richard F. 2005. In search of memory traces. *Annual Review of Psychology* 56: 1–23.

Tiger, Lionel. 2008. The human nature project. *Skeptical Inquirer* 32(3): 30–4.

Tiger, Lionel, and Robin Fox. 1971. *The imperial animal*. New York: Holt, Rinehart and Winston.

Tolman, Richard C. 1934. *Relativity, thermodynamics and cosmology*. Oxford: Oxford University Press.

Tomasello, Michael. 2006. Why don't apes point? In *Roots of human sociality*, ed. Enfield and Levinson, 506–24.

———. 2008. *Origins of human communication*. Cambridge, MA: MIT Press.

Tononi, Giulio, and Cristof Koch. 2008. The neural correlates of consciousness. *Annals of the New York Academy of Sciences* 1124: 239–61.

Tootell, Roger B. H., Martin S. Silverman, Eugene Switkes, and Russell L. de Valois. 1982. Deoxyglucose analysis of retinotopic organization in primate striate cortex. *Science* 218: 902–4.

Tootell, Roger B. H., Eugene Switkes, Martin B. Silverman, and Susan L. Hamilton. 1998. Functional anatomy of macaque striate cortex II. Retinotopic organization. *Journal of Neuroscience* 8: 1531–68.

Trigger, Bruce G. 2006. *A history of archaeological thought*, 2nd ed. Cambridge: Cambridge University Press.

———. 2003a. *Artifacts and ideas*. New Brunswick, NJ: Transaction Publishers.

———. 2003b. *Understanding early civilizations*. Cambridge: Cambridge University Press.

Truesdell, Clifford. 1984. *An idiot's fugitive essays on science: Method, criticism, training, circumstances*. New York: Springer-Verlag.

Tulving, Endel. 2002. Episodic memory: From mind to brain. *Annual Review of Psychology* 53: 1–25.

Uexküll, Jakob von. 1921. *Umwelt und Innenwelt der Tiere*, 2nd ed. Berlin: Springer-Verlag.

Van Fraassen, Bas C. 1980. *The scientific image*. Oxford: Clarendon Press.

Vaughan, Susan C., Randall D. Marshall, Roger A. McKinnon, Roger Vaughan, Lisa Mellman, and Steven P. Roose. 2000. Can we do psychoanalytic outcome research? A feasibiliy study. *International Journal of Psychoanalysis* 81: 513–27.

Vesey, G. N. A., ed. 1964. *Body and mind*. London: George Allen and Unwin.

Volchan, Sérgio B. 2002. What is a random sequence? *American Mathematical Monthly* 109: 46–62.

Vollmer, Gerhard. 1986. *Was können wir wissen?, vol. 2: Die Erkenntnis der Natur*. Stuttgart: S. Hirzel.

———. 1987. What evolutionary epistemology is not. In *Evolutionary epistemology*, ed. W. Callebaut and R. Pinxten, 203–21. Dordrecht, NL: D. Reidel.

Vygotsky, L[ev]. S[emyonovich]. 1978. *Mind in society: The development of higher psychological processes*. Cambridge, MA: Harvard University Press.

Weinberg, Steven. 1992. *Dreams of a final theory*. New York: Random House.

Westermarck, Edward. 1906–1908. *The origin and development of moral ideas*, 2 vols. London: Macmillan.

Wetter, Gustav A. 1958 [1952]. *Dialectical materialism: A historical and systematic survey of philosophy in the soviet union*. London: Routledge & Kegan Paul.

Whitehead, Alfred N. 1929. *Process and reality*. New York: Macmillan.

Whitson, Jennnifer A., and Adam D. Galinsky. 2008. Lacking control increases illusory pattern perception. *Science* 322: 115–7.

Wiener, Norbert. 1948. *Cybernetics: Or control and communication in the animal and the machine*. New York: Wiley; Paris: Hermann et Cie.

Wilczek, Frank. 2008. *The lightness of being: Mass, ether, and the unification of forces*. New York: Basic Books.

Wikström, Per-Olof H., and Robert J. Sampson, eds. 2006. *The explanation of crime: Context, mechanisms, and development*. Cambridge: Cambridge University Press.

Wikström, Per-Olof, and Kyle Treiber. 2009. Violence as situational action. *International Journal of Conflict and Violence* 3: 41–62.

Wilkinson, Richard, and Kate Pickett. 2009. *The spirit level: Why more equal societies almost always do better*. London: Allen Lane.

Wimsatt, William C. 2007. *Re-engineering philosophy for limited beings*. Cambridge, MA: Harvard University Press.

Wittgenstein, Ludwig. 1922. *Tractatus Logico-Philosophicus*. London: Routledge & Kegan Paul.

———. 1967. *Zettel*, ed. G. E. M. Anscomb and G. H. von Wright. Oxford: Basil Blackwell.

Wolf, Arthur P. 1995. *Sexual attraction and childhood association*. Stanford, CA: Stanford University Press.

Wolpert, Lewis. 1992. *The unnatural nature of science*. London: Faber & Faber.

Wright, Bradley R. Entner, Avshalom Caspi, Terrie E. Moffitt, and Phil A. Silva. 1999. Low self-control, social bonds, and crime: Social causation, social selection, or both? *Criminology* 37: 479–514.

Wright, John Paul, Kim N. Dietrich, M. Douglas Ris, Richard W. Hornung, Stephanie D. Wessel, Bruce P. Lanphear, Mona Ho, and Mary N. Rae. 2008. Association of prenatal and childhood blood lead concentrations with criminal arrests in early adulthood. *PLOS Medicine* 5: 0732–9.

Wright, Samantha B., Bryan W. Matlen, Carol L. Baym, Emilio Ferrer, and Silvia A. Bunge. 2008. Neural correlates of fluid reasoning in children and adults. *Frontiers in Human Neuroscience* 1: doi: 10.3389/neuro.09.0002007.

Zeki, Semir. 1993. *A vision of the brain*. Oxford: Blackwell.

Zimmer, Carl. 2004. *Soul made flesh: The discovery of the brain and how it changed the world*. New York: Free Press.

Index

C